COMPLETE DATA ANALYSIS USING

Sara Miller McCune founded SAGE Publishing in 1965 to support the dissemination of usable knowledge and educate a global community. SAGE publishes more than 1000 journals and over 800 new books each year, spanning a wide range of subject areas. Our growing selection of library products includes archives, data, case studies and video. SAGE remains majority owned by our founder and after her lifetime will become owned by a charitable trust that secures the company's continued independence.

Los Angeles | London | New Delhi | Singapore | Washington DC | Melbourne

MARCO LEHMANN

COMPLETE DATA ANALYSIS USING

YOUR APPLIED MANUAL

Los Angeles | London | New Delhi
Singapore | Washington DC | Melbourne

Los Angeles | London | New Delhi
Singapore | Washington DC | Melbourne

SAGE Publications Ltd
1 Oliver's Yard
55 City Road
London EC1Y 1SP

SAGE Publications Inc.
2455 Teller Road
Thousand Oaks, California 91320

SAGE Publications India Pvt Ltd
B 1/I 1 Mohan Cooperative Industrial Area
Mathura Road
New Delhi 110 044

SAGE Publications Asia-Pacific Pte Ltd
3 Church Street
#10-04 Samsung Hub
Singapore 049483

Editor: Jai Seaman
Editorial Assistant: Rhiannon Holt
Production editor: Victoria Nicholas
Marketing manager: Ben Griffin-Sherwood
Cover design: Shaun Mercier
Typeset by C&M Digitals (P) Ltd, Chennai, India
Printed in the UK

Library of Congress Control Number: 2022934375

British Library Cataloguing in Publication data

A catalogue record for this book is available from the
British Library

ISBN 978-1-4739-1364-6
ISBN 978-1-4739-1365-3 (pbk)

At SAGE we take sustainability seriously. Most of our products are printed in the UK using responsibly sourced
papers and boards. When we print overseas we ensure sustainable papers are used as measured by the PREPS
grading system. We undertake an annual audit to monitor our sustainability.

For Neele, Ida, and Mieke

TABLE OF CONTENTS

LIST OF R EXAMPLES

WBD – Web-based cognitive bias modification for depression

ITC – Independence of thinking bias and cognitive ability

UPS – Unpleasant sounds

SCB – Student conciliators against school bullying in German secondary schools

LIST OF FIGURES

LIST OF TABLES

LIST OF R SCRIPTS

PREFACE

As a psychologist, I am experienced in basic research in psychology, education, clinical research, and music psychology. Although grounded in empirical research, I fully endorse the idea that more knowledge about statistics and about computer programming will help you get along as a scientist. I am enthusiastic about R programming and how it helps with basic research. Nevertheless, I always keep an eye on how the empirical project progresses and subordinate statistics and programming to the aims of the project. This book is highly inspired by my experiences as a consultant for empirical methods and statistics of hundreds of research projects – big and small.

I am grateful to many friends and colleagues who supported my work on this book. Michael Großbach took part in the conception. Together, we devised what an R course for applied researchers should look like. It was planned that Michael would be co-author until he had to withdraw from the project. Pascal Jordan helped me with my statistical understanding in many issues relevant to this book, especially in Chapters 6 and 7. Ronald Böhm, my friend of more than 25 years, helped me find out that math and computer programming are not altogether bad leisure time activities. My friend René Gilster shared my enthusiasm for empirical science and gave me the initial idea to acquire programming skills as a psychologist. That was a good decision. Johannes Andres and Hartmut Fillbrandt were my mentors for methods and statistics at the University of Kiel. In Kiel I also took my first steps in command line programming under the supervision of Christian Kaernbach. Reinhard Kopiez was my indulgent mentor at the Hanover University of Music, Drama, and Media. He supported my early steps in empirical research. Friedrich Platz was my colleague in Hanover in empirical music research for several years. His fidelity in the planning and analysis of empirical research and his eagerness to conquer statistical programming promoted my will to prepare this course material. My supervisors Martin Spieß, Knut Schwippert, and Bernd Löwe provided me with the opportunity to work as a consultant for research methods and statistics for several years, which further strengthened the concept for this book. Simon Jebsen and Heiko Stüber invited me to the International Research Workshop for PhD students for many years and gave me plenty of opportunity to test my course material about R. Simon Blackwell, Christoph Reuter, and Michael Oehler contributed their original research data for the included examples. Denise Kästner and Natalie Uhlenbusch provided feedback on parts of the manuscript. Jai Seaman from SAGE promoted this project from the beginning with patience and understanding. Finally, I thank my wife Neele Lehmann for her enduring support and leniency in the many hours of writing.

INTRODUCTION

The aim of this book is to have fun with R and to practise the analysis of true research data. It is not organised according to statistical methods like a statistics book and not according to programming principles like a book for software developers. It is organised according to how a researcher would employ the R language in his or her daily labour in order to arrive at results ready for publication.

The content is organised around run-through examples from psychology, clinical research, and school research. The examples will be introduced in an early chapter. In subsequent chapters the examples serve for the illustration of statistics with R. The book welcomes the R novice and the experienced R user. We will start at level zero in Chapter 1 and gradually increase knowledge as we go along with the data analyses. We will, however, also deepen many aspects of the R programming language and its functions. I will highlight elements of the R language which are crucial for understanding and efficient programming. The aim is to equip you with all the knowledge that you need to carry on on your own. Furthermore, if we encounter some cool things that are possible with R, I will give my best to highlight them, for example, data imputation, HTML reports, 3D graphics, and publication-ready results tables. So, I hope that there is also a fun part in working through this book.

The structure of the book

Chapter 1 provides a quick start with R and introduces the basics to run the program. You will warm up with the R console and complete your first task: run the given lines of code at the console and carefully observe what they do. Programming features: vector, matrix, and data frame creation, calculations at the console, and function calls.

Chapter 2 introduces how to load, import, and save data in various formats. It stresses that teamwork requires agreement among scientists on the data format used and on data archiving. Since quality data sets are expensive, you should, as a researcher, document and maintain your data to be useful for the research community. You will learn how to set up plain text files as the format of choice for the secure storage of data and data documentation. Programming features: R objects, working directory and workspace, efficient variable access, file format for saving R objects, file formats for data frames.

Chapter 3 shows how to calculate and transform the variables of a data set. Usually, statistical analyses do not run with raw data but with aggregated variables, for example, a sum score of several items. So, the calculation of variables and the transformation of data are often required. The chapter further focuses on the formatting and annotation of variables of different types. Therein, we examine the preparation of factor variables for inferential statistics. Programming features: variable types, variable attributes, using code scripts to structure the analysis, calling scripts from other scripts, vector and matrix calculations, reshaping of data frames between wide and long format.

Chapter 4 introduces functions for descriptive and exploratory statistics in univariate, bivariate, and multivariate contexts. We practise the description of categorical and quantitative variables. We write the results in HTML reports to be shared with the research team. The programming technique of vectorisation further facilitates the calculation of statistics. Finally, we will visit more options to structure scripts and the whole analysis. Programming features: quick results report, vectorised operations, lists, search path.

Chapter 5 complements Chapter 4 as an integral part of exploratory analysis and describes R's graphical facilities to get the best overview of the data. For example, we implement response profiles of all subjects for the detection of response sets and also displays of multiple histograms for the evaluation of item characteristics. These are quick and convenient options to get the maximum immersion into the data. Programming features: high-level and low-level graphics, simultaneous output of multiple graphics with uniform layout.

Chapter 6 describes statistical analysis with one of the most common designs used in psychological research: the analysis of variance (ANOVA). It shows how to calculate the ANOVA model and how obtain the results. The chapter also includes factorial designs using more than one independent variable and the analysis of repeated measures. Programming features: formulas and statistical models, collection of information from complex R objects.

Chapter 7 broadens the scope of inferential statistics to linear regression. In regression analysis you will learn how to fit and test regression models. Furthermore, you will experience the ease of implementing interaction terms. The linear modelling approach is then extended to cover logistic regression. Programming features: simple and multiple linear regression, test of regression parameters, data imputation, logistic regression.

Chapter 8 shows how to produce tables for use in research publications. You will learn about R's facilities to export tables to be used in a word processor. So, you can finalise the display of your research results with R, which further enhances productivity. Programming features: file output of results, export of formatted tables, report generation.

Chapter 9 shows R's great power to create graphics from scratch. You will practise the standard graphical utilities and other graphical systems R offers, like grid graphics or ggplot2. Programming features: graphics programming, graphic systems, 3D graphics.

Chapter 10 introduces simulation as a powerful technique to create and evaluate experimental designs. R helps us to opt for the best design. It provides us with large numbers of simulation runs to compare the efficiency of different analyses as a supplement for statistical power analysis. Programming features: random number generation, simulation reproducibility, vector and matrix capabilities for the analysis of multiple data sets, display of simulation results.

Chapter 11 finds a remedy for going back to one of our statistical analyses with big worries as to how all these function calls fit together. We will review practical advice on programming style and on the structure of the analysis. This helps us to maintain the analysis and, further, to enhance and extend it after time has passed.

Throughout I maintain three levels of headings. The headings of level 1 reflect the overall approach to data analysis from data acquisition over exploratory and inferential statistics to publication-ready results output. If you read through the headings of level 2 you should notice many familiar topics that pertain to a reasonable data analysis with R. The headings on level 3 then announce the many particular issues of the R language, of statistics, and of programming that we need in the analysis.

DISCOVER THE ONLINE RESOURCES

Complete Data Analysis Using R is supported by a wealth of online resources for both students and lecturers to aid study and support teaching, which are available at https://study.sagepub.com/lehmann.

For students

- **Data sets** allow you to practise and apply the skills you've learned using real-life examples.
- **Scripts** ready to upload into R will generate meaningful information to help you master your data analysis skills.

For lecturers

- **Further reading** recommendations, including links to relevant publications, open data archives, and resources about statistical programming with R.
- **PowerPoint decks** can be customised for use in your own lectures and presentations.
- **Watch and learn! A video introduction** to the text by the author that discusses the advantages and disadvantages of using R.

1

QUICK START - BASIC TRAINING OF R FUNCTIONS

━━━━━━━━━━━━━━━━━━━━━━━━━━━━━━━━━ Chapter overview ━━━━━━━━━

This chapter provides a quick start with R and introduces the basics to run the program. You will warm up with the R console and complete your first task: run the given lines of code at the console and carefully observe what they do. Programming features: vector, matrix, and data frame creation, calculations at the console, and function calls.

1.1 Jump-starting R - Doing statistics at the R console

The first R session introduces functions as the building blocks of R programming. It uses <u>c()</u> and <u>rnorm()</u> to demonstrate how functions work.

EXAMPLE!

1.1.1 Warm up

Welcome to R! Let us go straight to statistics programming. All you need is an almost modern computer and an internet connection. R is open source, so it is freely available. *Download the installation file of the basic R program from the web page www.r-project.org/ and install the program.* After installation, double click on the R icon on your computer or select R in the start menu to start it. A plain window welcomes you with some information about the R installation, but then you are left with nothing more than a humble prompt: > | (Figure 1.1). This is the R console – this is where we begin.

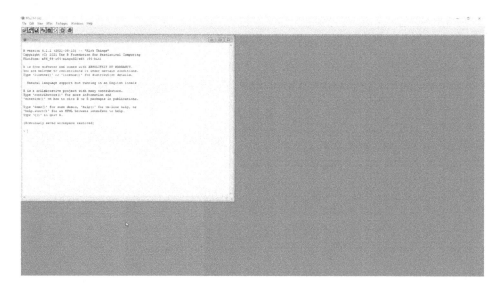

Figure 1.1   The R console.

Enter each line of the following R code into the console and press the return key at the end of each line. R immediately gives you the results. Closely inspect what happened after each line and what R did for you. You can find the script with the following file name in the repository that goes with this book.

Ch1_WarmUpRConsole.R

```
## Warm up with the R Console
# Basic operations
2 + 3
5 - 8
3 * 4
```

```
12 / 4
10 / 3
5^2
2^10
25^(1/2)
125^(1/3)
sqrt(36)
sqrt(4)
# Vectors
c(1,2,3) + 5
c(1,2,3) + c(4,5,6)
c(1,2,3) + c(4,5) # Error
c(1,2,3) * 3
c(10,20,25) / 5
x <- c(1,2,3)
x
x + 4
x
x * 3
y <- 2*x + 4
# Statistics with vectors
mean(x)
mean(y)
Means <- c(X = mean(x), Y = mean(y))
Means
sd(x)
sd(y)
y - mean(y)
(y - mean(y))^2
sum((y - mean(y))^2)
sum((y - mean(y))^2) / length(y)
sd(y)^2 * 2/3
# Statistics with data sets
X <- factor(rep(1:3, each=4))
X
Y <- c(rnorm(n = 4, mean = 10, sd = 2), rnorm(4,11,2), rnorm(4,10.5,2))
Y
d <- data.frame(X,Y)
d
str(d)
View(d)
tapply(d$Y, d$X, mean)
summary(d)
ANOVA <- aov(Y ~ X, d)
summary(ANOVA)
ls()
```

Analysis

In the first few lines R works as a calculator. It calculates plus and minus and even a square root with `25^(1/2)`. However, using an R function is easier to read, like `sqrt(36)`. In fact, **`sqrt()`** is an R function to carry out a certain operation; and `sqrt(36)` is a function call taking a number to produce the square root of. The 36 is called an argument for the function. This function call was easy, but it shows the general structure of all function calls in R: *all functions in R have a pair of parentheses where the function obtains information to use, the so-called arguments. Furthermore, the function has a so-called return value, which is the result of its calculations.*

As you can see, our lines at the console get more and more complex as more functions appear, for example, **`c()`**, **`mean()`**, or **`sd()`**. Notice that R is case-sensitive. For example, the functions **`sqrt()`** and **`SQRT()`** are different things. Some objects and function names use mixtures of uppercase and lowercase letters, so always observe the spelling of things.

Another important feature of the code is `<-`, which looks like an arrow to the left. This is the assignment operator in R and it works like this: to its right-hand side is an operation that R carries out. Without the assignment operator the result of this operation appears at the console. Using the operator, the result is stored in an object on the left-hand side. This is called assignment. The object now preserves the result of the operation for subsequent use. Notice in our code that some lines depend on the creation of objects in previous lines. For example, `mean(x)` only works if R knows what x is, which was created with the assignment operator some lines earlier in the code. In fact, some functions use several objects passed as arguments within one function call.

The last part of the script consists of an analysis of variance. We create an independent variable X with **`factor()`** and use three levels 1, 2, and 3. The colon operator **`:`** creates the sequence from 1 to 3. Each level is repeated four times for four subjects in each group. Then the dependent variable Y is randomly sampled from the normal distribution. The function **`aov()`**, which is the analysis of variance, requires a dependent variable and an independent variable, both linked with a **`~`** symbol. Using **`summary()`** with the ANOVA object produces the well known summary table for a one-way analysis of variance. In essence, all R programming involves this: the creation of objects, the application of statistical functions to objects, and the handling of output of these functions. *Learning R means learning about the use of functions and to acquire a big vocabulary of functions to use in every data analysis.*

R LANGUAGE! ## 1.1.2 Function calls at the R console

We already learned that the prompt `> |` suggests you to enter something. *But at the beginning, it is hard to find out what to enter.* By comparison, IBM® SPSS® Statistics software ('SPSS'), Excel, and OpenOffice Calc start with empty data spreadsheets and invite you to put some words or numbers in the data cells. In R there is nothing but the command prompt. We will discover that we achieve everything in R by writing lines of code. That is, the console receives the statistical commands like **`mean()`**, **`sd()`**, **`View()`**, **`anova()`**, or **`t.test()`**. To master it is very rewarding – it allows you to think in statistics and simply carry it out. For example, a factor analysis, display of missing data patterns, or diagrams of empirical distributions are only one line of code away with the right function at hand. However, if you start with R you need some endurance until you know enough functions to achieve your aims.

The general form of a function call is: ReturnValue <- FunctionName(Parameter = Argument) (Figure 1.2). Each function has a return value, which is the result of its operation. It also has one or more parameters which show the user the type of information the function needs to operate. Each parameter needs an argument, which is the concrete input for a particular function call. Commas separate the parameters. The <u>rnorm()</u> function is a good illustration for this. Type x <- rnorm(n = 10, mean = 100, sd = 10) and press return; in the next line type x and press return again. R shows that the object x contains 10 random numbers drawn from the normal distribution with mean 100 and standard deviation 10. The function <u>rnorm()</u> comprises the three parameters n, mean, and sd, which define the type of information that <u>rnorm()</u> needs: a sample size n, a theoretical mean, and a standard deviation. These parameters need concrete information – the arguments. That is, we want ten random numbers taken from the particular normal distribution with mean 100 and standard deviation 10.

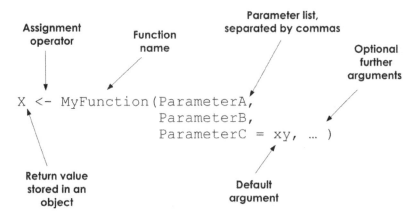

Figure 1.2 Functions, parameters, and arguments.

The console obtains many function calls in a usual R session. R knows a very large number of statistical and other functions. The scope of functions ranges from simple means to complete mixed-model or factor analyses. Your workflow will run smoothly if you know many functions. Your complex statistical tasks then divide as a sequence of function calls until you obtain your end result. Much of this book deals with how to set up the sequence of function calls to solve statistical problems. Each chapter will introduce new functions and their applications and thereby strengthen your vocabulary.

1.1.3 Everything in R is an object   R LANGUAGE!

In R, everything is an object and every object is an instance of a certain type or class. Like my glass of water on the table is an instance of the class glass of water, the particular vector x = (1, 2, 3) is an instance of the class vector. Some functions work with instances of the class vector. For example, scalar multiplication works with numeric vectors, but not with objects of the class string. In fact, classes have functions exclusively written for them by a computer programmer. To give a better impression, a list of some more types or classes available in R may help: vectors, matrices, strings, figures, tables, factor analyses, and many more. Proficient R users may even define their own object classes, relevant to their field of research, and also define the specific methods that apply to all objects of this class. In psycholinguistics, for example, some objects are of class text corpora, which allow for certain typical operations relevant in the research domain. *R itself and the add-on packages distinguish many different objects classes, and each class consists of a specific set of methods, which can be applied to its instances.* For example, there is a method which displays the means for the analysis of variance class; so, the method applies to each particular analysis of variance.

If you start R for the first time, there are no objects in the workspace. So, let us create one. Type, `myObject <- c(1, 2, 3, 4, 5)` and press return. Then type `ls()` and press return again. R returns `[1] "myObject"`. This is your first object in the so-called global environment. Display its content by typing `myObject` followed by the return key. This gives `[1] 1 2 3 4 5`. The function **mean()** applies to this object; that is, `mean(myObject)` returns 3. This is much of what using R is about: object creation and application of functions to objects. *The global environment hosts the statistical analyses done at the console.* It is like a desktop, where some objects like data sets or analyses of variance reside for further enquiry.

R contains several demos to show you how to start. Type in **demo()**, **help()**, or **help.start()** at the console. Calling **demo()** lists some of the available programming demos. Use `demo (graphics)`, for example, to produce a collection of nice diagrams. The console shows the corresponding programming code. Furthermore, `demo(persp)` shows some of the capabilities for three-dimensional displays and `demo(plotmath)` the possibilities to use mathematical annotation in graphics.

R FUNCTION! ## 1.1.4 c()

A most basic function in R is c(), which simply combines its arguments to a vector; it can combine numbers but also the content of other objects or strings. Let us try it at the console.

Ch1_c.R

```
## Combine everything with c()
a <- c(1, 2, 3)
a
b <- c(7, 8, a, "R is incredible")
b
c(47, 4, sample(5))
plot(1:10, ylim = c(1,15))
```

Output

```
> a <- c(1, 2, 3)
> a
[1] 1 2 3
> b <- c(7, 8, a, "R is incredible")
> b
```

```
[1]  "7"                    "8"                  "1"                 "2"
"3"              "R is incredible"
> c(47, 4, sample(5))
[1] 47  4  2  4  1  5  3
> plot(1:10, ylim = c(1,15))
```

Analysis

The first two uses of c() create the objects a and b. It combines everything within paren-theses into a vector. Vector a is made of numbers, vector b also includes a string. Strings are collections of letters or characters, usually to form words or text. A string inside c() turns all numbers into strings, too, as is indicated by "" in the output. We can also use function calls inside c() and it simply combines the result of the function call with the rest. The call of **plot()** shows how c() works to provide arguments to function parameters. The ylim parameter of **plot()** naturally requires two numbers – a minimum and a maximum. To hand over the two numbers requires c(). Here, ylim obtains c(1,15) as an argument, so that the minimum and the maximum of the y-axis become 1 and 15. The c() function is often used to supply arguments to functions calls. In consequence, *c() is one of the most commonly used functions and it is often embedded in other function calls to provide vectors as arguments to a func-tion. Later, we will practise how c() figures in indexing vectors, that is, in selecting only certain entries of a vector.*

1.1.5 The colon operator :   R FUNCTION!

The colon operator : defines sequences of numbers in R. Its simplest application is StartingNumber:FinalNumber. For example, 13:18 yields the six numbers 13, 14, 15, 16, 17, 18. However, R can also count backwards like in 154:17, or you can use decimals like 2.26:17. Later, we will also use **seq()** for an enhanced definition of sequences.

1.1.6 rnorm()   R FUNCTION!

The easiest way to obtain data is to simulate it. Simulation data is important in teaching sta-tistics, in statistical power analysis, and in testing statistical methods. The function **rnorm()** selects random data from the normal distribution. It needs a sample size, an expected mean value, and an expected standard deviation and then it returns a vector of numbers. Here is an example:

Ch1_HistogramsSamplesNormalDistribution.R

```
## Histograms of samples from the normal distribution
layout(matrix(1:4, 2))
Y <- rnorm(n = 100, mean = 50, sd = 5)
hist(Y, breaks = 20)
Y <- rnorm(n = 1000, mean = 50, sd = 5)
hist(Y, breaks = 20)
Y <- rnorm(n = 10000, mean = 50, sd = 5)
```

```
hist(Y, breaks = 20)
Y <- rnorm(n = 100000, mean = 50, sd = 5)
hist(Y, breaks = 20)
```

Output

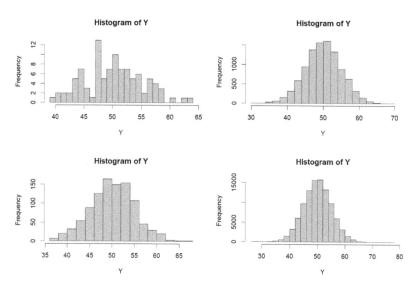

Figure 1.3 Histograms of data from the normal distribution with different sample sizes.

Analysis

The script uses four calls of **rnorm()** to generate the data. Each call is with a different sample size. The object Y stores the data of each call, and it is used, in turn, by **hist()** for the histogram of the frequency distribution. The graphics function **layout()**, to be introduced later, divides the page into four panels, so all diagrams find their place here. On the first glance, the data do not reveal that they have been selected from a normal distribution. With a large sample size the display comes very close to the ideal bell curve. It is fun to play around with **rnorm()**. Use different sample sizes and settings for mean and standard deviation to get a feeling for simulation data, which we will use frequently.

R LANGUAGE! ## 1.1.7 Using add-on R packages

You may have heard that R invites everybody to write his or her own functions or classes of objects and contribute these to the research community. This is the idea of add-on packages for R. You start your work with the basic installation of R, but then add several such packages to extend functionality in R. This way you may, for example, not only use one barplot function but five or six different versions from different packages and decide which one you like most. Packages come for free, and to obtain and manage them, use the packages menu in the R program (Figure 1.4). Select the menu item 'set cran mirror' to choose a download mirror near your location. Then select the menu item install packages. A list of all available packages appears and you can chose one or more packages and install them. Let us select the two packages **psych** and **Hmisc** for installation and let us use the **describe()** function provided by both packages.

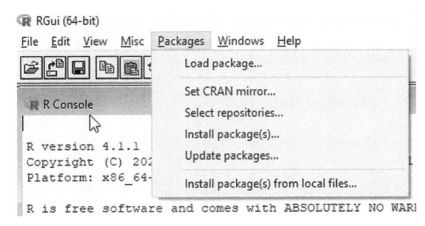

Figure 1.4 The packages menu in the R program.

Ch1_UseAddOnPackages.R

```
## Use add-on packages
x <- rnorm(100)
psych::describe(x)
Hmisc::describe(x)
```

Output

```
> x <- rnorm(100)

> psych::describe(x)
  vars   n  mean   sd median trimmed  mad   min  max range  skew
kurtosis  se
X1    1 100 -0.03 0.97   0.02   -0.02 1.12 -2.48 2.18  4.66 -0.07
-0.32 0.1

> Hmisc::describe(x)
x
        n  missing distinct     Info      Mean
Gmd      .05      .10      .25
      100        0      100        1 -0.02671    1.099  -1.5392
-1.1566  -0.7725
      .50      .75      .90      .95
   0.0151   0.7082   1.2504   1.5454


lowest : -2.481360 -2.356328 -2.178577 -1.848590 -1.602327
highest:  1.580611  1.682588  1.763949  1.879894  2.180808
```

Analysis

First, we create 100 random numbers from the normal distribution. Then we use the package name, the double colon operator, and the function name to calculate descriptive statistics for the data vector. Both function calls produce a descriptive analysis and you can decide which one you like. Note that describe(x) would not suffice, because R would not know where to look for the **describe()** function. Using the package name together with the function resolves the ambiguity. This way you can include functions from any package that you find interesting. Play also with the histogram functions **hist()**, **MASS::truehist()**, and **lattice::histogram()**. However, in everyday programming you will often use more than one function from a particular package. To include all functions of a package at once, call library(package name) and then use all functions without the preceding package names and double colon operator.

STATISTICAL
ISSUE!

1.1.8 Easily compare the means of two experimental conditions

You already know random sampling from the normal distribution with **rnorm()**. *A common application of this is to compare two experimental conditions with a t-test.* It determines whether the means of the data from the two conditions significantly differ. With your current knowledge in R you can already do this with the following strategy: initialise the parameters of two normal distributions. Then, draw random data from the two distributions and store it in a data vector; add a categorical variable to code two experimental conditions. Finally, compare the means of the two conditions and display the results. To program this we use two new functions **rep()** and **t.test()**, but they are harmless.

Ch1_MinimalSimulation_tTest.R

```
## Minimal simulation of a t-test
# Simulation
set.seed(12345)
N <- 100
Mu <- c(10, 11)
Sigma <- 5
Y <- rnorm(N, Mu, Sigma)
A <- rep(c("a1", "a2"), 50)

# Statistics and t-test
t.test(Y ~ A)
```

Output

```
> t.test(Y ~ A)

    Welch Two Sample t-test

data:  Y by A
```

```
t = -1.1617, df = 86.664, p-value = 0.2485
alternative hypothesis: true difference in means is not equal to 0
95 percent confidence interval:
 -3.5273162  0.9251337
sample estimates:
mean in group a1 mean in group a2
        11.07544        12.37653
```

Analysis

As can be seen, only a few lines of code at the console and only a handful of functions do the job. The objects Mu and Sigma contain the expected means and the standard deviations of two normal distributions. They figure in **rnorm()** as arguments. In effect, one call of **rnorm()** generates all the data for the two conditions. The sample size of $N = 100$ is automatically split in half: one half of the data belongs to one condition, the other half to the other condition. The vector Y then stores the random data. Note that in this vector it is not that the first fifty entries belong to one condition and the other fifty entries to the other condition; rather, the vector entries alternate between conditions. That is, the first, third, fifth and so on values belong to the first entries in Mu and Sigma, the second, fourth, sixth and so on values belong to the second entries in Mu and Sigma. Because in Y there is no information regarding which value belongs to which condition, the vector A holds the experimental condition for all data values. We then produce the *t*-test with a short function call of **t.test()** relating the random data Y to the conditions of A. The function displays group means, *t*-value, degrees of freedom, and the *p*-value.

1.2 **Programming statistics**

We assemble function calls in R scripts. These are plain text files comprising all the functions of an analysis in their logical order. Almost all examples in this book are scripts. Good scripting is a prerequisite for good teamwork and for reproducible science; therefore, we practise it from the beginning.

1.2.1 **Scripting the statistical analysis**

PROGRAMMING ISSUE!

One important feature of reproducible science is the documentation of all calculations of a statistical analysis. Whereas a point-click strategy of analysis is tempting to get the work done quickly, this method is dangerous because it is difficult to remember all steps of an analysis, that is, the sequence of clicks. Although it is possible to document all the clicks in normal language, scripting functions in plain text is more precise and is clearly mandatory in any scientific context. For example, for documentation it is better to store an analysis of variance as aov(Rating ~ Gender + Occupation, data = DataEmployeeQuestionnaire) than to click together the analysis in a statistics program and then document it: an ANOVA of the main influences of gender and occupation relating to the rating was calculated within the data set of the employee questionnaire. For a journal publication, the latter may suffice, but it leaves some questions open that the former and shorter form does not entail. For instance: were there other terms in the ANOVA model? Was the dependent variable transformed somehow?

How was the handling of the missing data? Repeatedly, you will find that this information is missing in journal publications. *So, the research documentation should keep the statistical operations in their most explicit form, which is the programming code that does the calculations.* R scripts hold this code and are actually plain text files. In the scripts all calculations follow each other in their logical order to produce the results of the study.

R LANGUAGE!

1.2.2 Create and run an R script

In R the menu command File – New script opens a new script. The script is an empty text file, which now receives all the R code. *Instead of solely using functions at the console, they all go into the script in their meaningful order.* That is, while at the console, usually trial and error helps. In the script there are only function calls that work and which often build on earlier function calls. The script is stored on the hard drive for later use and for further programming. In fact, doing the statistical analysis in a project ultimately amounts to creating the collection of scripts, which contain all calculations and all calls for results displays and diagrams. Now, write some of the first lines of the initial warm-up into the script file and save the script with the file extension .R. Then, select the function calls in the script with the mouse and use the key combination CTRL + R. The script should execute and you should see the results of all function calls at the console. Lines beginning with the hash tag key # are called code comments. You can use such comments to add human language to your scripts so that a reader could understand the script better and knows why the subsequent lines of code are needed. *All the further examples will either run at the console or as a script, depending on whether it is quick and dirty practice of R functioning at the console, or structured programming of a script.*

HELPER SOFTWARE!

1.2.3 Using RStudio

A good option to do statistics in R is to use the program RStudio (https://rstudio.com/). It is a nice looking program which hosts the R console and the scripting facilities of R. Moreover, RStudio comprises many useful panels which provide information about the current analysis. For example, there is a list view of all objects available for programming, a file browser for the current working directory, a viewer for figures, and a panel showing help pages. RStudio is not R; it is an interface to R. It does not bypass the requirement to program; however, it makes programming more convenient and helps you organise big analyses comprising several R scripts, data, and results files. It also supports report generation, be it as an HTML webpage or as slides for presentation. To do R programming, which is the intention of this book, we do not require RStudio; however, it is practical to use it.

HELPER SOFTWARE!

1.2.4 Using Tinn-R

Tinn-R is another program which helps with R programming (https://tinn-r.org/en/) (Figure 1.5). Like RStudio it bundles scripts as projects. Besides syntax highlighting and code completion it also offers the R console. Furthermore, a function catalogue offering templates of basic function calls and an explorer for objects available in the R environment are helpful additions. Best of all, it is open source and can be downloaded from https://nbc-gib.uesc.br/tinnr/en/ and http://sourceforge.net/projects/tinn-r. It has a long history dating back to 2001 and can help you to program your full analysis.

Figure 1.5 The Tinn-R program.

1.3 Creating data with a data frame or a matrix

Of course, we need data to analyse. Study data is stored in a data frame, which is one of the object types that R uses. We will introduce data frames and practise how to access the data.

1.3.1 data.frame()   R FUNCTION!

Regardless of the statistics program you use, the empirical data is stored in data frames. A data frame is a table with rows and columns, which typically consists of one row per subject and one column per empirical variable (Table 1.1). Most of the functions in R run on the rows or columns of such data frames. To handle data frames as objects in R is essential – to create, load, use, and save them.

Table 1.1 A data frame.

ID	A	B	X1	X2	Y1	Y2
1	...	...	...	...	...	...
2						
...						
n						

Let us practise how to work with this basic unit of every data analysis. R can import data frames or open a window for data entry by hand, which is part of Chapter 2. A data frame is stored as a data frame object in R and is created with **data.frame()**. The function aggregates variables to a data frame object. It is similar to the **cbind()** function which is, however, reserved for the combination of vectors to a matrix. Here is how **data.frame()** works.

Ch1_DataFrame.R

```
## The data.frame() function
set.seed(12345)
Data <- data.frame(ID = 1:20,
                IV = factor(rep(c("a", "b"),
                                times = 1,
                                each = 10)),
                y = c(rnorm(10,10,2),
                      rnorm(10,12,2)))
Data$y_standardised <- scale(Data$y)
View(Data)
```

Output

Table 1.2 A simple data frame with random numbers.

vp	IV	Y	y_standardised
1	a	11.171058	0.008299422
2	a	11.418932	0.122436813
3	a	9.781393	-0.631591801
4	a	9.093006	-0.948569987
5	a	11.211775	0.027048287
6	a	6.364088	-2.205139978
7	a	11.260197	0.049344993
8	a	9.447632	-0.785277199
9	a	9.431681	-0.792622198
10	a	8.161356	-1.377561690
11	b	11.767504	0.282942047
12	b	15.634624	2.063613837
13	b	12.741256	0.731320085
14	b	13.040433	0.869080594
15	b	10.498936	-0.301188804
16	b	13.633800	1.142304993
17	b	10.227285	-0.426274502
18	b	11.336845	0.084638555
19	b	14.241425	1.422095093
20	b	12.597447	0.665101438

Analysis

Because the data frame will consist of random data, we use **set.seed()** at the beginning. Next, the **data.frame()** function encloses the creation of three variables: the subject ID, the independent variable IV, and the dependent variable y. The function's output is the data frame object Data. After data creation we can add further variables with the $-operator. Here, we assign the standardised scores of y, created with **scale()**. Finally, **View()** shows us the data. Now that we know what a data frame in R is, why should we wait any longer to do real statistics? So, let us get to it and calculate an analysis of variance in only one line of code.

1.3.2 Calculate an analysis of variance in only one line of code

STATISTICAL ISSUE!

Working with a tight schedule and having to submit results the next day, just three lines of R code can produce an analysis of variance: data loading, calculation of results, and output of results. The combination of **data.frame()**, **aov()** *and* **summary()** *suffices.*

Ch1_ANOVA_OneLineOfCode.R

```
## ANOVA in one line of code
set.seed(12345)
Data <- data.frame(vp = 1:20,
                    UV = factor(rep(c("a", "b"),
                                times = 1,
                                each = 10)),
                    y = c(rnorm(10,10,2),
                          rnorm(10,12,2)))

summary(aov(y ~ UV, data = Data))
```

Output

```
> summary(aov(y ~ UV, data = Data))
            Df Sum Sq Mean Sq F value  Pr(>F)
UV           1  40.27   40.27   14.69 0.00122 **
Residuals   18  49.34    2.74
---
Signif. Codes:  0 '***' 0.001 '**' 0.01 '*' 0.05 '.' 0.1 ' ' 1
```

Analysis

As can be seen, *an ANOVA can be calculated in one line of code, given an available data frame.* The console extract below it shows the common results table. It shows information for the two sources of variation, the independent variable, and the error term. The information consists of degrees of freedom, sum squares, and the *F*-test, together with its *p*-value. As can be seen, the results do not reach significance, although a true effect is implemented in the simulation. The output could be your proposal for the next team meeting with your supervisor. Issues for discussion relating to the results which the team could discuss are: does the analysis include the correct variables? Which interaction terms are required? Is the amount of explained variance satisfactory? Answers from the team could lead you in good directions as to how to modify the analysis and where to efficiently spend more time for refinement. Then after another cycle of programming you could supply your team with results according to their suggestions.

So, your use of R is a success story already: you know how to inspect data at the console and then write a script to structure the analysis. Further, you programmed a statistical analysis yourself. The success definitely gives you the endurance to suffer a few more technical issues of R programming.

R FUNCTION! ### 1.3.3 matrix()

A similar concept to that of a data frame is that of a matrix. *Matrices are two-dimensional arrays of numbers, structured in rows and columns.* See below: **A** is a matrix with three rows and four columns of elements, usually numbers. All elements have indices that consist of their respective row and column number. For example, A_{23} = 7, that is, the element in the second row and the third column is 7. Many functions run on matrices. Some researchers would even prefer to operate only with data matrices when dealing with data, and many statistics books describe and justify the more complex statistical methods entirely in matrix notation. *R is exceptionally good in matrix calculations and contains lots of functions which are mostly efficient for use with matrices.* In fact, programming matrix calculations in R often comes close to the notation in statistics books. A matrix can be instantiated as a collection of vectors or it can be created with **matrix()**. For example, the function **cbind()** binds vectors of the same length and returns a matrix. Let us try to handle matrices.

$$A = \begin{pmatrix} 1 & 2 & 3 & 4 \\ 5 & 6 & 7 & 8 \\ 9 & 10 & 11 & 12 \end{pmatrix}$$

Ch1_SimpleMatrix.R

```
## Simple matrix containing random data
set.seed(12345)
x <- rnorm(10, 10, 2)
y <- rnorm(10, 50, 5)
Data <- cbind(x, y)
Data
```

Output

```
> Data
              x        y
[1,]  11.171058 49.41876
[2,]  11.418932 59.08656
[3,]   9.781393 51.85314
[4,]   9.093006 52.60108
[5,]  11.211775 46.24734
[6,]   6.364088 54.08450
[7,]  11.260197 45.56821
[8,]   9.447632 48.34211
```

```
 [9,]   9.431681 55.60356
[10,]   8.161356 51.49362
```

Analysis

At the console, we create two vectors x and y using the **rnorm()** function. Then, **cbind()** forms a matrix by simply binding the two vectors together. *Entering a matrix by hand looks like this*:

Ch1_ByHandMatrix.R

```
## By-hand entry of a matrix
A <- matrix(c(1, 2, 3, 4, 5, 6, 7, 8, 9, 10, 11, 12), nrow = 4,
ncol = 3)
A
```

Output

```
> A
     [,1] [,2] [,3]
[1,]    1    5    9
[2,]    2    6   10
[3,]    3    7   11
[4,]    4    8   12
```

Analysis

In the function call, the first argument gives the data entries and the second and third give the numbers of rows and columns. A matrix can be enhanced with names for rows and columns. The functions **rownames()** and **colnames()** allow you to query and set these names. The following lines illustrate how to use these functions at the console.

Ch1_MatrixRowAndColumnNames.R

```
## Setting row names and column names of a matrix
rownames(A) <- c("Bill", "Bob", "Anne", "Jill")
A
colnames(A) <- c("A", "B", "C")
A
rownames(A)
colnames(A)
rowSums(A)
colSums(A)
```

Output

```
> rownames(A) <- c("Bill", "Bob", "Anne", "Jill")
> A
```

```
       [,1] [,2] [,3]
Bill     1    5    9
Bob      2    6   10
Anne     3    7   11
Jill     4    8   12
> colnames(A) <- c("A", "B", "C")
> A
       A B  C
Bill 1 5  9
Bob  2 6 10
Anne 3 7 11
Jill 4 8 12
> rownames(A)
[1] "Bill" "Bob"  "Anne" "Jill"
> colnames(A)
[1] "A" "B" "C"
> rowSums(A)
Bill  Bob Anne Jill
  15   18   21   24
> colSums(A)
 A  B  C
10 26 42
```

Analysis

In the first line we use the assignment operator to declare some person names as row names. Notice that we call **rownames()** on the left side of the assignment operator. The same construction allows us to declare column names. Simply using **rownames()** and **colnames()** with the matrix queries the row and column names. The functions **rowSums()** and **colSums()** illustrate that names are preserved if functions operate on the matrix.

R LANGUAGE!

1.3.4 Indexing the elements of vectors, data frames, and matrices in R

Vectors, data frames, and matrices contain elements and the analysis often requires access to some of the elements for analysis or display. For example, a single element of a data vector or a single variable of a data frame may be required for an analysis. To get the information we call the data object holding the information and then use the operators [] and $ to locate the information inside the object.

Ch1_Indexing.R

```
## Vector, matrix, and data frame indexing
a <- c(21:32)
a
a[5]
a[c(3, 7, 11)]
```

```
A <- matrix(a, nrow = 4, ncol = 3)
A
A[1, 2]
A[c(1, 3), c(1, 2)]
A[1,]
A[,2]

dA <- as.data.frame(A)
dA
dA$V1
dA$V1[2]
dA[c("V1", "V2")]
dA[c(3, 4), c("V3")]
```

Output

```
> a <- 21:32
> a
 [1] 21 22 23 24 25 26 27 28 29 30 31 32
> a[5]
[1] 25
> a[c(3, 7, 11)]
[1] 23 27 31
>
> A <- matrix(a, nrow = 4, ncol = 3)
> A
     [,1] [,2] [,3]
[1,]   21   25   29
[2,]   22   26   30
[3,]   23   27   31
[4,]   24   28   32
> A[1, 2]
[1] 25
> A[c(1, 3), c(1, 2)]
     [,1] [,2]
[1,]   21   25
[2,]   23   27
> A[1,]
[1] 21 25 29
> A[,2]
[1] 25 26 27 28
>
> dA <- as.data.frame(A)
> dA
  V1 V2 V3
1 21 25 29
2 22 26 30
```

```
3 23 27 31
4 24 28 32
> dA$V1
[1] 21 22 23 24
> dA$V1[2]
[1] 22
> dA[c("V1", "V2")]
  V1 V2
1 21 25
2 22 26
3 23 27
4 24 28
> dA[c(3, 4), c("V3")]
[1] 31 32
```

Analysis

We first create vector a using, the colon operator. We use [5] to select the fifth element of it and then place three index numbers with c() inside [] to obtain three elements. Next we turn a into the matrix A. Matrix indexing requires two numbers between [] to specify rows and columns of the entries. The first position inside [] is for rows and the second for columns. In both positions, we can use c() to specify more than one row or column. If we put nothing behind or before the comma in matrix indexing we obtain the entries of all columns or all rows, respectively. Finally, we convert the matrix into a data frame with **as.data.frame()**. We call the columns of dA variables and we obtain single variables with the dollar sign operator $. The $ works even in combination with [] to get a certain entry from a variable. To obtain more than one variable from the data frame, use [] together with c() again, but now use the variable names with "". As can be seen, matrix indexing with two positions before and behind a comma can be combined with data frame indexing using the variable names. Find out for yourself which kind of indexing is most suitable for you and the given situation. Table 1.3 gives an overview of the options we just practised.

Table 1.3 Index operators for vectors, matrices, and data frames.

Index	Meaning
x[i]	Select element i from vector x
x[i, j]	Select element in row number i and column number j from matrix x
x[i,]	Select row number i from matrix x
x[,j]	Select column number j from matrix x
x$a	Select element a from list x by its name
x$a[i]	Select element i from element a from list x

1.3.5 names()

We have just seen that we can select variables of a data frame by name using the $-operator. However, in large data frames we often need an overview of the available variables. For data frames and other complex objects in R, **names()** returns the names of its elements. The $-operator then provides access to these elements. In the case of data frames, **names()** returns the variable names; in the case of a regression analysis, it returns information relevant for the analysis. Let us inspect some examples.

Ch1_names.R

```
## Inspect the elements of an R object with names()
set.seed(12345)
x <- rep(1:5, each = 20)
y <- x + rnorm(100)
d <- data.frame(x, y)
m <- lm(formula = y ~ x, data = d)

names(d)
names(m)
m$coefficients
names(summary(m))
```

Output

```
> names(d)
[1] "x" "y"

> names(m)
 [1] "coefficients"  "residuals"     "effects"       "rank"
"fitted.values"
 [6] "assign"        "qr"            "df.residual"   "xlevels"
"call"
[11] "terms"         "model"

> m$coefficients
(Intercept)           x
 0.09763797  1.04918641

> names(summary(m))
 [1] "call"          "terms"         "residuals"     "coefficients"
"aliased"
 [6] "sigma"         "df"            "r.squared"     "adj.r.squared"
"fstatistic"
[11] "cov.unscaled"
```

Analysis

First, we create two data vectors x and y and then aggregate them in data frame d. We further add the linear model object m consisting of the regression analysis of the outcome y on the predictor x. The three subsequent calls of **names()** show diverse results, depending on the type of object the function uses. For data frame d, **names()** returns the variable names, for the linear model m it returns a list of information important for the regression analysis. Use the **$**-operator to access the different items of this list, for example, m$coefficients. The last line shows how the available information changes if a call of **summary()** is nested inside **names()**. Whenever you encounter an unfamiliar object in R, use **names()** with the object and check whether it offers you information. Usually it will.

1.4 Getting help in R

To handle all R functions and add-on packages we need the help system. It is our everyday companion – even as experienced R users. Let us develop a strategy for how to use it.

R LANGUAGE! ### 1.4.1 Where to get help for R functions

With experience you will quickly increase your vocabulary of R functions. However, R comprises so many functions that you cannot expect to know them all. *For a given statistical problem you will often need help with how to use the functions.* Some statistical problems may even allow for different solutions, so that before each statistical analysis you would inquire about the different options. Let us practise different options to get help.

Ch1_FindingHelp.R

```
## Finding help in R
?mean
?Mean
help(plot)

??boxplot
??Mean
??"error bar"
```

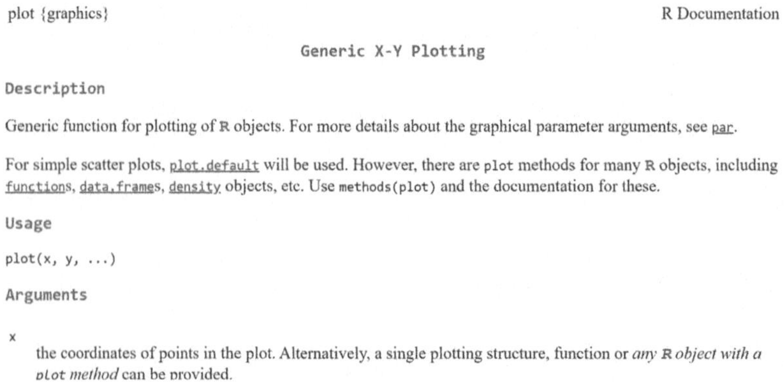

plot {graphics}   R Documentation

Generic X-Y Plotting

Description

Generic function for plotting of R objects. For more details about the graphical parameter arguments, see par.

For simple scatter plots, plot.default will be used. However, there are plot methods for many R objects, including functions, data.frames, density objects, etc. Use methods(plot) and the documentation for these.

Usage

plot(x, y, ...)

Arguments

x   the coordinates of points in the plot. Alternatively, a single plotting structure, function or *any R object with a plot method* can be provided.

Figure 1.6 Online help page for the function **plot()**.

The most straightforward situation which requires help: *you want to calculate something with a given function you know, but you do not remember the exact list of parameters.* In this situation, `?` together with the function name at the console helps. R or a web browser will then open the help page for the function. However, you must know the exact spelling of the function. For example, `?mean` works but `?Mean` does not. Next, a call of `help(plot)` gives details of the **plot()** function. Let us inspect this help page more closely. The description section gives an overview of the function and about situations when it is used. Below is the proper usage of the function. Between parentheses are the already familiar `x` and `y` parameters. The arguments section tells us that the first parameter, `x`, goes on the x-axis the second parameter, `y`, on the y-axis. *The three dots mean that* **plot()** *takes more arguments than are actually shown here and hands them down to further functions, which will be explained in detail later.* With this help page we found out that **plot()** usually requires two arguments `x` and `y` to be passed to the function.

To learn the details of an R function you should study the details of the help page. There are always some examples at the bottom of each help page. *Each R function has a help page with a standardised structure. Even experienced R programmers need the help pages because one cannot have all the details of a function in memory.* However, to use R efficiently you need to know the names of many functions from memory so that you can readily call their help pages.

Sometimes, the exact function name is not available to you, for example, **boxplot.matrix()***; then the double question marks ?? together with a single descriptive term may do it.* R then searches all functions in all installed packages for this keyword and displays the results in a new window. `??Mean` also works. The result is a list of function names and short descriptions. On the left, package and function name stand separated with the double colon operator, on the right is a short description of the function. Notice in the search results that the construction using the double colon operator also tells you how to quickly use the discovered functions. Simply use them at the console the same way as they are presented here. Now the correct function name is available to use with the `?` operator. Look also at some of the other results of `??Mean`. For example, the function **weighted.mean()** sounds interesting, which calculates the mean with use of the frequencies of the measurements. If a search term comprises two words with a blank in between, quotation marks work, like `??"error bar"`.

Other options to get help are to search the R website (www.r-project.org/) for relevant functions or to visit https://rseek.org/. *If you do not search for particular functions but rather for how to deal with a statistical problem, then the so-called task views on the R homepage offer information about which packages deal with particular topics, for example, automatic linguistic analyses or multivariate statistics.* The link CRAN on the main R page links to so-called cran mirrors, which host the task views and other R resources. A click on task views shows the list of statistical topics included. A click on multivariate gives much information about multivariate statistics in R.

The examples given on the help pages vary in quality. Some are quite comprehensive and demonstrate all options of the respective function. The help pages of the **plotrix** package are a good example. Simply copy and paste the example code to the console and hit return. However, some help pages offer only basic examples. With more experience you will find your favourite functions that you repeatedly use for your work. Functions, perhaps, which offer good examples, which you modify slightly to apply them in your project. Functions which do not offer comprehensive help you will simply avoid or neglect, because you will not have the time to figure it out yourself. That's perfect. Find your preferred functions and do not waste your time with the bad ones.

1.4.2 Finding help on the web

In your daily work with R you will spend much time searching for useful functions or packages. To be most successful, combine all search strategies. Several websites help you with the learning of R or with specific programming problems. R also offers mailing lists and special interest groups on www.r-project.org/mail.html. Finally, if you start to use R for your statistics, then you may cite it appropriately in your publication. Use the `citation()` function to view the reference information in bibtex format, which you can copy-paste into your bibliography software.

1.5 Teamwork and project work - Our lives as researchers

This chapter closes with a look at how project work with R may enrich your scientific endeavours. We look at how researchers work together and how R can foster your team productivity.

1.5.1 Share your R vocabulary and read code

R consists of a great number of powerful functions and it is impossible to know them all. The add-on packages supply a multitude of extra functions, each tailored to a specific problem. However, the parameter lists of functions for the same purpose may differ although the output of the functions is almost the same. So, how do you learn about the best functions for your statistical problems, their applicability, their merits, and weaknesses? *Talk about all the functions with your team members. Ask them which function they know is suitable for a given problem. Further, ask which functions they find difficult to use. Such talk about functions strengthens your vocabulary.* A good way to acquire vocabulary and to learn how others structure their analyses is to do code reading in the team. In a reading session you walk through the code together with your team without immediately executing the code. It helps you to assess whether your code is intelligible. *A practical and vast vocabulary of R functions, that every team member knows, enhances comprehensibility of shared code scripts.* You may even maintain a code repository for the team to establish a collective ownership of the code (Goodliffe, 2007).

1.5.2 Plan the analysis

Statistical programming leads you into unexpected complexities and consumes work time. Sometimes data frames have to be transformed in order to apply a certain function for analysis, which is often the case in repeated measures designs. Variables of the data frame sometimes have to be converted to another type to match the parameters requirements of a function. All these troubles arise with statistical programming in R. Even worse, the results that you eventually produce may not look like much effort to your professor or supervisor, if it is only a results table or a simple *p*-value. In fact, experience shows that supervisors are eager to promote you to get only the best out of the data and make suggestions for some further calculations, which may take another two weeks of your time to accomplish. *The point is, statistical programming requires a planning dialogue with your supervisor or with your team. It is best to agree on some programming tasks which cover all required statistical tasks and all desired results.* For example, you can initially agree about the tables or figures in a certain format for publication or presentation. Afterwards you can use your time to create only the required work products. This keeps you from playing around and optimising code where it is not necessary. *Finish your initial analysis quickly and present it to your supervisor or your team.* Usually, they will suggest changes to the calculations and you will be in charge of programming them. If you spend too much time with your initial presentation of the results, much of the time may appear wasted afterwards, because of changes or complete deletions of parts of the analysis. *In planning the analysis, put the research hypothesis or the research question at the beginning of the discussion and agree it with your supervisor or team. Rephrase the hypothesis or question until there is no room for divergent interpretation.* So, be most explicit, for example, about the inclusion criteria for your experimental subjects, about the dependent variables to analyse, and about the comparison groups and factors of influence. *Without such agreement you would start programming without the assurance that your time investment is reasonable.* So, avoid stopping the discussion before consensus.

1.5.3 How statistical software supports your statistical judgment

PLANNING ISSUE!

The use of statistical software like R is not the same as doing statistics. In the social sciences, scholars know about the statistical issues of data analysis. Consequently, the curricula nowadays consist not only of basic but also of elaborated statistical methods. In fact, data analysis not only requires selection of a statistic, but also arranging all calculations in a meaningful sequence. Statistical software is not involved in these decisions. The software only does the calculations. For example, you have to judge whether the suggested procedure of a significance test for statistical assumptions is tenable or if it would have been better to use some sort of graphical analysis. In fact, it is not really possible to select a statistical procedure and do adequate adjustments to your project if you have no knowledge of statistics. Harrell (2015) offers good advice and checklists, which includes your judgment as a researcher, before reasonable statistical modelling can happen.

Here are some examples of statistical judgement which rely on your personal statistical intuitions:

- Can you use Likert scale ratings with statistical methods for interval data?
- Can you calculate statistics if your test for non-normality in the data is significant?
- In a project with several variables and hypotheses what is the conceptual unit to use for alpha adjustment?
- Which evidence is acceptable as a truly unbiased estimation of a treatment effect?
- Is it better to use the .05 or .01 significance level?

You learn nothing about these issues when you learn a computer program. However, on your way to get a feeling for these statistical issues your knowledge of good computer programs can help you. *R supports your statistical judgement, because you can explore your data on all levels from raw data to the statistical test on a high level of data aggregation.* R can easily display views of the raw data to discover response sets in subjects' responses or bad item distributions. Furthermore, you can compare between different versions of a statistical analysis, for example, regarding different parameter settings in factor analysis. In summary, statistical software supports, but does not replace, your statistical judgment. Although it lets you quickly calculate analyses, you still have to plan it.

What you have learned so far

1   To start R and complete a first programming session at the R console
2   To use R functions and supply function parameters with arguments
3   To use add-on packages
4   To create and handle data frames and matrices
5   To calculate an analysis of variance in one line of code
6   To use R's help pages
7   To use R scripts for the functions of an analysis

Exercises

Explore functions

1   Use **rnorm()** to select data from the normal distribution. Use different arguments for the function parameters. How can you easily select from the standard normal distribution?

2   Use **rnorm()** to generate 100 data values. Supply a vector of two or three greatly varying numbers to the mean parameter, for example, c(5, 50, 500). What does this illustrate?

3   Use **matrix()** with 1:16 and parameter nrow = 4 to create a matrix containing the numbers from 1 to 16. Do this two times and switch the function parameter byrow between TRUE and FALSE. Observe the difference between the results.

4   Create two matrices A <- matrix(1:16, 4) and B <- matrix(1:8, 4). Use these two matrices to explore the functions **cbind()** and **rbind()**. Also use matrix indexing with **[]** to select subsets of the matrices and bind them together.

Solve tasks

1   Create a data frame d with **data.frame()**. The data frame should consist of a subject ID from 1 to 20, two experimental conditions a1 and a2, and random data from the normal distribution. Use the colon operator **:** to create the ID variable. Use **c()** to create a balanced experimental factor A with the levels a1 and a2. Remember to use "" when declaring the factor levels. Finally, use **rnorm()** and **rep()** to generate data from the normal distributions with means 5 and 7 for the two levels of A, respectively, and standard deviation of 2.

2   Turn the data frame of the previous task into a matrix with **as.matrix()**. Also use **data.matrix()** for this and observe the difference.

3   Create a 4 × 3 matrix **C** with entries of 0 or 1 in all of the twelve fields. Then assign as row names: "Condition 1", "Condition 2", "Condition 3", "Condition 4". Assign as column names: "Comparison 1", "Comparison 2", "Comparison 3". In analysis of variance, the matrices consisting of contrast coefficients look like this.

Read code

1   Fill in the gaps of the following code fragments. Use **rep()**, **seq()**, **()**, and **:** where appropriate.

```
> 1___15
 [1]  1  2  3  4  5  6  7  8  9 10 11 12 13 14 15
> ___(1,5,.3)
 [1] 1.0 1.3 1.6 1.9 2.2 2.5 2.8 3.1 3.4 3.7 4.0 4.3 4.6 4.9
> rep(1:2,5, ___)
 [1] 1 1 2 2 1 1 2 2 1 1 2 2 1 1 2 2 1 1 2 2 1 1 2 2
> (1 : 4 ^ 2
 [1]  1  4  9 16
>  1:4___
 [1]  1  2  3  4  5  6  7  8  9 10 11 12 13 14 15 16
```

2   Create the following vectors. Use **rep()** and **seq()** where appropriate.

```
[1] 1 2 1 2 1 2 1 2 1 2
[1] 1 1 1 1 1 2 2 2 2 2
[1] 1 3 5 7 9
[1] 0.0 0.2 0.4 0.6 0.8 1.0
[1] 1.00 1.25 1.50 1.75 2.00 1.00 1.25 1.50 1.75 2.00
[1] "a" "a" "a" "a" "b" "b" "b" "b" "c" "c" "c" "c"
```

Analyse data

1   Use the following code to create a data frame. Analyse the variables in the data frame with **summary()** and **table()**. Observe the differences between the outputs for categorical and quantitative variables.

```
d <- data.frame(A = rep(c("a1", "a2", "a3"), each = 20),
                B = rep(c("b1", "b2"), times = 30),
                Y = round(rnorm(60, 10, 2)))
```

Become a statistics programmer

1   Translate some statistical terms from your language into English. Search for a list of R functions associated with these terms using the **??** function. Then open the help file for some of the discovered functions. Suggestions: mean, standard deviation, factor analysis, sum of squares ...

2

GETTING YOUR DATA IN AND OUT OF R

━━━━━━━━━━━━━━━━━━━━━━━━━ Chapter overview ━━━━━━━━━━━

This chapter introduces how to load, import, and save data in various formats. It stresses that teamwork requires agreement among scientists on the data format used and on data archiving. Since quality data sets are expensive, you should, as a researcher, document and maintain your data to be useful for the research community. You will learn how to set up plain text files as the format of choice for the secure storage of data and data documentation. Programming features: R objects, working directory and workspace, efficient variable access, file format for saving R objects, file formats for data frames.

In your empirical research, you will mainly use R for data analysis. The research yields data sets comprising the experimental units (e.g., subjects) displayed in rows and the variables in columns. There are *multiple ways to load such data sets in R*. Especially difficult for beginners is that even if one loads the data it is still invisible unless one uses special functions for display. Therefore, we will first learn to import and then inspect the data.

2.1 Examples in this book

Let us now visit the run-through examples in this book – their research questions and analysis plans. The examples originate from different fields. All examples appear in the subsequent chapters to illustrate R programming. This application of our R knowledge in the real world will yield comprehensive analyses towards the end of the book.

EXAMPLE! ### 2.1.1 WBD - Web-based cognitive bias modification for depression

The first example is from clinical psychology and it is about the effectiveness of depression therapy. Depression is a disease with severe suffering of patients and an enhanced risk of suicide. The US Preventive Services Task Force recommends depression screening in the general adult population (U.S. Preventive Services Task Force, 2009), because it is one of the most prevalent diseases. Simon E. Blackwell and colleagues report evidence about the effectiveness of a web-based therapy approach to modify thinking biases in depressive patients (Blackwell et al., 2015). The study builds on the idea that depressive patients show a tendency to expect negative outcomes in daily events. So, promoting positive mental imagery in these patients may help to overcome this thinking bias. In a randomised controlled experiment the researchers compared an experimental condition of web-based cognitive bias modification with a control group with the following manipulation: in repeated sessions all patients were presented with ambiguous stories or pictures of everyday situations, which always resolved positively for the experimental group and positively or negatively for the control group. The results of the experiment, however, do not indicate a significant effect of cognitive bias modification on depression symptoms. Instead, a treatment effect on anhedonia was found. The study is an example of a randomised controlled trial, which is a standard procedure in clinical research to prove the effectiveness of treatments in all kinds of diseases. Its main methodological aim is to create an adequate pair of intervention group and control group to secure the internal validity of conclusions (see Kirk, 2012 for more about threats to valid inference making).

EXAMPLE! ### 2.1.2 ITC - Independence of thinking bias and cognitive ability

The next example is from general psychology. There is a temptation to assume that rational thinking is associated with cognitive ability; that is, smarter people think more rationally and show no thinking biases. Stanovich and West (2008) in a series of experiments all consisting of cognitive tasks demonstrated, however, that this association does not hold across the different tasks. The thinking biases tested in the tasks consisted of framing effects, anchoring effects, base-rate neglect, or certainty effects. One of the tasks employed was the famous *Wason card selection task* (Wason, 1966) which requires the participant to select two of four presented cards, which show a letter on one side and a number on the other, respectively (Figure 2.1). The selection is done in order to validate a given rule, say "if a vowel is printed on one side an even number must be printed on the other side". The Wason task is very popular in general and social psychology; its original results show that people are not especially good at the task, although it has an objective solution. In the given example, the rule can be validated by turning over the A and the 7 card. If the A shows a consonant on the other side, the rule is invalidated, and also if the 7 shows a vowel on the other side. Although experimental subjects consistently turn over the A, they usually do not turn over the 7 but the 4. It is as if subjects somehow prefer to find examples which validate a rule rather than those to invalidate it.

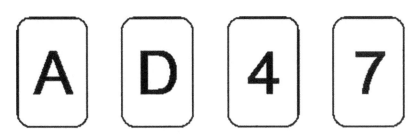

Figure 2.1 Example of the Wason card selection task.

The experiment by Stanovich et al. relating to the Wason task was part of the reproducibility project instigated in the psychology community (Open Science Collaboration, 2015a). The idea of the reproducibility project was to demonstrate whether the well known effects of psychological science truly reproduce, or if they are just random findings which do not originate from any true effect. Within the reproducibility project, Baranski reproduced the Wason experiment and could not fully confirm the previous findings. However, the author reports that there were some aspects relating to the measurement of cognitive ability which differed between studies. The data set of the experimental replication by Baranski is one of the run-through examples here. It is publicly available for download at the open science framework (www.osf.io) and, thus, invites us to replicate the statistical calculations.

2.1.3 UPS - Unpleasant sounds   EXAMPLE!

Scratching your finger nails on a chalk board is one of the most unpleasant sounds. It sends cold shivers down the spine and leaves some people in pure horror. In an experiment, Reuter and Oehler (2011; Reuter, Oehler, & Mühlhans, 2014) compared different candidates of unpleasant sounds and asked their participants to judge several different variations of the sounds for unpleasantness. The experiment consisted of the factor 'quality of the sound', like scratching on a chalk board, vomiting, or styrofoam. Another factor was the applied sound filter for the sound, like different band-pass filters. Further variables were age and sex of the participants. The experiment included presenting all participants repeatedly with all conditions, thus creating 28 unpleasantness ratings per person. As a run-through example, we will recalculate and extend the results of this study.

2.1.4 SCB - Conflict reconciliators against school bullying

Before we go to the example, think about how it would be (1) to conjure up an experiment first, then (2) to simulate the experiment with random data and some good guesses about design parameters, then (3) to see how the experimental design behaves in terms of statistics, then (4) to adjust the experiment to make it better, for example, more sensitive, and finally (5) to conduct the experiment in the real world. In the simulation you can familiarise yourself with the prospective data, discuss all options for data analysis, and fine-tune the analysis before a true commitment to the execution of the project and investment of plenty of resources is made. Let us see how this works in our next example. Although the example is hypothetical, its topic is serious.

The problem of bullying in schoolyards challenges our vision of how we can offer secure environments for children and adolescents. Provocations as well as bodily injuries are common (Ahmed & Braithwaite, 2006). Enhancing attendance by teachers is not a viable solution due to scarce budget. Another option involves school students themselves becoming conflict reconciliators or mediators and thus providing conflict reconciliation for their peers. The training curriculum would consist of negotiation skills for interpersonal or intergroup conflicts. For example, students could learn how to appropriately convey the negative consequences of violence to aggressors in a conflict situation.

A randomised experiment would prove the effectiveness of the training. Schools would be randomly allocated to three conditions (N = 180): 60 schools act as a control condition and receive no treatment, 60 schools double the attendance of teachers in the schoolyards, and 60 schools appoint trained student conflict reconciliators. A subdivision of the 60 schools of each condition yields groups of 20 primary schools, secondary schools, and schools of further education, respectively. The project duration is four months. The first two months consist of no intervention and act as a baseline for the measurement of bullying incidents among school students. The primary outcome of the experiment is the count of incidents in a given month. It may be that conflict reconciliators are more effective in one type of bullying than in others. Therefore, the design distinguishes three types of incidents: (1) one person against one other person, (2) several persons against one person, and (3) several persons against several other persons. The aim of the statistical analysis will be to test whether schools employing conflict reconciliators show smaller counts of bullying incidents than schools without intervention, and are comparable to schools with higher teacher attendance in the schoolyard.

2.2 Load and import research data

We now proceed to data import. The data sets in real life come in diverse formats. So, we study different import functions and see how parameter variations in these functions allow for fine-tuning.

2.2.1 R working directory and workspace

Before going to the data import, we need to distinguish two types of things: objects and files. Objects in R are not files but instances of classes, as we saw earlier. In contrast, files can store R objects or data and they reside on the hard drive or some other storage medium.

Usually, you create a so-called working directory associated with your analysis. *A working directory is a directory in the file system, for example, an ordinary folder in Windows, which is designated in R to hold all required files and store all results files for a given statistical analysis.* Any directory in the file system can become the working directory and different analyses or projects can have different working directories. At the beginning of each R session you should select a directory with the menu option 'change directory' in the file menu. The function **setwd()** for *set working directory* does the same. The function **getwd()** asks for the current working directory.

To prepare for a statistical analysis it is helpful to create a project directory with an ordinary file explorer and move all required files to that directory. Most often, you would put a data frame in it, but sometimes also plain text files holding variable names, variable labels, and value labels exist here. If this directory is then declared as the working directory in the analysis session, R can easily load all files stored here. Functions like **load()**, **read.csv()**, **foreign::read.spss()**, or **read.lines()** operate in the directory to bring the files to the R session. Other functions like **save()**, **write.csv()**, **capture.output()**, **pdf()**, **png()** write results into files in the working directory.

Furthermore, it is a good idea to use a certain fixed structure of subfolders and folder names within the working directory. The research team can agree about it and maintain it across projects. This makes it easier to get an overview about the available files, and every co-worker in the team can make good guesses about where to find a certain file. One structure of subfolders could be: (1) Data, (2) Scripts, (3) Results, (4) Figures. Every file easily finds its place in the right folder and the structure is quite simple. All file operations that we will explore would then access the correct subfolder. In sum, a usual R session sets the working directory at the beginning, loads data frames and other objects from files in the directory, and finally writes back the results to other files. Establishing a fixed folder structure for all projects is helpful and you should think about your own preferences for this.

2.2.2 Options to load and import data into R R LANGUAGE!

There are at least *six ways to load, import, or acquire data.*

- load data frame stored in an .RData file with function **load()**,
- enter data by hand in a spread sheet with Data <- edit(as.data.frame(NULL)),
- import data from an SPSS .sav file with **foreign::read.spss()**,
- import data in comma-separated values from a .csv file with **read.csv()**,
- read a data frame from the clipboard with **psych::read.clipboard()**,
- use an installed data set with **data()**.

The most common options are to *import a data set from a csv file or to import from another statistics program*. Both options are highly useful in team projects because one team member can enter and edit data in his or her preferred file format and all other team members can load the data for analysis. Not for research projects but for quick and dirty work with small data frames from lectures or textbooks is the **read.clipboard()** function in the **psych** package. Simply copy the data table from any application to the clipboard and then use this function to paste it in a data frame in R.

2.2.3 read.csv()

Files in the csv format are plain text files and can be edited with any text editor. However, they have a characteristic structure which is the only difference from txt-files. That is, csv-files are structured like a data set: each experimental unit occupies one line in the file and all its characteristics or measurements are stored in this line, all separated with commas or semicolons. In fact, csv means character-separated values, which means that a certain designated character, usually a comma, stands between adjacent measurements. Statistics programs can easily handle this structure: they only have to be told what the character for separation of the measurements is and can then import the data. In R we use **read.csv()** for this purpose, and the function consists of a parameter sep which is used to designate the character used for separation of the measurements in the file.

Let us try a small example. Open the text editor of your operating system and type in the following small data set. Be careful to store nothing else in the file except the data. Save it with the file name MyData.csv. Then start R and change the working directory to the folder containing the csv file. Finally, run the R code at the console.

MyData.csv

```
Subject, FactorA, FactorB, Response
1, 1, 1, 5
2, 1, 2, 4
3, 2, 1, 3
4, 2, 2, 5
```

Ch2_readcsv.R

```
## Read character-separated values
d <- read.csv("MyData.csv")
View(d)
str(d)
attributes(d)
```

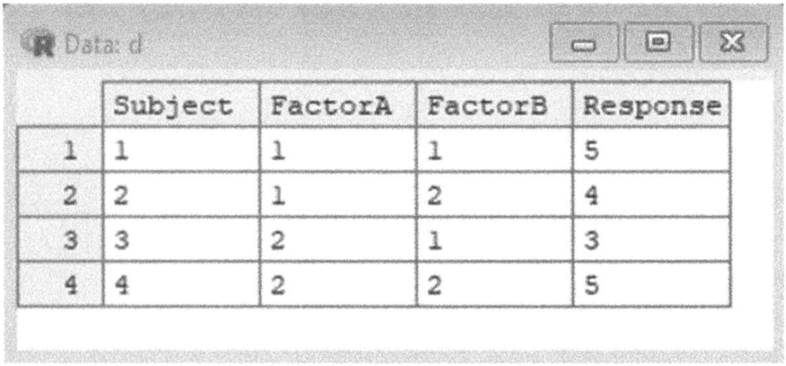

	Subject	FactorA	FactorB	Response
1	1	1	1	5
2	2	1	2	4
3	3	2	1	3
4	4	2	2	5

Figure 2.2 View of the data frame after import with **read.csv()**.

Output

```
> str(d)
'data.frame':   4 obs. of 4 variables:
 $ Subject : int  1 2 3 4
 $ FactorA : int  1 1 2 2
 $ FactorB : int  1 2 1 2
 $ Response: int  5 4 3 5

> attributes(d)
$names
[1] "Subject" "FactorA" "FactorB" "Response"

$class
[1] "data.frame"

$row.names
[1] 1 2 3 4
```

Analysis

After data import with **read.csv()**, the data set is ready for statistics in R. We first inspect the data and then see that there are no attributes stored with the data frame. In fact, everything that is available was stored in the csv file. Other helpful functions to import data are **read.table()** and **scan()**. Chapter 3 demonstrates how to enrich the data frame with additional information like variable labels or value labels.

2.2.4 SCB – Data import from a csv file   EXAMPLE!

The simulation data for the project about student reconciliators against school bullying comes as a csv file. Any text editor can display the file (Figure 2.3). So, the following shows

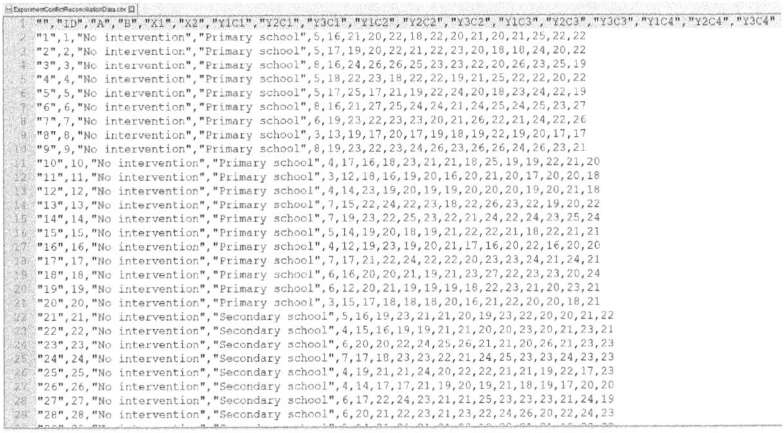

Figure 2.3 Data set for the SCB project in csv format as viewed with a text editor.

Source: Screenshot from the Notepad++ Software

the script to import this data in R. Note that it is OK that the script executes only one task: data import. Subsequent scripts for statistical analyses will all include this script. For the analysis with this example, create a folder SCB on your computer system. Set this folder as the working directory using the menu option "Change dir..." in the R program or **setwd()** together with the folder path. Further create four subfolders in SCB: Data, Script, Figures, and Results. Put the data set **ExperimentConflictReconciliationData.csv** in the Data subfolder, so it can be found by the **read.csv()** function. Place the import script and all subsequent scripts of this example in the Script subfolder.

SCB_PRE_DataImport.R

```
## SCB - data import from a csv file
if(!exists('SCB'))
{
  SCB                                                                    <-
read.csv("./Data/ExperimentConflictReconciliationData.csv",
header = TRUE)
}
```

Output

```
View(SCB)
str(SCB)
attributes(SCB)
```

	X	ID	A	B	X1	X2	Y1C1	Y2C1	Y3C1	Y1C2	Y2C2	Y3C2	Y1C3	Y2C3	Y3C3
1	1	1	No intervention	Primary school	5	16	21	20	22	18	22	20	21	20	
2	2	2	No intervention	Primary school	5	17	19	20	22	21	22	23	20	18	
3	3	3	No intervention	Primary school	8	16	24	26	26	25	23	23	22	20	
4	4	4	No intervention	Primary school	5	18	22	23	18	22	22	19	21	25	
5	5	5	No intervention	Primary school	5	17	25	17	21	19	22	24	20	18	
6	6	6	No intervention	Primary school	8	16	21	27	25	24	24	21	24	25	
7	7	7	No intervention	Primary school	6	19	23	22	23	23	20	21	26	22	
8	8	8	No intervention	Primary school	3	13	19	17	20	17	19	18	19	22	
9	9	9	No intervention	Primary school	8	19	23	22	23	24	26	23	26	26	
10	10	10	No intervention	Primary school	4	17	16	18	23	21	21	18	25	19	
11	11	11	No intervention	Primary school	3	12	18	16	19	20	16	20	21	20	
12	12	12	No intervention	Primary school	4	14	23	19	20	19	19	20	20	20	
13	13	13	No intervention	Primary school	7	15	22	24	22	23	18	22	26	23	
14	14	14	No intervention	Primary school	7	19	23	22	25	23	22	21	24	22	
15	15	15	No intervention	Primary school	5	14	19	20	18	19	21	22	22	21	
16	16	16	No intervention	Primary school	4	12	19	23	19	20	21	17	16	20	
17	17	17	No intervention	Primary school	7	17	21	22	24	22	22	20	23	23	

Showing 1 to 17 of 180 entries, 18 total columns

Figure 2.4 Data set of the SCB project after data import in R.

Source: RStudio and Shiny are trademarks of RStudio, PBC

```
> str(SCB)
'data.frame':   180 obs. of   18 variables:
 $ X   : int  1 2 3 4 5 6 7 8 9 10 ...
 $ ID  : int  1 2 3 4 5 6 7 8 9 10 ...
 $ A   : Factor w/ 3 levels "Conflict reconciliators",..: 3 3 3
3 3 3 3 3 3 3 ...
 $ B   : Factor w/ 3 levels "Further education",..: 2 2 2 2 2 2
2 2 2 2 ...
 $ X1  : int  5 5 8 5 5 8 6 3 8 4 ...
 $ X2  : int  16 17 16 18 17 16 19 13 19 17 ...
 $ Y1C1: int  21 19 24 22 25 21 23 19 23 16 ...
 $ Y2C1: int  20 20 26 23 17 27 22 17 22 18 ...
 $ Y3C1: int  22 22 26 18 21 25 23 20 23 23 ...
 $ Y1C2: int  18 21 25 22 19 24 23 17 24 21 ...
 $ Y2C2: int  22 22 23 22 22 24 20 19 26 21 ...
 $ Y3C2: int  20 23 23 19 24 21 21 18 23 18 ...
 $ Y1C3: int  21 20 22 21 20 24 26 19 26 25 ...
 $ Y2C3: int  20 18 20 25 18 25 22 22 26 19 ...
 $ Y3C3: int  21 18 26 22 23 24 21 19 24 19 ...
 $ Y1C4: int  25 24 23 22 24 25 24 20 26 22 ...
 $ Y2C4: int  22 20 25 20 22 23 22 17 23 21 ...
 $ Y3C4: int  22 22 19 22 19 27 26 17 21 20 ...
> attributes(SCB)
$names
 [1] "X"    "ID"    "A"    "B"    "X1"    "X2"    "Y1C1" "Y2C1" "Y3C1"
[10] "Y1C2" "Y2C2" "Y3C2" "Y1C3" "Y2C3" "Y3C3" "Y1C4" "Y2C4" "Y3C4"

$class
[1] "data.frame"

$row.names
  [1]   1   2   3   4   5   6   7   8   9  10  11  12  13  14  15  16
 [17]  17  18  19  20  21  22  23  24  25  26  27  28  29  30  31  32
 [33]  33  34  35  36  37  38  39  40  41  42  43  44  45  46  47  48
 [49]  49  50  51  52  53  54  55  56  57  58  59  60  61  62  63  64
 [65]  65  66  67  68  69  70  71  72  73  74  75  76  77  78  79  80
 [81]  81  82  83  84  85  86  87  88  89  90  91  92  93  94  95  96
 [97]  97  98  99 100 101 102 103 104 105 106 107 108 109 110 111 112
[113] 113 114 115 116 117 118 119 120 121 122 123 124 125 126 127 128
[129] 129 130 131 132 133 134 135 136 137 138 139 140 141 142 143 144
[145] 145 146 147 148 149 150 151 152 153 154 155 156 157 158 159 160
[161] 161 162 163 164 165 166 167 168 169 170 171 172 173 174 175 176
[177] 177 178 179 180
```

Analysis

The script only imports the data set and stores it in the data frame object SCB, our abbreviation for the project. At the beginning of the script is a so-called inclusion guard, which ensures that the data frame is only imported once in an analysis session. The logic behind this is: if the object SCB already exists in the global environment (i.e. your workspace), then skip the whole script from execution. If SCB is not already available, the script executes. After data import, a few lines at the console help us inspect the data (Figure 2.4). As was the case before, the data frame consists only of the naked data set – no beautiful variable names and no value labels.

R LANGUAGE! ## 2.2.5 Data import into R from other statistics programs

Many data sets in psychology and other social or clinical sciences come in the sav format of SPSS®. Import of data sets from the SPSS format into R is easy, although there are some traps along the way, which will bother us. Let us import the following small SPSS data set and test some of the available options.

SPSS_Data.sav

Name	Type	Width	Dec...	Label	Values	Missing	Columns	Align	Measure	Role
A	String	8	0	Factor A	None	None	8	Left	Nominal	Input
B	Numeric	8	2	Factor B	{1,00, b1}...	None	8	Right	Nominal	Input
X	Numeric	8	2	Variable X	None	None	8	Right	Scale	Input
Y	Numeric	8	2	Variable Y	{1,00, definitely no}...	None	8	Right	Scale	Input

		A	B	X	Y
1		a1	1,00	1,00	2,00
2		a1	2,00	3,00	1,00
3		a2	1,00	5,00	6,00
4		a2	2,00	6,00	4,00
5					

Figure 2.5 SPSS data set.

Source: Reprint courtesy of IBM Corporation ©

Ch2_SPSS_DataImport.R

```
## SPSS data import
library(foreign)

# Import data
d1 <- read.spss("SPSS_Data.sav",
                use.value.labels = TRUE,
                to.data.frame = TRUE)
d2 <- read.spss("SPSS_Data.sav",
                use.value.labels = FALSE,
                to.data.frame = TRUE)
```

```
# View data
d1
d2

# View additional information
str(d1)
str(d2)
```

Output

```
> # View data
> d1
        A  B X              Y
1 a1      b1 1             no
2 a1      b2 3  definitely no
3 a2      b1 5 definitely yes
4 a2      b2 6      rather yes

> d2
        A B X Y
1 a1      1 1 2
2 a1      2 3 1
3 a2      1 5 6
4 a2      2 6 4

> # View additional information
> str(d1)
'data.frame':  4 obs. of  4 variables:
 $ A: Factor w/ 2 levels "a1       ","a2       ": 1 1 2 2
 $ B: Factor w/ 2 levels "b1","b2": 1 2 1 2
 $ X: num  1 3 5 6
 $ Y: Factor w/ 6 levels "definitely no",..: 2 1 6 4
 - attr(*, "variable.labels")= Named chr  "Factor A" "Factor B"
"Variable X" "Variable Y"
   ..- attr(*, "names")= chr  "A" "B" "X" "Y"
 - attr(*, "codepage")= int 65001

> str(d2)
'data.frame':  4 obs. of  4 variables:
 $ A: Factor w/ 2 levels "a1       ","a2       ": 1 1 2 2
 $ B: num  1 2 1 2
   ..- attr(*, "value.labels")= Named chr  "2" "1"
   .. ..- attr(*, "names")= chr  "b2" "b1"
 $ X: num  1 3 5 6
 $ Y: num  2 1 6 4
```

```
    ..- attr(*, "value.labels")= Named chr  "6" "5" "4" "3" ...
    .. ..- attr(*, "names")= chr  "definitely yes" "yes" "rather
yes" "rather no" ...
  - attr(*, "variable.labels")= Named chr  "Factor A" "Factor B"
"Variable X" "Variable Y"
    ..- attr(*, "names")= chr  "A" "B" "X" "Y"
 - attr(*, "codepage")= int 65001
```

Analysis

Use SPSS to create the small data set shown in Figure 2.5. Save it with the file name SPSS_Data. sav in the working directory of your R session. Then run the code lines at the console. We use **read.spss()** from the **foreign** package and apply different arguments for the parameter use.value.labels. In the first call we set this parameter TRUE; in the second call we set it FALSE. After data import we inspect the two data frames at the console. As can be seen in d1, the value labels set in SPSS for the variables B and Y are used as data entries in R. If use. value.labels is set FALSE as in d2, R uses the numeric entries for the two variables. Variable A is a string in SPSS and is not affected by the parameter setting. The two calls of **str()** with the data frames reveals that d2, nevertheless, comprises the value labels of B and Y. They are attached to B and Y as the value.labels attribute. In sum, data import from SPSS is easy. Beware, however, of how you want to handle value labels for numeric variables.

EXAMPLE! ## 2.2.6 UPS - Data import from an SPSS file

The data set of the project unpleasant sounds comes in the sav format and requires data import with **foreign::read.spss()**. To start with this example, create a folder, UPS, in your file system. Use the menu option "Change dir..." or **setwd()** to make this folder the current directory in R. In this directory, create the following subfolders: Data, Script, Figures, and Results. Put the data set UnpleasantSounds.sav in the Data subfolder and the following script in the Script subfolder.

UPS_PRE_DataImport.R

```
## UPS - Data import
if(!exists('UPS'))
{
  UPS <- foreign::read.spss("./Data/UnpleasantSounds.sav",
                            to.data.frame = TRUE,
                            use.value.labels = TRUE)
}
```

Analysis

As shown, the script calls **foreign::read.spss()**, which includes **read.spss()** from the **foreign** package with the double colon operator. Although the package is already installed with the R program, it has to be explicitly included in the code. The script also uses an inclusion guard made of **if()** and **exists()** to protect against repeated execution of the script if the data frame has already been created. The line reads like this: If the object UPS does not (!) exist in the workspace, execute the following lines of code. And these following lines actually create the object UPS.

If you have access to SPSS, look at the data in this program. There, the data frame shows that it contains numbers and each number has a value label shown in the variable view. In contrast, after data import in R, and a call of View(UPS), R uses the value labels in the data display.

2.2.7 WBD - Data import from an SPSS file

EXAMPLE!

Before the data import, we look at the paper of Blackwell et al. (2015) to get an idea of what the data structure is about. From the paper we can already learn many aspects of the data structure and what to expect in the raw data set. Then, we use **foreign::read.spss()** to import the data set from the SPSS sav format to R. There are a few options for parameter settings, which change the representation of the data set in R as a data frame, and a little experimentation reveals what suits best for the actual purpose. The first thing to try is the simple use of the function with the only parameter to.data.frame = TRUE. To run the following script, create a project folder WBD in your file system. Set the folder as working directory in R with the menu option "Change dir..." or **setwd()** using the location where the folder is stored on your system. In the WBD folder, create four subfolders to structure the analysis: Data, Scripts, Results, and Figures. Place the data files Blackwelletal2015BDIallitemsalltimepoints.sav and Blackwelletal2015CPSDatabase_20160927.sav for this example in the Data folder. Place the following script and all subsequent scripts for this example in the Scripts folder.

WBD_PRE_DataImport.R

```
## WBD - Data import from SPSS
if(!exists("WBD"))
{
  library(foreign)
  WBD <- list(BDI = read.spss("./Data/Blackwelletal2015BDIallitemsalltimepoints.
sav",
                              to.data.frame = TRUE),
              CPS = read.spss("./Data/Blackwelletal2015CPSDatabase_20160927.
sav",
                              to.data.frame = TRUE))
}
```

Analysis

To import an sav file from SPSS, the script includes the **foreign** package with the **library()** function. The script creates the list WBD collecting the two available data sets from the project. We call **read.spss()** two times to import and store them in the data frame objects BDI and CPS. Do not worry too much about the expression if(!exists("WBD")); it is the inclusion guard that prevents repeated execution of the data import by other scripts.

EXAMPLE! ## 2.2.8 ITC - Data import

Finally, we import the data for the independence of thinking bias study. The project offers three data sets in different formats and each containing different information. Let us import these data sets in one list of three data frames. Before you run the following script, create a folder ITC as the working directory for this example. Use the menu option "Change dir..." in R to declare ITC as working directory. In this folder, create four subfolders to structure the analysis: Data, Scripts, Figures, and Results. Place the data files `stanovich_data_used.sav`, `Stanovich_raw_data.csv`, and `Stanovich_clean_data.xls` in the Data subfolder. Place the following script in the Scripts subfolder.

ITC_PRE_LoadData.R

```
## ITC - Data import
## Imports three versions of the data set: raw data,
## clean data, and used data

if(!exists('ITC'))
{
  # Data import
  ITC <- list(stvu = foreign::read.spss("./Data/stanovich_data_used.sav",
                                 to.data.frame = TRUE),
          stvr = read.csv("./Data/Stanovich_raw_data.csv",
                          skip = 1),
          stvc = readxl::read_excel("./Data/Stanovich_clean_data.xls"))
}
```

Analysis

Three data sets are available. We import them in data frames. In turn, we put these data frames in a list ITC. I favour this option instead of having three independent data frames because it does not clutter the workspace with too many objects. However, each access to the data frame `stvu` requires to index it in the list ITC, for example `ITC$stvu`.

2.3 Data scrutiny

After data import, the data is hidden in an R object. We need some functions to obtain a data display and access the information that accompanies the data.

R LANGUAGE! ### 2.3.1 Quick and dirty overview of the data at the R console

After data import, it is often a good idea to explore many aspects of the data at the console. This gives a feeling for the data, the variables, and their distributions. If I want to figure out a thing relating to the inner life of a complex R object, usually a data frame, I fire ten or twenty information functions at the thing and see how it behaves; some of these are names(), str(), attributes(), View(), class(), is(), unlist(), print(), plot(), summary(), and some others. The following is an example of a console scripting session, to get a grip on the data. The session does not figure in the final analysis; it is a live performance, which is fun if done in a team session. The following console session shows with dummy data how to simply apply many functions to gain insight into the properties of a data frame.

Ch2_QuickDataOverview.R

```
## Quick data overview
d <- data.frame(ID = c("s1", "s2", "s3", "s4"),
                A = c("a1", "a2", "a3", NA),
                X1 = c(1, 2, 3, 4),
                X2 = c(5, 6, "a", 8),
                X3 = c(9, 10, 11, 12))

ls()
View(d)
names(d)
str(d)
attributes(d)
attributes(d$X2)
plot(d$X1, d$X2)

summary(d)
Hmisc::describe(d)
Hmisc::contents(d)
psych::describe(d)
```

	ID	A	X1	X2	X3
1	s1	a1	1	5	9
2	s2	a2	2	6	10
3	s3	a3	3	a	11
4	s4	NA	4	8	12

Figure 2.6 Quick data overview with View().

Source: RStudio and Shiny are trademarks of RStudio, PBC

Output

```
> ls()
[1] "d"
```

```
> View(d)

> names(d)
[1] "ID" "A"  "X1" "X2" "X3"

> str(d)
'data.frame':  4 obs. of  5 variables:
 $ ID: Factor w/ 4 levels "s1","s2","s3",..: 1 2 3 4
 $ A : Factor w/ 3 levels "a1","a2","a3": 1 2 3 NA
 $ X1: num  1 2 3 4
 $ X2: Factor w/ 4 levels "5","6","8","a": 1 2 4 3
 $ X3: num  9 10 11 12

> attributes(d)
$names
[1] "ID" "A"  "X1" "X2" "X3"

$class
[1] "data.frame"

$row.names
[1] 1 2 3 4


> attributes(d$X2)
$levels
[1] "5" "6" "8" "a"

$class
[1] "factor"


> plot(d$X1, d$X2)

> summary(d)
   ID        A            X1         X2          X3
 s1:1    a1  :1    Min.   :1.00    5:1    Min.   : 9.00
 s2:1    a2  :1    1st Qu.:1.75    6:1    1st Qu.: 9.75
 s3:1    a3  :1    Median :2.50    8:1    Median :10.50
 s4:1    NA's:1    Mean   :2.50    a:1    Mean   :10.50
                   3rd Qu.:3.25           3rd Qu.:11.25
                   Max.   :4.00           Max.   :12.00

> Hmisc::describe(d)
d
```

```
5  Variables     4  Observations
---------------------------------------------------------------------------
--------------------
ID
      n  missing distinct
      4        0        4

Value         s1    s2    s3    s4
Frequency      1     1     1     1
Proportion  0.25  0.25  0.25  0.25
---------------------------------------------------------------------------
--------------------
A
      n  missing distinct
      3        1        3

Value          a1     a2     a3
Frequency       1      1      1
Proportion  0.333  0.333  0.333
---------------------------------------------------------------------------
--------------------
X1
      n  missing distinct    Info    Mean     Gmd
      4        0        4       1     2.5   1.667

Value          1     2     3     4
Frequency      1     1     1     1
Proportion  0.25  0.25  0.25  0.25
---------------------------------------------------------------------------
--------------------
X2
      n  missing distinct
      4        0        4

Value          5     6     8     a
Frequency      1     1     1     1
Proportion  0.25  0.25  0.25  0.25
---------------------------------------------------------------------------
--------------------
X3
      n  missing distinct    Info    Mean     Gmd
      4        0        4       1    10.5   1.667

Value          9    10    11    12
Frequency      1     1     1     1
```

```
Proportion 0.25 0.25 0.25 0.25
--------------------------------------------------------------------------------
--------------------

> Hmisc::contents(d)

Data frame:d   4 observations and 5 variables   Maximum # NAs:1


     Levels Storage NAs
ID        4 integer   0
A         3 integer   1
X1          double   0
X2        4 integer   0
X3          double   0


+--------+-----------+
|Variable|Levels     |
+--------+-----------+
|   ID   |s1,s2,s3,s4|
+--------+-----------+
|   A    |a1,a2,a3   |
+--------+-----------+
|   X2   |5,6,8,a    |
+--------+-----------+

> psych::describe(d)
      vars n mean    sd median trimmed  mad min max range skew kurtosis   se
ID*     1 4  2.5 1.29    2.5     2.5 1.48   1   4     3    0    -2.08 0.65
A*      2 3  2.0 1.00    2.0     2.0 1.48   1   3     2    0    -2.33 0.58
X1      3 4  2.5 1.29    2.5     2.5 1.48   1   4     3    0    -2.08 0.65
X2*     4 4  2.5 1.29    2.5     2.5 1.48   1   4     3    0    -2.08 0.65
X3      5 4 10.5 1.29   10.5    10.5 1.48   9  12     3    0    -2.08 0.65
```

Analysis

After data creation, we apply several functions to the data frame d. A certain order of the functions is not necessary at this point. We simply apply functions that we feel would give us interesting information. However, we should cover some topics of particular interest. First, **ls()** shows that the data frame d is indeed available for us. It is natural to display d with **View()** and its variable names with **names()**. Then, **str()** shows us detailed information about the variables in the data, for example, the location of the independent and dependent variables. Here we notice that although X1, X2, and X3 were all meant to be numeric variables, X2 is characterised as a factor. Sure, this is because of the "a" value in the data vector, which lets R treat all entries as strings. Next, **attributes()** also provides information about the data frame, which we will discuss in more detail in Chapter 3. Notice that the attributes of a data frame are different from the attributes of a variable comprised in the data frame. For quantitative variables, also call **plot()** to display the scatter plot of their joint distribution. Finally, the script uses some functions for a quick statistical overview of the data. Besides **summary()**, the functions **describe()** from the **Hmisc** and the **psych** packages show helpful information.

2.3.2 Variable access for a quick data overview in R R LANGUAGE!

To access variables in a data frame, we use the **$**-operator or the square brackets **[]**. Now, we add three handy methods to make access even more efficient. These are (1) to use **attach()** to make all variables of the data frame directly accessible at the console, (2) to use **paste()** to access many variables by constructing a string vector of their names, and (3) to use **grep()** to access all variables which consist of a particular string pattern in their names. Let us try it with a short script to create some data and a subsequent console session to practise variable access.

Ch2_VariableAccess_DataCreation.R

```
## Variable Access - Data creation
d <- data.frame(ID = c(1, 2, 3, 4),
                t0_ANX_1 = c(3, 4, 3, 2),
                t0_ANX_2 = c(3, 3, 5, 4),
                t0_DEP_1 = c(2, 4, 1, 5),
                t0_DEP_2 = c(4, 4, 4, 2),
                t1_ANX_1 = c(4, 3, 2, 2),
                t1_ANX_2 = c(5, 3, 4, 5),
                t1_DEP_1 = c(1, 3, 4, 3),
                t1_DEP_2 = c(3, 4, 4, 5))
```

Ch2_VariableAccess.R

```
## Access to variables in a data.frame
# Available variables in data frame
names(d)

# Access to single variables
d$t0_ANX_1
d["t0_ANX_1"]

# Direct access with attach()
t0_ANX_1 ## Error
attach(d)
t0_ANX_1
detach(d)
t0_ANX_1 ## Error
```

```
# Access to multiple variable with paste() and grep()
d[paste("t0_ANX", 1:2, sep="_")]
d[grep(pattern = "DEP", names(d))]
```

Output

```
> # Available variables in data frame
> names(d)
[1] "ID"       "t0_ANX_1" "t0_ANX_2" "t0_DEP_1" "t0_DEP_2"
"t1_ANX_1" "t1_ANX_2" "t1_DEP_1" "t1_DEP_2"
>
> # Access to single variables
> d$t0_ANX_1
[1] 3 4 3 2
> d["t0_ANX_1"]
  t0_ANX_1
1        3
2        4
3        3
4        2
>
> # Direct access with attach()
> t0_ANX_1 ## Error
Error: object 't0_ANX_1' not found
> attach(d)
> t0_ANX_1
[1] 3 4 3 2
> detach(d)
> t0_ANX_1 ## Error
Error: object 't0_ANX_1' not found
>
> # Access to multiple variable with paste() and grep()
> d[paste("t0_ANX", 1:2, sep="_")]
  t0_ANX_1 t0_ANX_2
1        3        3
2        4        3
3        3        5
4        2        4
> d[grep(pattern = "DEP", names(d))]
  t0_DEP_1 t0_DEP_2 t1_DEP_1 t1_DEP_2
1        2        4        1        3
2        4        4        3        4
3        1        4        4        4
4        5        2        3        5
```

Analysis

The script creates some data in a data frame d. The data frame consists of an ID variable for the participants and eight clinical variables as they could be used in real research. Notice that the naming of the variables follows rules. That is, the abbreviations ANX and DEP could stand for anxiety and depression variables, respectively, t0 and t1 could indicate time point of the measurement, and the number in each variable name could indicate that the study uses more than one anxiety or depression measurement at each time point, as is common research practice. We will now take advantage of these naming rules when we access the variables in the console session.

The console session starts with **names()** to remind us about the available variables. We then use **$** and **[]** to access one variable the way you already know. Next, we enter a variable name directly in the console and confirm that this produces an error, because R does not know where to find the variable – it is hidden inside d. In the next step, calling attach(d) makes all variables of d directly accessible; that is, we can simply use their names at the console to get them. After we do that once with t0_ANX_1, we use **detach()** to hide the variable again and, indeed, t0_ANX_1 is not accessible any more. Next, we use **paste()** and **grep()** in the session with the aim of easily returning more than one variable. The **paste()** function concatenates the common first part of the two anxiety variables at t0 with the numbering of the variables. As a result, both variables are addressed. Note that the **paste()** construction goes inside the square brackets **[]**, so it is, in fact, variable access using **[]**. Although **paste()** creates strings useful in indexing, **grep()** matches a string fragment with all variable names contained in d. It returns all index numbers where there is a match, and these numbers are then used inside the square brackets **[]** as indices.

You may wonder why it could be good to use such complicated constructions for variable access where one could directly use the variable names. Imagine handling hundreds of variables in one data set, as is often the case in research, and that the data consists of 50 anxiety and depression measures. In this case, using **paste()** with the rule governing all variable names for indexing is shorter, easier to read, and safer than listing all 50 variables inside the **[]**. Or, in another case, where you know that there are several depression variables in the data, but you do not bother with the particular naming rule, then simply guess that all depression variables are very likely to have "dep" in their name; so, to deploy **grep()** with this is an obvious option.

I suggest never using **attach()** in a programmed script. It creates confusion about where variables in the script belong: whether they exist at the console or in a data frame. I use **attach()** only if I have obtained a new data set and want familiarise myself with it at the console. In this case I use **attach()** with the data set and then juggle with all the variables. However, when it comes to the programming of my published results I use **$** and **[]** together with **paste()** and **grep()** to always make clear where the variables belong.

2.3.3 paste()

We used **paste()** to make variable access in a data frame more efficient. However, it is useful in many other situations where we need strings; that is, letters enclosed in quotation marks. Although **c()** can also concatenate string vectors like c("Max", "Fips", "Tups"), string creation according to a rule uses **paste()** or **paste0()**. String vectors occur frequently in programming, for example, in the creation of filenames from building blocks of file path, content, timestamp, and file extension. Let us practise at the console how **paste()** works.

Ch2_paste.R

```
## Using paste() to create strings
paste("Strings", "linked", "together")


MyFriends <- c("Max", "Fips", "Tups")
paste("I like", MyFriends)


DepressionVariables <- paste("t0_DEP", 1:15, sep = "_")
DepressionVariables


today <- format(Sys.time(), "%Y%m%d")
MyResultsFile <- paste("Results", today, ".txt", sep = "")
MyResultsFile
```

Output

```
> paste("Strings", "linked", "together")
[1] "Strings linked together"

> MyFriends <- c("Max", "Fips", "Tups")

> paste("I like", MyFriends)
[1] "I like Max"  "I like Fips" "I like Tups"

> DepressionVariables <- paste("t0_DEP", 1:15, sep = "_")

> DepressionVariables
 [1] "t0_DEP_1"  "t0_DEP_2"  "t0_DEP_3"  "t0_DEP_4"  "t0_DEP_5"  "t0_DEP_6"
 [7] "t0_DEP_7"  "t0_DEP_8"  "t0_DEP_9"  "t0_DEP_10" "t0_DEP_11" "t0_DEP_12"
[13] "t0_DEP_13" "t0_DEP_14" "t0_DEP_15"

> today <- format(Sys.time(), "%Y%m%d")

> MyResultsFile <- paste("Results", today, ".txt", sep = "")

> MyResultsFile
[1] "Results20210111.txt"
```

Analysis

In its first use, **paste()** takes three strings and simply links them to one long string. In its second use, paste takes a string as a first argument and a string vector consisting of three names as a second argument. The result shows that **paste()** linked the first string to all elements of the string vector, thus creating another string vector. The next example uses the same construction in another context. That is, **paste()** creates 15 names of depression variables from a prefix indicating the content of the variables and the vector `1:15` to number all variables. This example also uses the `sep` parameter to set the character that goes between linked elements. In the final example, **paste()** creates a file name from a term indicating content, creation date, and the file extension. With **format()** and **Sys.time()** we create the object today and then use this in **paste()**. To create a file name in a script this way is very handy if you use functions which automatically store the results in files like **capture.output()** or **write.csv()**. Notice also the alternative **paste0()**, which does the same as **paste()** except that the default argument for `sep` is `""`. In sum, many situations involving strings require **paste()**. However, some add-on packages like **stringr** (Wickham, 2019) and **stringi** (Gagolewski, 2020) enhance string handling even more.

2.3.4 grep()

R FUNCTION!

The **paste()** function has a good friend – **grep()**. It searches string vectors for a given string pattern and returns the vector indices where the string pattern occurs. Sometimes you expect a certain sequence of characters in a string or in a vector of strings. For example, the keyword "anxiety" should be retrieved in a list of interview responses from ten interviewees. Or, in a list of 1000 variable names, all variables containing the pattern "dep" for depression should be used for descriptive statistics. The function **grep()** can specify such patterns of characters and returns the index numbers of the vector elements which contain it. Let us practise at the console how it works.

Ch2_grep_DataCreation.R

```
## grep() function - Data creation
d <- data.frame(t0_dep01 = c(1, 2, 1),
                t0_dep02 = c(2, 2, 2),
                t0_dep03 = c(1, 2, 2),
                t0_anx01 = c(2, 2, 1),
                t0_anx02 = c(1, 1, 2),
                t1_dep01 = c(1, 1, 1),
                t1_dep02 = c(1, 2, 1),
                t1_dep03 = c(2, 2, 1),
                t1_anx01 = c(2, 2, 2),
                t1_anx02 = c(1, 2, 1))
```

Ch2_grep.R

```
## Using the grep() function for indexing
# Statistics for selected columns
apply(d[grep("dep", names(d))], 2, mean)
apply(d[grep("t0", names(d))], 2, mean)
apply(d[grep("t1_dep", names(d))], 2, mean)
```

```
# Aggregation of selected columns
t0_Mean_dep <- rowMeans(d[grep("t0_dep", names(d))])
t0_Mean_dep
```

Output

```
> # Statistics for selected columns
> apply(d[grep("dep", names(d))], 2, mean)
t0_dep01 t0_dep02 t0_dep03 t1_dep01 t1_dep02 t1_dep03
1.333333 2.000000 1.666667 1.000000 1.333333 1.666667
> apply(d[grep("t0", names(d))], 2, mean)
t0_dep01 t0_dep02 t0_dep03 t0_anx01 t0_anx02
1.333333 2.000000 1.666667 1.666667 1.333333
> apply(d[grep("t1_dep", names(d))], 2, mean)
t1_dep01 t1_dep02 t1_dep03
1.000000 1.333333 1.666667
> # Aggregation of selected columns
> t0_Mean_dep <- rowMeans(d[grep("t0_dep", names(d))])
> t0_Mean_dep
[1] 1.333333 2.000000 1.666667
```

Analysis

This example shows the power of **grep()** to address several related variables with a small function call. We create a data frame in the first script. It contains short variable names which indicate time point of measurement, concept, and item index. Apparently, the variable names have been generated according to a certain rule, which produced the usual appearance of variable names in data frames in empirical studies. The **grep()** function now uses this information and allows for the selection of certain variables and the omission of others. In this example, a combination with **apply()** yields means of selected groups of variables. The first combination of **apply()** and **grep()** returns the means for all depression variables in the data, the second for all variables of the t0 measurement, and the third for all depression variables at time point 1. Furthermore, **grep()** is useful in the aggregation of data columns to new variables. In this case, the combination of **grep()** with **rowMeans()** produces a new variable which contains the means per subject of the depression items at time point 0.

This was only a small example to illustrate the power of **grep()**. Imagine an analysis with hundreds of variables. Instead of the specification of all variable names explicitly between the indexing operator [], a simple call of **grep()** saves much typing. However, if the rule for the generation of the variable names is known, using **paste()** with this rule is more precise. But if different rules of name generation apply for different sets of variables, even a **paste()** construction becomes clumsy. Then, it is easiest to make a good guess of what all the desired variable names have in common and then use **grep()**. For example, if we have hundreds of clinical psychological variables with different kinds of naming, a good guess would be that all the depression items or scales somewhere contain the string "dep" in their name, or that all anxiety variables somewhere contain "anx". So, to get a feeling for the data and the variable names, repeated uses of **grep()** at the console returns you variable groups available for further analysis.

2.3.5 UPS - Data inspection

Let us inspect the unpleasant sounds data after the import from an SPSS file. We use a few handy functions for this, which are related to the attributes of the data frame. Attributes are information units stored along with an R object. Now, we inspect the attributes of a data frame; later, we will practise how to manipulate them.

UPS_CON_DataInspection.R

```
## UPS - Data inspection
source("Scripts/UPS_PRE_DataImport.R")

class(UPS)
View(UPS)
names(UPS)
str(UPS)

# Query variable labels
attr(UPS, "variable.labels")
```

Output

```
> class(UPS)
[1] "data.frame"
> View(UPS)
> names(UPS)
[1] "subject"   "group"     "sound"     "variation" "age"
"gender"    "rating"
> str(UPS)
'data.frame':   2688 obs. of  7 variables:
 $ subject  : num  28 28 28 28 28 28 28 28 28 28 ...
 $ group    : Factor w/ 2 levels "Origin known",..: 1 1 1 1 1 1 1 1 1 1 ...
 $ sound    : Factor w/ 4 levels "Fingernails",..: 3 3 3 2 2 3 2 3 1 1 ...
 $ variation: Factor w/ 7 levels "Original","High-pass",..: 6 5 4 4 3 2 2 1 5 3 ...
 $ age      : num  30 30 30 30 30 30 30 30 28 28 ...
 $ gender   : Factor w/ 2 levels "female","male": 2 2 2 2 2 2 2 2 2 2 ...
 $ rating   : num  1 1 1 1 1 1 1 1 1 1 ...
 - attr(*, "variable.labels")= Named chr  "Subject" "Group" "Sound"
"Variation" ...
  ..- attr(*, "names")= chr  "subject" "group" "sound" "variation" ...
 - attr(*, "codepage")= int 1252
>
> # Query variable labels
> attr(UPS, "variable.labels")
    subject       group       sound    variation          age      gender       rating
  "Subject"     "Group"     "Sound" "Variation"        "Age"    "Gender"     "Rating"
```

Analysis

The function **class()** confirms that UPS is indeed a data frame; **View()** shows us the data frame. Then, **names()** shows us the variable names of the seven contained variables. I use **names()** very often in each R session to remind myself of the available variables. The function **str()** then returns an overview of the structure and attributes of the imported data. For example, it tells us which variables are categorical factor variables and which are numeric variables. It also retrieves the variable labels. Finally, **attr()** with the argument "variable. labels" returns the labels for all variables. There can be long and elaborate variable labels, as is the case in other data sets. However, here the difference between variable names and variable labels is that the labels start with capital letters. Beautiful variable labels, instead of sometimes ugly variable names, can be useful for graphics annotation and to better understand the content of variables.

2.4 Save, export, and archive the data

This section shows how to save and export the data. Nowadays, besides storing the data for yourself and your team, you make it publicly available for the research community. This section reviews some important options for such teamwork and how to archive the data appropriately.

R LANGUAGE! ## 2.4.1 Saving and exporting the data in R

In an R session, you usually create several objects. Some of these you will even send to colleagues for further use. So, we now discuss ways to store and export data frames and other R objects. The options are to use the proprietary RData format or to export in another file format. Using the RData format, you can store data frames which retain all their attributes. Furthermore, not only data frames can be stored but any R object. However, only R can load this file format, so, before data sharing with colleagues, you should ensure that all are willing to use R for their analyses using the objects you supply. The alternative is data export in another file format. The csv format is recommended. It is a plain text format which only uses a designated character, usually the comma or semicolon, to separate all the data fields from each other. In my own analyses I never store R objects but only data frames and scripts. All intermediate objects, like regression models or summary tables, I create anew with the corresponding script when I need the object. This way I can always access the particular calculations which create the required object, and it is often necessary to tinker with this code even a month after I wrote it. So, let us practise data storage in the next example. It also demonstrates export to SPSS and Excel.

EXAMPLE! ## 2.4.2 SCB - Data storage

We imported the data for the student conciliators data from the csv format, which is basically plain text. To store more information than mere numbers with the data frame, **save()** uses the RData file format. The format retains all changes in attributes or other characteristics of the data frame. In the following script we practise export in different file formats. Notice that the script requires installation of the **XLConnect** package.

SCB_PRE_DataStorageAndExport.R

```
## SCB - Data storage and export
# Save in R format
save(SCB, file = "./Data/SCB_Data.RData")


# Export for archivation and publication
write.csv(SCB, file = "./Data/SCB_Data.csv")


# Export in SPSS and Excel-format
foreign::write.foreign(SCB,
                    datafile = "./Data/SCB_Data.csv",
                    codefile = "./Data/SCB_Codefile.sps",
                    package = "SPSS")

XLConnect::xlcDump("SCB", file = "./Data/SCB_Data.xlsx")
```

Analysis

This script shows examples of how to save and export data. Look at your working directory in the file browser and open the inspected files. The function **save()** preserves all information about the data frame and is a good choice if you exclusively analyse the data in R and do not expect to pass on the data to others. For data sharing, **write.csv()** is the best choice since the output can be read in any statistics or text software. If you wanted to make life a little easier for your colleagues using SPSS or Excel you would use **write.foreign()** from the **foreign** package, or **xlcDump()** from the **XLConnect** package.

2.4.3 Standards for data sharing   TEAMWORK!

R can store the data frames in various formats, and add-on packages even extend this functionality. Moreover, R can format and process a data frame and then store it for further use by other researchers. In fact, *the idea of reproducible science treats data as a separate entity for publication*, because others may recalculate your results, calculate their own analyses, or even integrate the data into a bigger data frame, which furnishes more test power for the establishment of associations between variables. There are standards for data storage and publication to enhance the workflow of data sharing. Defossez and colleagues (2020) propose an elaborate rule set for naming and storing data to make it useful for others and to allow the merging of data frames from different sources. Hrynaszkiewicz and colleagues (2010) demonstrate how to prepare clinical data for publication. Furthermore, the UK Data Archive offers detailed advice on how to prepare data for dissemination in the archive (www.data-archive.ac.uk). Finally, Harvard Dataverse offers deposition of research data for diverse scientific fields (dataverse.harvard.edu). *Following such guidelines is beneficial. If you know them you can guess many things about the structure of data supplied by other researchers. Thus, it becomes easier to familiarise yourself with this data.* Most guidelines recommend using a plain text file format for data storage and archiving such as the csv format. The letters csv mean character-separated values. The **write.csv()** function creates such files.

The strategy to create a data set and prepare it for the statistical analysis should allow for long-term maintenance of the data. Your colleagues should have no problem importing the data into their preferred statistical program and finding all things in the data as they expect them. It should also be accessible from the internet. So, better to spend more time initially on data preparation according to guidelines and have a professional data set afterwards. Finally, in addition to the sharing of data, the sharing of R code may further boost productivity. *Team colleagues could share the code for standard tasks in their analyses, so everybody can increase speed.*

R LANGUAGE! ## 2.4.4 Data subsets

In psychological, social, and clinical research it is common practise to divide large data frames into smaller sub data frames, each comprising only those variables necessary to answer a particular research question. R offers different options to create subsets. One option is to use indexing of the data frame to select certain rows and columns. Another option is to use **subset()** as a more explicit way of subsetting. Using **subset()** may be more comprehensible when reading the code after some time. Let us try both options and also note their drawbacks.

Ch2_Subset.R

```
## Subset of a data frame
# Data creation
set.seed(12345)
d <- data.frame(A = rep(c("a1", "a2", "a3"), each = 20),
                B = rep(c("b1", "b2"), times = 3, each = 10),
                Y = rnorm(60,
                          mean = c(10, 10, 11, 9, 9, 11),
                          sd = 2))
attr(d, "variable.labels") <- c(A = "Variable A",
                                B = "Variable B",
                                Y = "Variable Y")
comment(d) <- "My first data set"
attr(d, "Date of data recording") <- "2020/11/02"


# Data subsets
d1 <- d[d$A == "a1",]
d2 <- d[d$A %in% c("a1", "a2"),]
d3 <- d[d$A == "a1" & d$B == "b1",]
d4 <- d[d$A == "a1", c("A", "B")]
d5 <- subset(d,
             subset = A == "a1")
d6 <- subset(d,
             subset = A == "a1",
             select = c("A", "Y"))
attributes(d1)
```

```
attributes(d4)
attributes(d6)
```

Output

```
> attributes(d1)
$names
[1] "A" "B" "Y"

$variable.labels
          A           B           Y
"Variable A" "Variable B" "Variable Y"

$comment
[1] "My first data set"

$`Date of data recording`
[1] "2020/11/02"

$row.names
 [1]  1  2  3  4  5  6  7  8  9 10 11 12 13 14 15 16 17 18 19 20

$class
[1] "data.frame"

> attributes(d4)
$names
[1] "A" "B"

$row.names
 [1]  1  2  3  4  5  6  7  8  9 10 11 12 13 14 15 16 17 18 19 20

$class
[1] "data.frame"

> attributes(d6)
$names
[1] "A" "Y"

$row.names
 [1]  1  2  3  4  5  6  7  8  9 10 11 12 13 14 15 16 17 18 19 20

$class
[1] "data.frame"
```

Analysis

The script first creates a small data frame d consisting of three variables A, B, and Y. It then assigns a `variable.labels` attribute to the data frame holding more beautiful variable names. To use variable labels in results reports or graphical displays is usually better than cryptical variable names. The script also adds a comment and `"Date of data recording"` as another arbitrary attribute. After this, the script shows six different expressions to subset the data frame into six smaller data frames. The first four expressions use data indexing with **[]** to select the subsets: d1 consists of all cases where A = a1. Note that testing for equality uses **==**, that is, a double equals sign. Further note that the comma before the closing bracket indicates that the subset should include all of the original variables. Next, d2 selects cases belonging either to a1 or a2 and also uses all variables as indicated with the comma. Then, d3 combines two conditions using the & operator; that is, cases of the subset will belong to conditions a1 and b1. Data frame d4, again, consists only of cases showing category a1, but this time we select only the variables A and B for the subset, using the expression behind the comma inside **[]**. The data frames d5 and d6 do not use indexing but **subset()** to create the data subsets. The parameter subset declares the cases to select; the parameter select declares the variables. The script closes with three calls of **attributes()** for d1, d4, and d6, respectively. Naturally, we would expect that data selection should not change the data attributes, but it does. While subsetting cases using indexing for d1 preserves the attributes, subsetting variables with **subset()** does not. If you want to create a subset of your data for a colleague, you may reassign the attributes using **attr()**. If the task is not really subsetting a portion of the data but only omitting a few variables, use `DataFrame$ToBeOmittedVariable <- NULL` to remove variables.

R FUNCTION! ## 2.4.5 merge()

Subsetting allows us to partition the data frame into smaller data frames. Sometimes, however, you receive data from different sources and want to create a master data frame containing all the information. For example, a clinical study may consist of data from different time points or different study sites. *After data collection, the project statistician gathers all data from different sources and merges them in one data frame for the statistical analysis.* The **merge()** function easily accomplishes this using one or more columns which allow matching of the data rows.

Ch2_merge.R

```
## Aggregate data sets with merge()
d1 <- data.frame(ID = c("s1", "s2", "s3", "s4"),
                 A = c("a1", "a2", "a3", NA),
                 X1 = c(1, 2, 3, 4),
                 X2 = c(5, 6, "a", 8),
                 X3 = c(9, 10, 11, 12))
d2 <- data.frame(ID = c("s1", "s2", "s3", "s4"),
                 A = c("a1", "a4", NA, NA),
                 Y1 = c(13, 14, 15, 16),
                 Y2 = c(17, 18, 19, 20),
                 Y4 = c(21, 22, 23, 24))
```

```
d3 <- merge(d1, d2, by = "ID")
d4 <- merge(d1, d2, by = "A")
d5 <- merge(d1, d2, by = "A", all = TRUE)
d6 <- merge(d1, d2, by = "A", incomparables = NA)
```

Output

```
> d3
  ID  A.x X1 X2 X3  A.y Y1 Y2 Y4
1 s1   a1  1  5  9   a1 13 17 21
2 s2   a2  2  6 10   a4 14 18 22
3 s3   a3  3  a 11 <NA> 15 19 23
4 s4 <NA>  4  8 12 <NA> 16 20 24
> d4
     A ID.x X1 X2 X3 ID.y Y1 Y2 Y4
1   a1   s1  1  5  9   s1 13 17 21
2 <NA>   s4  4  8 12   s3 15 19 23
3 <NA>   s4  4  8 12   s4 16 20 24
> d5
     A ID.x X1   X2 X3 ID.y Y1 Y2 Y4
1   a1   s1  1    5  9   s1 13 17 21
2   a2   s2  2    6 10 <NA> NA NA NA
3   a3   s3  3    a 11 <NA> NA NA NA
4   a4 <NA> NA <NA> NA   s2 14 18 22
5 <NA>   s4  4    8 12   s3 15 19 23
6 <NA>   s4  4    8 12   s4 16 20 24
> d6
   A ID.x X1 X2 X3 ID.y Y1 Y2 Y4
1 a1   s1  1  5  9   s1 13 17 21
```

Analysis

The script creates two data frames, each with an ID variable for subject IDs and with a few more variables. Let us say that the A is a qualitative variable, the X variables are pretest, and the Y variables are posttest variables. The subsequent calls of **merge()** show how the function behaves. It always receives the two data frames to be merged. The by parameter then obtains the name of the ID variable. The result of this function call is d3 with four rows corresponding with the subject IDs and with all other data columns assembled. This would be the normal use of **merge()**; however, merging by A lets us inspect further parameters of the function. As can be seen, A consists of one common entry between d1 and d2; it also consists of one NA in d1 and two NA in d2. Merging by A produces d4, consisting of one row for the a1 entry and two rows with NA entries in A. Inspecting d4 closer reveals that the NA from d1 was merged with both rows consisting of NA in d2. In fact, the ID and X entries from d1 are used twice in d4, respectively. In many situations this behaviour is not wanted and results in an ID variable that contains double entries, although it was supposed to contain unique entries. Using **duplicated()** with the ID variable before merging identifies double entries beforehand. Furthermore, d4 does not contain data rows for non-matching entries of A. To include all data, the parameter all is used to create d5, which shows several missing values for cases where no match in variable A was possible between d1 and d2. Finally, to avoid the problem of matching one entry multiple times, the parameter incomparables can be set to those entries, which should be omitted from the matching. This results in a data frame d6, which omits all entries set as incomparables.

What you have learned so far

1 Several methods to load and import data in R
2 To explore the raw data and its attributes
3 To efficiently access the variables in data frames
4 To save and export data frames

Exercises

Explore functions

1 Search for data import functions. Start with **read.csv()**, **read.table()**, and **data()**. Also try functions from add-on packages like **gdata::read.xls()**, **readxl::read_excel()**, **Hmisc::spss.get()**, or **sjlabelled::read_spss()**.
2 Use **setwd()** and **getwd()** at the console to navigate through the file system of your computer. Enter the complete path of the folder where you stored the example data sets **MusicData.csv** and **MusicPreferencesSchool.csv** in **setwd()** to set this folder as a working directory. Check the contents of this folder with **dir()**.
3 Use **paste()** to create a system of variable names for a survey. Create three time points and 10 measurements for each of the variables depression, somatic symptoms, and anxiety. For example, a variable name could look like this: t1_anx10 or t2_dep3.

Solve tasks

1 Create a small SPSS data set and use different combinations of variable settings, for example, a numeric variable with value labels, string entries in the data frame, numeric entries in the data frame with or without value labels. Then import the data into R. Use different parameter settings in **foreign::read.spss()** and observe how the data frame turns out in R. What suits you best and your subsequent analysis?
2 Create a small table with four rows and three columns in a word processor and copy this table to the clipboard. Then, use **read.clipboard()** from the **psychTools** package to paste the table in an R object. Does the table in R look much like the original?
3 Many data sets come with the installation of R and the installation of add-on packages. Call data(package = .packages(all.available = TRUE)) to get a list of all these data sets. Then, use **data()** with the name of one data set and have the data frame in your workspace, for example, data(morley), which provides the speed of light data of the famous Michelson–Morley experiment.

4   Read in the data sets `MusicData.csv` and `MusicPreferencesSchool.csv` used in this book and explore their structure. Use `str()` and `attributes()` for this.
5   Create a data subset of males only from the `MusicData.csv` data set. Use indexing with `[]` and `subset()` to complete the task.

Read code

1   Read the following code fragments. The first line of each pair of lines creates variable names; the second line indexes variable names. Decide which pairs create and index corresponding variable names. Check the code with dummy data, for example, `d <- data.frame(mvtnorm::rmvnorm(10, rep(5, 10)))`.

```
names(d)       <-      paste("dep",      1:10,      sep     =      "_")
d[paste("dep", 1:10, sep = "_")]
```

```
names(d)      <-      paste("t0",     "anx",     1:5,     sep     =      "_")
d[grep("anx", names(d))]
```

```
names(d)      <-      c(paste0("dep",      1:5),      paste0("anx",      1:5))
d[paste0("depr", 1:5)]
```

```
names(d) <- c(paste("dep", 1:5, sep = "_"), paste("anx", 1:5, sep = "_"))
d[grep("_", names(d))]
```

```
names(d)      <-     c("t0_dep_1",      "t0_dep_2",      "t0_anx",      "t0_int")
d[grep("t0", names(d))]
```

Analyse data

1   Obtain the data set `MusicData.csv` from the book's repository. In R, change the working directory to where you stored the data set.
2   Open `MusicData.csv` in a text editor and inspect its structure. The data consists of music evaluations by 40 persons for 20 musical examples. Pay attention to how many data rows and columns there are and which character separates the data entries.
3   Create a script for the following analysis of the music data and save it in your working directory. Create an appropriate header for your script, for example, Date and Analysis music data.
4   Import the data set `MusicData.csv` with read.csv as data frame `MusicData`.
5   Get an overview about the attributes of the music data with `names()`, `str()`, and `attributes()`.
6   Obtain summary statistics with `summary()` for the music evaluations only. Use `[]` together with `paste()` or `grep()` to index the 20 music evaluations.
7   Save the `MusicData` object as an .RData file for further use.
8   Use `MusicData.csv`. One subject is 17 years old and not 15 or 16 like all other subjects. Analyses which use age group as a factor will not work properly if one group consists only of one subject. An option to solve this problem may be to replace the

(Continued)

age for this subject with NA. This subject will be excluded in all analyses using `Age`. Another option might be to set `Age` as a factor and declare a contrast about the ages of 15 and 16. A third option might be to recode all values of the `Age` variable and declare the 16- and 17-year-olds as the older subjects in contrast to the 15-year-olds as the younger subjects. Explore all three options and reflect about their advantages and disadvantages.

9   Read the data set **MusicPreferencesSchool.csv** into a data frame in R. The data set contains preference ratings from school students of different schools for two musical examples. A measure of creativity was also collected for each student. Check the variable types that R extracts from the data set with **str()**.

Apply in the real world

1   Import your own data in R. Create a script for data import and search for the right import function.

2   After data import get a quick data overview with **View()**, **names()**, **str()**, and **attributes()**.

3   If possible, try to create a systematic way to access many of your variables with only a little code. Use **paste()** or **grep()** for this.

Become a statistics programmer

1   Search for the standards for data sharing in your research domain. Which file formats are standard? Are there online repositories, which may store your data and which provide you with the data of others?

2   Establish a standard format for research data in your team. Agree on the spelling of commonly used variable names and on how to arrange variables in the data frame.

EXAMPLE! **WBD**

1   Read the publication by Blackwell and colleagues (2015). Create a list of the variables of the study design. Familiarise yourself with the planned efficacy analysis.

2   Use **spss.get()** from the **Hmisc** package to import the data. Is it easier or more difficult to use than **read.spss()** from the **foreign** package? Import the two SPSS sav files in R. Vary the parameters for data import and inspect how the data is then represented in R.

3   Explore the data at the console. Use **View()**, **str()**, **attributes()**, and **names()**. Match the available variables in the data frame with the variables from the study design. Which variables serve as potential grouping variables? Which variables comprise the primary and secondary outcomes?

EXAMPLE! **ITC**

1   Go to the website osf.io and locate the reproducibility project in psychology. In the project, search for the replication study of Stanovich & West (2008) to find out more about the background of the study. The codebook in particular helps you to understand the data.

2   Import the data file **stanovich_data_used.sav** and familiarise yourself with the data in a console session. Retrieve the variable names, information about the structure, and attributes of the data frame.

UPS

1 Read the conference poster by Reuter and colleagues (2014) about the Physical and   EXAMPLE!
 Acoustical Correlates of Unpleasant Sounds. Sketch the experimental design including
 all available factor variables.
2 Import the data set from the SPSS sav file UnpleasantSounds.sav. Familiarise yourself
 with the data. Use `str()`, `names()`, or `attributes()` to learn about its structure.
 Note which variables are imported as factors and which variables as numeric variables.
 Also try `contents()` from the `Hmisc` package, or `codebook()` from `memisc`.

SCB

1 Create a working directory with your file explorer and store the file   EXAMPLE!
 `ExperimentConflictReconciliationData.csv` in this directory. Then import the
 data set with `read.csv()` in a data.frame object SCB.
2 After data import of the SCB data frame, store the object in an .RData file. In this file,
 the object will keep all its attributes and other declarations. Use `save()` to store the
 object and `load()` to load it in your next R session.
3 Explore the data set with `View()`, `names()`, `str()`, and `attributes()`.

3

PREPARING THE DATA FOR ANALYSIS

━━━━━━━━━━━━━━━━━━━━━━━━━━ Chapter overview ━━━━━━━━━━━━━━━━━━━━━━━━━━

This chapter shows how to calculate and transform the variables of a data set. Usually, statistical analyses do not run with raw data but with aggregated variables, for example, a sum score of several items. So, the calculation of variables and the transformation of data are often required. The chapter further focuses on the formatting and annotation of variables of different types. Therein, we examine the preparation of factor variables for inferential statistics. Programming features: variable types, variable attributes, using code scripts to structure the analysis, calling scripts from other scripts, vector and matrix calculations, reshaping of data frames between wide and long format.

3.1 **Data transformations**

This section demonstrates how to create new variables and add them to the existing variables in a data frame. We also focus on matrix calculations to increase efficiency and on how to modify the format of a data frame to suit different analyses.

R LANGUAGE! ### 3.1.1 **Calculation of new variables in R**

You already know about data frames and how to select variables. Now we add variables with the help of the [] and the $-operator. Say a data frame d contains variables X1, X2, and X3. We plan to calculate data columns containing the sum and mean of the three variables. The **+** operator or the function **rowSums()** do this. The + looks more intuitive, but **rowSums()** is usually better because it has a parameter about how to treat missing data. The console helps us to see how it works.

Ch3_AddingVariablesToADataFrame.R

```
## Adding a variable to a data frame
d <- data.frame(X1 = c(3, 4, 4, 3, 5),
                X2 = c(2, 3, 3, 4, 3),
                X3 = c(4, 3, 4, 5, 3))
d$X_Sum <- rowSums(d)
d["X_Mean"] <- rowMeans(d)
d[paste0("X_roots", 1:3)] <- sqrt(d[c("X1", "X2", "X3")])
d
```

Output

```
> d
  X1 X2 X3   X_Sum   X_Mean X_roots1 X_roots2 X_roots3
1  3  2  4 27.64626 5.786566 1.732051 1.414214 2.000000
2  4  3  3 30.46410 6.366025 2.000000 1.732051 1.732051
3  4  3  4 33.23205 6.933013 2.000000 1.732051 2.000000
4  3  4  5 35.96812 7.492030 1.732051 2.000000 2.236068
5  5  3  3 33.20017 6.925042 2.236068 1.732051 1.732051
```

Analysis

These lines show different ways to add variables to a data frame. The top lines create a small data frame, d, for the demonstration. Then $ and [] help us to add the variables X_Sum and X_Mean to the data frame. Although both operators do the same here – add a variable – they work differently. The operator, $, can only add a single variable, which is convenient in many cases. The other construction is more complicated and requires a string between []; it is, however, more flexible. It can generate several variables in one line of code. Adding **paste0()** to the call, the names of the new variables can even be generated according to a rule. The example creates three variables containing the square-roots of other variables and then adds these to the data frame. This way of variable creation is powerful: it can create several variables with one function call.

3.1.2 SCB - Aggregation of variables

EXAMPLE!

The student reconciliator study consists of several time points for measurement of bullying in the schoolyard. We aggregate the pre-intervention and the post-intervention measurements for the three types of bullying, respectively. Moreover, we calculate total numbers of bullying pre-intervention and post-intervention. Notice that the script first runs the script for data import that we created earlier. The **source()** function allows us to hook required earlier scripts into the current script. It uses the file path of the data import script relative to the current working directory.

SCB_PRE_DataAggregation.R

```
## SCB - Data aggregation
source("./Script/SCB_PRE_DataImport.R")

SCB$PersonPerson_PRE <- SCB$Y1C1 + SCB$Y1C2
SCB$PersonGroup_PRE <- SCB$Y2C1 + SCB$Y2C2
SCB$GroupGroup_PRE <- SCB$Y3C1 + SCB$Y3C2

SCB$PersonPerson_POST <- SCB$Y1C3 + SCB$Y1C4
SCB$PersonGroup_POST <- SCB$Y2C3 + SCB$Y2C4
SCB$GroupGroup_POST <- SCB$Y3C3 + SCB$Y3C4

SCB$PersonPerson_PREPOST <- SCB$PersonPerson_PRE - SCB$PersonPerson_POST
SCB$PersonGroup_PREPOST <- SCB$PersonGroup_PRE - SCB$PersonGroup_POST
SCB$GroupGroup_PREPOST <- SCB$GroupGroup_PRE - SCB$GroupGroup_POST

SCB$SumPreIntervention <- rowSums(SCB[grep("PRE", names(SCB))])
SCB$SumPostIntervention <- rowSums(SCB[grep("POST", names(SCB))])
```

Analysis

The script first calls the import script to make the data frame SCB available. It then adds six variables to the data frame using the **$**-operator, one for each combination of type of bullying and time point. The plus sign executes vector addition, that is, element-wise addition to produce a sum vector with the same number of elements as the original vectors. Subsequently, the use of **rowSums()** illustrates another option to aggregate data vectors. The two code lines create the total scores of bullying pre-intervention and post-intervention. Also, in this code **grep()** creates index vectors by finding all variables containing the strings PRE and POST in their variable name, respectively. This use of **grep()** was introduced in Chapter 2 already and it is a convenient option to index many variables with little code. However, the plus sign is easier to comprehend than the construction using **rowSums()** and **grep()**.

3.1.3 Naming conventions

TEAMWORK!

Your analyses in R will rapidly grow and comprise many objects and variables. Each object and variable needs a name and you should early on agree with your team about rules how to create these names – you should agree for a naming convention. "To know the name is to know the object" (Goodliffe, 2007, p. 41) as a most general rule means that a reader of your code should be able to guess from the name the type of data an object consists and how it behaves. Good naming of objects and variables entails better intelligibility of your whole analysis. Some teams prefer abbreviations of concepts, like `anx_i5` for a variable holding the fifth item of an anxiety questionnaire, or `t0_dep_score` for the depression sum score at timepoint t0 (usually baseline). Other teams prefer full variable names which express the content of a variable and its transformations, like `IntelligenceCategorised`. Full names are sometimes easier to read and understand than abbreviations; however, if hundreds of such names are involved, they may clutter your scripts. However, "favor clarity over brevity" (Goodliffe, 2007, p. 43) in naming. In the name of a variable or object you could include how it figures in the empirical design or analysis. For example, an object name could include whether the object is a factor variable, an item of a certain scale, or a measure at a certain time point. With consistent names, your code is easier to read for you and others who use it. In agile software programming, good naming conventions permit programming with little use of redundant code comments because the code comments itself (Martin, 2009).

Because every analysis consists of so many objects, you should apply a practical naming convention straight away. Let us now visit some good rules for naming objects and variables.

- Choose descriptive and unambiguous names (Horstmann, 2007; Martin, 2009); for example, depression, anxiety, dep, or anx may immediately suggest the right content to the reader.
- Variables in classical empirical designs could be assigned the common abbreviations used in statistics textbooks. Usually, the letters A and B stand for qualitative factors in the design. An X usually represents a predictor or covariate in regression analysis; Y usually denotes the dependent variable.
- To analyse longitudinal data, use the time point of measurement in each variable name; for example t0, t1, t2.
- If variables or objects belongs to a certain group, number them, like items 1 to 10 of a certain scale.
- Objects as results of a sequence of operations may reflect this sequence in their names. For example, `ModelIntelligenceMultipleImputations` expresses that the given model was calculated with imputed data.
- Agree on how to connect the parts of each name. To use the underscore _ between parts is easy to read, like in t0_depression. Another good option is the camel case, using uppercase letters to connect parts of a name, like in `DepScoreStandardised`, resembling humps of a camel (Goodliffe, 2007). Baath (2012) also suggests period-separated names.

A naming convention allows for easy reference of all variables and objects in the analysis. One could correctly guess a variable name by simply adhering to the convention. For example, if we know that `t0_dep_1` is the first depression item at baseline, we may correctly guess the name of the fifth anxiety item at the first follow-up to be `t1_anx_5`.

The convention to express the transformations which produced a variable in the variable name has a drawback. Usually, the main analysis uses such derived variables in many places, in fact, more often than the variable consisting of the raw data. Consequently, the naming convention unfortunately entails that long variable names are more frequent in the analysis than short ones. In contrast, programming is easier and intelligibility improves with short variable names. So, a remedy would be to turn the naming convention upside down, that is, give the derived variable the short name and the raw data variable the longer name. For example, employment status in the analysis could become EmploymentStatus and in the raw data EmploymentStatusRawData. Using this strategy enables that the variables in high use obtain short names. A variable `t1_DEP` is easier to use than `t1_DEP_ItemSum_MissingsReplaced` in the analysis. So, we could use the names of raw variables and intermediate variables to mirror the data transformations. This would reserve the shortest names for the variables in high use.

Naming variables and objects not only concerns the naming of a single entity; it requires one to find a strategy to consistently name numerous things, while still letting you create names quickly and keeping the names handy for your use and the use of others. Baath (2012) asserts that there is no convention in R of how to name things. However, there are some style guides by Google http://google-styleguide.googlecode.com/svn/trunk/google-r-style.html or Hadley Wickham http://stat405.had.co.nz/r-style.html. At www.r-bloggers.com Robin Lovelace (2014) suggests that naming should be addressed more in textbooks about R. In sum, most important is to be consistent with names throughout. Play with the suggested options and discuss with your team.

3.1.4 Vector and matrix calculations

STATISTICAL ISSUE!

We may treat data frames as matrices and apply the different functions of matrix algebra to modify them. We know already that matrices are two-dimensional arrays of numbers. We denote matrices with bold capital letters, for example, matrix **A**. In the following, we show what a little programming with matrix functions can accomplish. The example uses random multivariate sampling from the **mvtnorm** package.

Ch3_MatrixOperations.R

```
## Matrix operations
set.seed(12345)
D <- mvtnorm::rmvnorm(n = 12,
                      mean = c(10, 10, 10),
                      sigma = matrix(c(2, .3, .2,
                                       .3, 2, .2,
                                       .2, .2, 1),
                                     nrow = 3))
D <- round(D)
D

# Sum per subject or row
D %*% c(1, 1, 1)

# z-transformation
scale(D)

# Sums of cross-products
t(D) %*% D

solve(t(D) %*% D)
det(t(D) %*% D)
```

Output

```
> D
      [,1] [,2] [,3]
 [1,]   11   11   10
 [2,]    9   11    8
 [3,]   11   10   10
 [4,]    9   10   12
 [5,]   11   11    9
 [6,]   11    9   10
 [7,]   12   11   11
 [8,]   12    9    9
 [9,]    8   12   10
[10,]   11   11   10
[11,]   12   13   12
[12,]   12   11   11
> # Sum per subject or row
> D %*% c(1, 1, 1)
      [,1]
 [1,]   32
 [2,]   28
 [3,]   31
 [4,]   31
 [5,]   31
 [6,]   30
 [7,]   34
 [8,]   30
 [9,]   30
[10,]   32
[11,]   37
[12,]   34
> # z-transformation
> scale(D)
           [,1]        [,2]        [,3]
 [1,]  0.1842569  0.2196488 -0.1396551
 [2,] -1.2897985  0.2196488 -1.8155163
 [3,]  0.1842569 -0.6589465 -0.1396551
 [4,] -1.2897985 -0.6589465  1.5362061
 [5,]  0.1842569  0.2196488 -0.9775857
 [6,]  0.1842569 -1.5375419 -0.1396551
 [7,]  0.9212847  0.2196488  0.6982755
 [8,]  0.9212847 -1.5375419 -0.9775857
 [9,] -2.0268263  1.0982442 -0.1396551
[10,]  0.1842569  0.2196488 -0.1396551
[11,]  0.9212847  1.9768396  1.5362061
[12,]  0.9212847  0.2196488  0.6982755
```

```
attr(,"scaled:center")
[1] 10.75000 10.75000 10.16667
attr(,"scaled:scale")
[1] 1.356801 1.138180 1.193416
> # Sums of cross-products
> t(D) %*% D
     [,1] [,2] [,3]
[1,] 1407 1385 1315
[2,] 1385 1401 1316
[3,] 1315 1316 1256
> solve(t(D) %*% D)
             [,1]         [,2]         [,3]
[1,]  0.03584437 -0.01163008 -0.02534249
[2,] -0.01163008  0.04895336 -0.03911549
[3,] -0.02534249 -0.03911549  0.06831319
> det(t(D) %*% D)
[1] 775575
```

Analysis

The most important functions are **t()**, which returns the transposition of a matrix, and the **%*%** operator, which is matrix multiplication. Given the correct vectors and matrices, they accomplish many statistically relevant calculations. Furthermore, **solve()** and **det()** calculate the inverse of a matrix and the determinant of a square matrix. Table 3.1 shows a basic inventory of some common matrix operations. In the statistics literature there are in-depth treatments of matrix algebra (Timm, 1975) and the topic is often included in appendices to statistics books (Hays, 1994; Kirk, 2012; Winer, Brown & Michels, 1991). In fact, there are formulations of commutative, associative, and distributive laws for matrices widely available, which help in the day to day work.

Table 3.1 Common matrix operations in R.

Function	Notation			R function call
Matrix creation	$A=\begin{pmatrix} 1 & 2 \\ 3 & 4 \\ 5 & 6 \end{pmatrix}$,	$B=\begin{pmatrix} 1 & 2 \\ 2 & 1 \end{pmatrix}$,	$C=\begin{pmatrix} 1 & 2 \\ 1 & 2 \\ 1 & 2 \end{pmatrix}$	`A <- matrix(c(1, 2, 3, 4, 5, 6), nrow = 3, byrow = TRUE)` `B <- matrix(c(1, 2, 2, 1), nrow = 2, byrow = TRUE)`
Matrix transposition	$A'=\begin{pmatrix} 1 & 3 & 5 \\ 2 & 4 & 6 \end{pmatrix}$			`t(A)`
Scalar addition	$A+2=\begin{pmatrix} 1 & 2 \\ 3 & 4 \\ 5 & 6 \end{pmatrix}+2=\begin{pmatrix} 3 & 4 \\ 5 & 6 \\ 7 & 8 \end{pmatrix}$			`A + 2`

(Continued)

Table 3.1 (Continued)

Function	Notation	R function call
Scalar multiplication	$A \times 5 = \begin{pmatrix} 1 & 2 \\ 3 & 4 \\ 5 & 6 \end{pmatrix} \times 5 = \begin{pmatrix} 5 & 10 \\ 15 & 20 \\ 25 & 30 \end{pmatrix}$	A * 5
Matrix addition	$A + C = \begin{pmatrix} 1 & 2 \\ 3 & 4 \\ 5 & 6 \end{pmatrix} + \begin{pmatrix} 1 & 2 \\ 1 & 2 \\ 1 & 2 \end{pmatrix} = \begin{pmatrix} 2 & 4 \\ 4 & 6 \\ 6 & 8 \end{pmatrix}$	A + C
Matrix multiplication	$AB = \begin{pmatrix} 1 & 2 \\ 3 & 4 \\ 5 & 6 \end{pmatrix} \begin{pmatrix} 1 & 2 \\ 2 & 1 \end{pmatrix}$ $= \begin{pmatrix} 1*1+2*2 & 1*2+2*1 \\ 3*1+4*2 & 3*2+4*1 \\ 5*1+6*2 & 5*2+6*1 \end{pmatrix}$ $= \begin{pmatrix} 5 & 4 \\ 11 & 10 \\ 17 & 16 \end{pmatrix}$	A %*% B
Combination of matrices by column	$(AC) = \begin{pmatrix} \begin{pmatrix} 1 & 2 \\ 3 & 4 \\ 5 & 6 \end{pmatrix} \begin{pmatrix} 1 & 2 \\ 1 & 2 \\ 1 & 2 \end{pmatrix} \end{pmatrix} = \begin{pmatrix} 1 & 2 & 1 & 2 \\ 3 & 4 & 1 & 2 \\ 5 & 6 & 1 & 2 \end{pmatrix}$	cbind(A, C)
Combination of matrices by row	$\begin{pmatrix} A \\ B \end{pmatrix} = \begin{pmatrix} 1 & 2 \\ 3 & 4 \\ 5 & 6 \\ 1 & 2 \\ 2 & 1 \end{pmatrix}$	rbind(A, B)

Usual R sessions do not need many matrix calculations because for almost all statistical purposes, there are statistical functions which take the required matrices and vectors as arguments. However, some occasions of by-hand calculations of regression weights or model predictions grossly profit from these basic formulas.

R FUNCTION! **3.1.5 reshape()**

We need the __reshape()__ *function to switch between different layouts of a data frame. This is important in the analyses of repeated measures. In many data sets, repeated measures occupy adjacent data columns; for example, three measurements per subject result in three data columns. This is the so-called wide format of a data frame. In contrast, repeated measures in R stand in only one column stacked one on top of the other. So, each subject occupies more than one data row, and an index variable marks each measurement within a subject. This is the long format* of a data frame. This may be counter intuitive to IBM SPSS Statistics software ('SPSS') users who always think that one subject occupies one data row. But on closer inspection the solution of stacked measurements more closely resembles the terminology in the statistics books. That is, repeated measurements ANOVA traditionally belongs to the univariate linear modelling approach, which splits the variance of one variable into different effects, one of these being a random effect of the subject. So, before adjacent repeated measurements can be used for linear modelling in R, they need to be switched from wide to long format with **reshape()**.

In the following code demonstration we create dummy data in a factorial design with two between-subject factors A *and* B, *one within-subject factor* C, *and two covariates* X1 *and* X2. *At first, we create the data with three measurements per subject corresponding to three levels of factor* C. The design could also be termed split-plot factorial (Kirk, 2012), which combines between-subject and within-subject factors. In the terminology of randomised-block or repeated-measures designs, it could also be termed a mixed model, because in the subsequent statistical analysis, the random effect of the subjects would be incorporated in the linear model.

Ch3_reshape.R

```
## reshape() data frame between wide and long format
# Data creation
set.seed(12345)
DataWide <- data.frame(VP = 1:20,
                       A = gl(n = 2, k = 10, labels = c("a1", "a2")),
                       B = gl(n = 2, k = 5, length = 20,
                              labels = c("b1", "b2")),
                       X1 = 101:120,
                       X2 = 201:220,
                       Y1 = 10 + rnorm(20),
                       Y2 = 20 + rnorm(20),
                       Y3 = 30 + rnorm(20))

# Reshape to long format
DataLong <- reshape(DataWide,
                    direction = "long",
                    varying = c("Y1", "Y2", "Y3"),
                    v.names = "Y",
                    timevar = "C",
                    times = c("c1", "c2", "c3"))
```

Output

```
> DataWide
  VP  A  B  X1  X2        Y1       Y2       Y3
1  1 a1 b1 101 201 10.585529 20.77962 31.12851
2  2 a1 b1 102 202 10.709466 21.45579 27.61964
3  3 a1 b1 103 203  9.890697 19.35567 28.93973
4  4 a1 b1 104 204  9.546503 18.44686 30.93714
5  5 a1 b1 105 205 10.605887 18.40229 30.85445
```

```
6    6 a1 b2 106 206  8.182044 21.80510 31.46073
7    7 a1 b2 107 207 10.630099 19.51835 28.58690
8    8 a1 b2 108 208  9.723816 20.62038 30.56740
9    9 a1 b2 109 209  9.715840 20.61212 30.58319
10  10 a1 b2 110 210  9.080678 19.83769 28.69320
11  11 a2 b1 111 211  9.883752 20.81187 29.45961
12  12 a2 b1 112 212 11.817312 22.19683 31.94769
13  13 a2 b1 113 213 10.370628 22.04919 30.05359
14  14 a2 b1 114 214 10.520216 21.63245 30.35166
15  15 a2 b1 115 215  9.249468 20.25427 29.32902
16  16 a2 b2 116 216 10.816900 20.49119 30.27795
17  17 a2 b2 117 217  9.113642 19.67591 30.69117
18  18 a2 b2 118 218  9.668422 18.33795 30.82380
19  19 a2 b2 119 219 11.120713 21.76773 32.14507
20  20 a2 b2 120 220 10.298724 20.02580 27.65306
> DataLong
        VP  A  B  X1  X2  C         Y id
1.c1     1 a1 b1 101 201 c1 10.585529  1
2.c1     2 a1 b1 102 202 c1 10.709466  2
3.c1     3 a1 b1 103 203 c1  9.890697  3
4.c1     4 a1 b1 104 204 c1  9.546503  4
5.c1     5 a1 b1 105 205 c1 10.605887  5
6.c1     6 a1 b2 106 206 c1  8.182044  6
7.c1     7 a1 b2 107 207 c1 10.630099  7
8.c1     8 a1 b2 108 208 c1  9.723816  8
9.c1     9 a1 b2 109 209 c1  9.715840  9
10.c1   10 a1 b2 110 210 c1  9.080678 10
11.c1   11 a2 b1 111 211 c1  9.883752 11
12.c1   12 a2 b1 112 212 c1 11.817312 12
13.c1   13 a2 b1 113 213 c1 10.370628 13
14.c1   14 a2 b1 114 214 c1 10.520216 14
15.c1   15 a2 b1 115 215 c1  9.249468 15
16.c1   16 a2 b2 116 216 c1 10.816900 16
17.c1   17 a2 b2 117 217 c1  9.113642 17
18.c1   18 a2 b2 118 218 c1  9.668422 18
19.c1   19 a2 b2 119 219 c1 11.120713 19
20.c1   20 a2 b2 120 220 c1 10.298724 20
1.c2     1 a1 b1 101 201 c2 20.779622  1
2.c2     2 a1 b1 102 202 c2 21.455785  2
3.c2     3 a1 b1 103 203 c2 19.355672  3
4.c2     4 a1 b1 104 204 c2 18.446863  4
5.c2     5 a1 b1 105 205 c2 18.402290  5
6.c2     6 a1 b2 106 206 c2 21.805098  6
7.c2     7 a1 b2 107 207 c2 19.518353  7
8.c2     8 a1 b2 108 208 c2 20.620380  8
9.c2     9 a1 b2 109 209 c2 20.612123  9
10.c2   10 a1 b2 110 210 c2 19.837689 10
```

```
11.c2 11 a2 b1 111 211 c2 20.811873 11
12.c2 12 a2 b1 112 212 c2 22.196834 12
13.c2 13 a2 b1 113 213 c2 22.049190 13
14.c2 14 a2 b1 114 214 c2 21.632446 14
15.c2 15 a2 b1 115 215 c2 20.254271 15
16.c2 16 a2 b2 116 216 c2 20.491188 16
17.c2 17 a2 b2 117 217 c2 19.675913 17
18.c2 18 a2 b2 118 218 c2 18.337950 18
19.c2 19 a2 b2 119 219 c2 21.767734 19
20.c2 20 a2 b2 120 220 c2 20.025801 20
1.c3    1 a1 b1 101 201 c3 31.128511  1
2.c3    2 a1 b1 102 202 c3 27.619642  2
3.c3    3 a1 b1 103 203 c3 28.939734  3
4.c3    4 a1 b1 104 204 c3 30.937141  4
5.c3    5 a1 b1 105 205 c3 30.854452  5
6.c3    6 a1 b2 106 206 c3 31.460729  6
7.c3    7 a1 b2 107 207 c3 28.586901  7
8.c3    8 a1 b2 108 208 c3 30.567403  8
9.c3    9 a1 b2 109 209 c3 30.583188  9
10.c3 10 a1 b2 110 210 c3 28.693201 10
11.c3 11 a2 b1 111 211 c3 29.459614 11
12.c3 12 a2 b1 112 212 c3 31.947693 12
13.c3 13 a2 b1 113 213 c3 30.053590 13
14.c3 14 a2 b1 114 214 c3 30.351663 14
15.c3 15 a2 b1 115 215 c3 29.329023 15
16.c3 16 a2 b2 116 216 c3 30.277954 16
17.c3 17 a2 b2 117 217 c3 30.691171 17
18.c3 18 a2 b2 118 218 c3 30.823795 18
19.c3 19 a2 b2 119 219 c3 32.145065 19
20.c3 20 a2 b2 120 220 c3 27.653056 20
```

Analysis

In the upper part of the script, data is created in the wide format, which corresponds to the usual data layout where each subject occupies one row. We first create a vector VP containing the subject ids. Then two experimental factors A and B are created with gl(). Two covariates X1 and X2 are added with dummy values for better visual comparison between wide and long format. In reality these would be measurements. Finally, three columns of measurements are created, Y1 to Y3, which will eventually be turned into measures on the different levels of factor C.

3.1.6 WBD – Reshaping the data frame before the analysis   EXAMPLE!

We need the **reshape()** function for many real-world data sets, because the data layout we obtain from others often does not correspond with the layout we need for the analysis. The web-based depression therapy study is such an example. Using **reshape()** we transform the data from wide to long format, which we need in subsequent chapters for the analysis.

WBD_PRE_ReshapeToLongFormat.R

```
## WBD - Convert data frame to long format
source("Scripts/WBD_PRE_DataImport.R")

if(!exists("CPS_LongFormat", where = WBD))
{
  WBD$CPS_LongFormat <- reshape(WBD$CPS, varying = list(BDI =
paste0("BDI_", 1:5),
                                                ANH =
paste0("ANH_", 1:5),
                                                PITPV =
paste0("PITPV_", 1:5)),
                        timevar = "Measurement",
                        v.names = c("BDI", "ANH", "PITPV"),
                        idvar = c("Participant"),
                        direction = "long")
  WBD$CPS_LongFormat$Measurement <-
factor(WBD$CPS_LongFormat$Measurement,
                        labels =
c("Baseline",
                                        "Post-treatment",
                                        "1-month",
                                        "3-month",
                                        "6-month"))
}
```

Analysis

The output of the script is a data frame consisting of the depression measurements BDI 1 to 5 stacked on top of each other in the new variable BDI. To identify the time point of each variable, the parameter timevar declares the new variable Measurement. Because in the long format more than one row belongs to a single person, an id variable is required keeping measurements of one person identifiable. The last line of the script transforms the new variable Measurement into a factor, which is useful for subsequent analyses in the ANOVA or Regression context. Note that some variables in wide format are not included in the **reshape()** call. However, these variables also appear in the long format.

3.2 Types of variables

Scripts and functions work better if you correctly manage different types of variables. Be it numbers, characters, dates, or times – depending on the type of variable, some R functions are available and others are not.

STATISTICAL
ISSUE!

3.2.1 Scale levels of variables

From our education as researchers, we know that there are different types of empirical variables: nominal, ordinal, interval, and ratio scales. Nominal scales distinguish mere identities of measurement objects; ordinal scales distinguish rank orders between objects; interval scales capture measured distances between objects; and ratio scales introduce an origin to the scale to allow for the measurement of ratios between objects. The distinction originated with Stevens (1946) and has been subject to much dispute in the literature (Thomas, 2019). In R, all scale levels can be captured in numeric variables, that is, by assigning different numbers to objects with different measurements, and equal numbers to objects with the same measurements. Strings usually capture nominal scales because no numeric relationship is needed but only different identities. Experimental conditions like control group and treatment group are good examples for nominal scales captured in a string variable. In the following we learn how to use numeric variables and strings for the different types of measurements we encounter in our research.

3.2.2 Numeric variables in R

R LANGUAGE!

Most variables in R are numeric. Usually they consist of measurements and reside in data frames. We encounter measurements as integers, like 1, 2, 3; and we encounter measurements with decimals, like 1.2 or 3.83. In special circumstances, scientific expression of numbers occurs like $2 \cdot 10^{12}$ or $5 \cdot 10^{-5}$. The functions **numeric()**, **as.numeric()**, **double()**, **as.double()**, **integer()**, or **as.integer()** help us to handle numeric variables.

Suppose the following vector b <- c(5, 3, "2", 4). Returning the vector at the console gives "5" "3" "2" "4" and mean(b) gives an error because the vector is not numeric. In fact, we meant numbers and wonder why b is a character vector now, containing the numbers as strings. Upon creation of the vector, accidentally, the 2 is marked as a character with quotation marks, and this turns all content of the vector into characters. Calling mean(b) does not work for this vector, nor any other numerically based function. We may now convert b to a numeric vector with as.numeric(b). Now, the strings are converted to numbers again and we can recalculate the mean with this.

R uses scientific notation with very small numbers. For example, a very small p-value of a t-test of $p = 0.000000037$ could be written as $3.7 \cdot 10^{-8}$. In R this would be 3.7e-08.

Infinity is useful for categorisation of quantitative variables into a fixed number of categories. Usually, the lowest and highest categories are assumed to cover the value range from minus infinity to the lowest category break and from the highest category break to plus infinity. So, the reserved R terms -Inf and +Inf apply here. For example, a vector of measurements a <- c(-27, 3, 5, 5, 7, 8, 8, 13, 17, 30) could be categorised into four categories with cut(a, breaks = c(-Inf, 5, 10, Inf)).

A word on rounding: the display of numbers with ten decimal places is clumsy and does not serve to get a grasp on the data. So, numbers can be rounded with **round()**, **ceiling()**, or **floor()**. However, the position of the rounding needs consideration. Usually, do not apply rounding in the course of statistical calculations unless there is a good reason. Use rounding for data display only. For numbers with many decimal places, R internally retains a certain number of places even if the numbers are rounded for display. However, using rounding in assignments of numbers to a variable would drop decimal places and only store the rounded numbers in the variable. Subsequent calculations would suffer from avoidable imprecision.

3.2.3 String variables in R

Strings are sequences of characters. Words and sentences are sequences of strings, but also paragraphs, essays, and books. R offers functions to handle them all. Factor levels are often coded with strings, like in "intervention group 1", "intervention group 2", and "control group". Furthermore, in many studies open responses of subjects are gathered and entered in the data frame in string variables. Usually, within the same variable, responses range from short expressions "no idea" to multi-sentence statements "When I entered this troublesome situation I incidentally experienced pure fear. I thought my heart would stop and I would die here and now." As a first distinction, on the one hand, sometimes strings come with a fixed small number of different levels, and these are usually declared as a factor variable in R. On the other hand, strings can infinitely vary, like in open responses. Matloff (2011) provides an in-depth treatment of strings and there are R packages for advanced string manipulation, for example, **stringi** (Gagolewski, 2020) or **stringr** (Wickham, 2019).

3.2.4 Date and time variables in R

Date and time variables are often used in data sets if time points for measurement or other relevant dates, durations in the experiment, or between measurements are needed. They require special treatment and, thus, a special type of variable. Although time points could easily be recorded as strings, as in "20th August 2014", the use of a specialised class for this bears some advantages. These are that differences between dates can be calculated as durations, dates can be converted in different notations, durations can be converted between different units, for example, days, hours, or seconds. Functions that we need are **Date()**, **Sys.Date()**, **Sys.time()**, **difftime()**, **ISOdatetime()**. Let us practise some uses of dates and times.

Ch3_DateTimeVariables.R

```
## Use of date time variables
today <- date()
today
tomorrow <- as.Date("14.1.2021", format = "%d.%m.%Y")
tomorrow

date_baseline <- c("15.3.2011", "15.3.2011", "16.3.2011", "20.4.2011")
date_followup <- c("13.9.2011", "14.9.2011", "7.12.2011", "22.11.2011")
class(date_baseline)

date_baseline <- as.Date(date_baseline, format = "%d.%m.%Y")
date_followup <- as.Date(date_followup, format = "%d.%m.%Y")
class(date_baseline)
date_baseline
methods(class = "Date")
months(date_baseline)
quarters(date_baseline)
format(date_baseline, "%m.%Y")
as.POSIXct(date_baseline)
```

```
date_followup - date_baseline
difftime(date_followup, date_baseline, units = "weeks")
date_baseline + 30

# Today string for use in time stamps
timestamp <- format(Sys.Date(), "%Y%m%d")
timestamp
```

Output

```
> today <- date()
> today
[1] "Fri Jan 15 20:43:00 2021"
> tomorrow <- as.Date("14.1.2021", format = "%d.%m.%Y")
> tomorrow
[1] "2021-01-14"
> date_baseline <- c("15.3.2011", "15.3.2011", "16.3.2011", "20.4.2011")
> date_followup <- c("13.9.2011", "14.9.2011", "7.12.2011", "22.11.2011")
> class(date_baseline)
[1] "character"
> date_baseline <- as.Date(date_baseline, format = "%d.%m.%Y")
> date_followup <- as.Date(date_followup, format = "%d.%m.%Y")
> class(date_baseline)
[1] "Date"
> date_baseline
[1] "2011-03-15" "2011-03-15" "2011-03-16" "2011-04-20"
> methods(class = "Date")
 [1] -              [               [[             [<-            +
 [6] as.character   as.data.frame  as.list        as.POSIXct     as.POSIXlt
[11] Axis           c              coerce         cut            diff
[16] format         hist           initialize     is.numeric     julian
[21] length<-       Math           mean           months         Ops
[26] pretty         print          quarters       rep            round
[31] seq            show           slotsFromS3    split          str
[36] summary        Summary        trunc          weekdays       weighted.mean
[41] xtfrm
see '?methods' for accessing help and source code
> months(date_baseline)
[1] "März"  "März"  "März"  "April"
> quarters(date_baseline)
[1] "Q1" "Q1" "Q1" "Q2"
> format(date_baseline, "%m.%Y")
[1] "03.2011" "03.2011" "03.2011" "04.2011"
> as.POSIXct(date_baseline)
[1] "2011-03-15 01:00:00 CET"  "2011-03-15 01:00:00 CET"
"2011-03-16 01:00:00 CET"
```

```
[4] "2011-04-20 02:00:00 CEST"
> date_followup - date_baseline
Time differences in days
[1] 182 183 266 216
> difftime(date_followup, date_baseline, units = "weeks")
Time differences in weeks
[1] 26.00000 26.14286 38.00000 30.85714
> date_baseline + 30
[1] "2011-04-14" "2011-04-14" "2011-04-15" "2011-05-20"
> # Today string for use in time stamps
> timestamp <- format(Sys.Date(), "%Y%m%d")
> timestamp
[1] "20210115"
```

Analysis

The **date()** function asks the computer what today is and returns a character object; in contrast, **as.date()** takes a user input in any format and stores it as a Date object. There is an ugly argument to the format parameter of **as.date()**: "%d.%m.%Y". This tells R what the days, month, and year information in the first argument is. Around the world, some write their dates as 1.7.2017, others as 7-1-2017. So, is it the first day of the seventh month or the seventh day of the first month? The format string %d declares the day, %m declares the month, and %Y the year. The points in between correspond to the points that we used to write the date. In sum, the format string takes the declarations of special fields in a string and other characters that surround these specifications. The help page of function **strptime()** gives information about all format options.

Next, we create the two vectors date_baseline and date_followup containing strings of dates. As the names suggest, these may be the dates when experimental subjects supplied their data at two different time points in the experimental design. For illustration, we created the vectors as character vectors as shown with class(date_baseline). The point is: for a human reader the entries in these vectors already look like dates, but for R they are only characters. So, a function is required that tells R that the entries are, in fact, dates and what the year, month, and day in the given strings are. The function **as.Date()** does this with the two vectors. A call of class(date_baseline), then, shows that R treats date_baseline as a date, now. Simply printing date_baseline shows that R stripped off our notation Day. Month.Year and simply used its own for the display.

R now treats our dates as dates and we can do many things with them. A call of methods (class = "Date") shows several useful functions applicable to date objects. The function **months()** returns only the names of the months of the dates; **quarters()** returns the quarter of the year of a given date; parts of the date can be extracted with **format()** and be returned as a character vector; finally, the dates can be converted to class POSIXct with as.POSIXct, adding time information to each vector element.

We can not only use methods to return information about the dates, we can also calculate with them. For example, the elapsed time between baseline and follow-up simply is date_followup - date_baseline counted in days. The function **difftime()** gives even more control about time differences. In many experiments using repeated measurements, this is important information and will be controlled by the experimenter. Finally, the script shows how to create a time stamp for the statistical analysis. It returns the current date in the nice YearMonthDay form, which I often use in the names of automatically created files. For example, such a name could be `paste0("ResultsLinearModel", today, ".pdf")`, which uses the today time stamp and secures that repeated runs of the scripts at different days would not overwrite results files created some days earlier.

In sum, R has specialised functions to handle dates and times. Use these functions to store dates as dates and not strings in your data frame before you give it to your team. Your colleagues can then easily extract and display dates any way they want. Another package relating to the handling of dates is **chron** (James & Hornik, 2020).

3.2.5 Type conversion in R R LANGUAGE!

The variable types visited so far can all be converted back and forth between types. Numbers can become strings and the other way around, dates can become strings, data frames can become matrices, lists can become vectors and so on. Many conversion functions have the prefix "as.", like in, **as.data.frame()** or **as.matrix()**. Let us now try some of the functions for type conversion at the console.

Ch3_TypeConversion.R

```
## Type conversion
A <- 1:10
A
B <- as.character(A)
B
C <- matrix(as.numeric(B), 5)
C
D <- as.data.frame(C)
D
E <- as.matrix(D)
E
```

Output

```
> A <- 1:10
> A
 [1]  1  2  3  4  5  6  7  8  9 10
> B <- as.character(A)
> B
 [1] "1"  "2"  "3"  "4"  "5"  "6"  "7"  "8"  "9"  "10"
> C <- matrix(as.numeric(B), 5)
> C
     [,1] [,2]
```

```
[1,]    1    6
[2,]    2    7
[3,]    3    8
[4,]    4    9
[5,]    5   10
> D <- as.data.frame(C)
> D
  V1 V2
1  1  6
2  2  7
3  3  8
4  4  9
5  5 10
> E <- as.matrix(D)
> E
      V1 V2
[1,]   1  6
[2,]   2  7
[3,]   3  8
[4,]   4  9
[5,]   5 10
```

Analysis

First, we store the numbers from 1 to 10 in A. Then we convert A to character values. Their display now uses "" appropriate for strings. Then, we convert the characters back to numbers and aggregate them in a matrix. This matrix is then converted in a data frame. Finally, we convert the data frame back in a matrix. Note how the display of rows and numbers vary between matrix and data frame. The matrix display already suggests using [] for indexing.

The conversion functions have many uses in statistical analyses and are often placed unsuspiciously in the midst of code for other purposes. For example, there are functions which require a matrix with data as an argument, but you only have your data frame. Then a quick and dirty solution is to hand over a conversion of the data frame in the function call; that is, function(as.matrix(DataFrame)). If the code works now – fine. So, help yourself with type conversion if you want to proceed quickly, but do not completely rely on your code. Remember to check your results when you have more time.

3.3 Attributes – Adjunct information for variables

In this section we enrich objects with attributes to make them more meaningful in the analysis and to allow for better R programming. Attributes are basically information adjuncts for objects or variables to make them easier to use for you and your colleagues.

3.3.1 **attributes() and attr()**

Attributes hold additional object information and they can be arbitrarily assigned to any object. The attributes of an object are enquired with **attributes()** and are set with **attr()**. For example, an attribute could be a memo you wrote about the data collection, which you would assign with `attr(data, which = memo) <- MemoString`. Whether or not to use attributes with data frames and variables is a matter of personal preference. You could program a complete statistical analysis without touching attributes. However, we now want to explore some options where attributes are useful and allow for more beautiful, more robust, and more flexible analyses. Now, *we test functions to set and access object attributes.* Generally, **attr()** can do all jobs with attributes, but some other functions offer more convenience.

Ch3_attributes.R

```
## Using attributes of a matrix or data frame
d <- cbind(1:5,
           c(21, 22, 20, 21, 21),
           c(25, 33, 26, 23, 27),
           c(7, 9, 15, 8, 9))
d

# Get attributes
attributes(d)
str(d)
class(d)

# Set attributes
comment(d) <- "A 5 x 4 matrix of numbers"
dimnames(d) <- list(paste("Subject", 1:5), paste("Variable", 1:4))
d

# Get attributes again
attributes(d)
str(d)

# Convert d to a data frame
d_df <- as.data.frame(d)
d_df

# Get attributes
attributes(d_df)
attributes(d_df$`Variable 1`)
str(d_df)
class(d_df)

# Set attributes
names(d_df) <- c("ID", "ASO", "ASR", "DEP")
row.names(d_df) <- c("Susi", "John", "Mary", "Jane", "Dave")
comment(d_df) <- "Data from my first five subjects"
```

```
comment(d_df$ID) <- "Subject ID"
comment(d_df$ASO) <- "Age at symptom onset"
comment(d_df$ASR) <- "Age of symptom remission"
comment(d_df$DEP) <- "Depression score"

# Get attributes again
attributes(d_df)
str(d_df)

# Attributes of a data subset
d_df_HighDepression <- d_df[d_df$DEP > 8,]
d_df_HighDepression
attributes(d_df_HighDepression)
str(d_df_HighDepression)
```

Output

```
> d
     [,1] [,2] [,3] [,4]
[1,]   1   21   25    7
[2,]   2   22   33    9
[3,]   3   20   26   15
[4,]   4   21   23    8
[5,]   5   21   27    9
> # Get attributes
> attributes(d)
$dim
[1] 5 4

> str(d)
 num [1:5, 1:4] 1 2 3 4 5 21 22 20 21 21 ...
> class(d)
[1] "matrix"
> # Set attributes
> comment(d) <- "A 5 x 4 matrix of numbers"
> dimnames(d) <- list(paste("Subject", 1:5), paste("Variable", 1:4))
> d
          Variable 1 Variable 2 Variable 3 Variable 4
Subject 1          1         21         25          7
Subject 2          2         22         33          9
Subject 3          3         20         26         15
Subject 4          4         21         23          8
Subject 5          5         21         27          9
> # Get attributes again
> attributes(d)
$dim
[1] 5 4
```

```
$comment
[1] "A 5 x 4 matrix of numbers"

$dimnames
$dimnames[[1]]
[1] "Subject 1" "Subject 2" "Subject 3" "Subject 4" "Subject 5"

$dimnames[[2]]
[1] "Variable 1" "Variable 2" "Variable 3" "Variable 4"

> str(d)
 num [1:5, 1:4] 1 2 3 4 5 21 22 20 21 21 ...
 - attr(*, "comment")= chr "A 5 x 4 matrix of numbers"
 - attr(*, "dimnames")=List of 2
  ..$ : chr [1:5] "Subject 1" "Subject 2" "Subject 3" "Subject 4" ...
  ..$ : chr [1:4] "Variable 1" "Variable 2" "Variable 3" "Variable 4"
> # Convert d to a data frame
> d_df <- as.data.frame(d)
> d_df
          Variable 1 Variable 2 Variable 3 Variable 4
Subject 1          1         21         25          7
Subject 2          2         22         33          9
Subject 3          3         20         26         15
Subject 4          4         21         23          8
Subject 5          5         21         27          9
> # Get attributes
> attributes(d_df)
$names
[1] "Variable 1" "Variable 2" "Variable 3" "Variable 4"

$class
[1] "data.frame"

$row.names
[1] "Subject 1" "Subject 2" "Subject 3" "Subject 4" "Subject 5"

> attributes(d_df$`Variable 1`)
NULL
> str(d_df)
'data.frame':   5 obs. of  4 variables:
 $ Variable 1: num  1 2 3 4 5
 $ Variable 2: num  21 22 20 21 21
 $ Variable 3: num  25 33 26 23 27
 $ Variable 4: num  7 9 15 8 9
> class(d_df)
[1] "data.frame"
> # Set attributes
> names(d_df) <- c("ID", "ASO", "ASR", "DEP")
```

```
> row.names(d_df) <- c("Susi", "John", "Mary", "Jane", "Dave")
> comment(d_df) <- "Data from my first five subjects"
> comment(d_df$ID) <- "Subject ID"
> comment(d_df$ASO) <- "Age at symptom onset"
> comment(d_df$ASR) <- "Age of symptom remission"
> comment(d_df$DEP) <- "Depression score"
> # Get attributes again
> attributes(d_df)
$names
[1] "ID"  "ASO" "ASR" "DEP"

$row.names
[1] "Susi" "John" "Mary" "Jane" "Dave"

$comment
[1] "Data from my first five subjects"

$class
[1] "data.frame"

> str(d_df)
'data.frame':   5 obs. of  4 variables:
 $ ID : num  1 2 3 4 5
  ..- attr(*, "comment")= chr "Subject ID"
 $ ASO: num  21 22 20 21 21
  ..- attr(*, "comment")= chr "Age at symptom onset"
 $ ASR: num  25 33 26 23 27
  ..- attr(*, "comment")= chr "Age of symptom remission"
 $ DEP: num  7 9 15 8 9
  ..- attr(*, "comment")= chr "Depression score"
 - attr(*, "comment")= chr "Data from my first five subjects"
> # Attributes of a data subset
> d_df_HighDepression <- d_df[d_df$DEP > 8,]
> d_df_HighDepression
      ID ASO ASR DEP
John  2  22  33   9
Mary  3  20  26  15
Dave  5  21  27   9
> attributes(d_df_HighDepression)
$names
[1] "ID"  "ASO" "ASR" "DEP"

$comment
[1] "Data from my first five subjects"

$row.names
[1] "John" "Mary" "Dave"
```

```
$class
[1] "data.frame"

> str(d_df_HighDepression)
'data.frame':   3 obs. of  4 variables:
 $ ID : num  2 3 5
 $ ASO: num  22 20 21
 $ ASR: num  33 26 27
 $ DEP: num  9 15 9
 - attr(*, "comment")= chr "Data from my first five subjects"
```

Analysis

In this code example we create a matrix d *with numbers, first, and then inspect it at the console.*
Then, **attributes()**, **str()**, and **class()** show that there is nothing special about d at
this point. However, with **comment()** and **dimnames()** we add some useful information,
and another call of **attributes()** and **str()** shows the information that we just added.
Next, we convert d to the object d_df of class data.frame and inspect it at the console.
Interestingly, a call of **attributes()** with d_df reveals that the former matrix attribute
dimnames, which was a list of two elements for subjects and variables, is now stored as the
two attributes names and row.names. In data frames, not only attributes of the whole data
frame, but of any variable contained can be enquired and manipulated. For example, the call
attributes(d_df$'Variable 1') enquires the attributes of the first variable. Next, observe
how the two calls of **str()** for the matrix d and for the data frame d_df differ. For the matrix
str() shows num [1:5, 1:4] which means that it is a numeric vector arranged in a matrix
of five rows and four columns, for the data frame it shows that it is, in fact, a data frame of
five observations and four variables. Further, notice that the former dimnames attribute has
been translated to appropriate data frame attributes.

Let us move on to the assignment of attributes to our data frame. In the script, we first change
the names and row.names of d_df to make the data frame prettier. Then, we add a comment
to the data frame about the origin of the data and further add variable labels as comments
to the single variables in the data frame. Note how the five calls of **comment()** differ here:
the first one only uses the data frame, but the other four use indexing of the variables
with the **$**-operator. Afterwards, **attributes()** and **str()** reveal our changes to the attri-
butes. Here, a difference between the two functions becomes obvious: **attributes()** only
returns the attributes of the data frame, whereas **str()** also return attributes of the contained variables, namely the variable labels stored in comments. You have to get used to the
output format of **str()** used with complex objects to decipher which attributes belong to
which element of the object. The data frame comment is preceded by - attr(*, "comment");
the variable comments are preceded by ..- attr(*, "comment"). The two dots in the case
of the variables indicate that there is an attribute further down in the structure of the data
frame. It is helpful to know how to read the information **str()** returns because in the analysis you sometimes want to access a single piece of information nested deep in the structure
of an object and it costs your time to figure out how to do it.

EXAMPLE! ## 3.3.2 SCB – Setting variable labels

In the student reconciliator example we can add the available information to the data frame SCB. We use the **attr()** function to add variable labels. However, before running the following script use str(SCB) to inspect the attributes of the data frame.

SCB_PRE_VariableLabels.R

```
## SCB – Variable labels
source("./Script/SCB_PRE_DataImport.R")

attr(SCB, "variable.labels") <- c(ID = "School ID",
                                  A = "Intervention program",
                                  B = "Type of school",
                                  X1 = "Need of renovation of the school",
                                  X2 = "Average pocket money of school students
per week",
                                  Y1C1 = "Person x Person M1",
                                  Y2C1 = "Group x Person M1",
                                  Y3C1 = "Group x Group M1",
                                  Y1C2 = "Person x Person M2",
                                  Y2C2 = "Group x Person M2",
                                  Y3C2 = "Group x Group M2",
                                  Y1C3 = "Person x Person M3",
                                  Y2C3 = "Group x Person M3",
                                  Y3C3 = "Group x Group M3",
                                  Y1C4 = "Person x Person M4",
                                  Y2C4 = "Group x Person M4",
                                  Y3C4 = "Group x Group M4")
```

Output

```
> str(SCB)
'data.frame':   180 obs. of  18 variables:
 $ X   : int  1 2 3 4 5 6 7 8 9 10 ...
 $ ID  : int  1 2 3 4 5 6 7 8 9 10 ...
 $ A   : chr  "No intervention" "No intervention" "No intervention"
"No intervention" ...
 $ B   : chr  "Primary school" "Primary school" "Primary school"
"Primary school" ...
 $ X1  : int  5 5 8 5 5 8 6 3 8 4 ...
 $ X2  : int  16 17 16 18 17 16 19 13 19 17 ...
 $ Y1C1: int  21 19 24 22 25 21 23 19 23 16 ...
 $ Y2C1: int  20 20 26 23 17 27 22 17 22 18 ...
 $ Y3C1: int  22 22 26 18 21 25 23 20 23 23 ...
 $ Y1C2: int  18 21 25 22 19 24 23 17 24 21 ...
 $ Y2C2: int  22 22 23 22 22 24 20 19 26 21 ...
```

```
$ Y3C2: int  20 23 23 19 24 21 21 18 23 18 ...
$ Y1C3: int  21 20 22 21 20 24 26 19 26 25 ...
$ Y2C3: int  20 18 20 25 18 25 22 22 26 19 ...
$ Y3C3: int  21 18 26 22 23 24 21 19 24 19 ...
$ Y1C4: int  25 24 23 22 24 25 24 20 26 22 ...
$ Y2C4: int  22 20 25 20 22 23 22 17 23 21 ...
$ Y3C4: int  22 22 19 22 19 27 26 17 21 20 ...
- attr(*, "variable.labels")= Named chr [1:17] "School ID" "Intervention
program" "Type of school" "Need of renovation of the school" ...
 ..- attr(*, "names")= chr [1:17] "ID" "A" "B" "X1" ...
```

Analysis

Calling `str(SCB)` reveals that the data frame obtained the new attribute `variable.labels`, which is a string vector consisting of the actual labels. The attribute itself has a names attribute to associate the labels with the variable names of the data frame. Further analyses of the data can now use the good-looking variable labels in results tables or graphics annotation.

3.3.3 The class attribute of an R object   R LANGUAGE!

Every object in R has a class attribute, which is used for mapping functions to the specific content of an object. We noticed at the very beginning that everything in R is an object of some type or class. Some essential characteristics of R programming go with this notion. As it is true that objects of our natural environment which belong to the same class share some characteristics, so do objects which belong to the same class in R. Two cars of different brands, however, have four wheels, a steering wheel, seatbelts, and other characteristics in common. We could say that the two particular cars belong to one general class named car. Similarly, the data frames of two different studies have the characteristics of data rows, data columns, variable names, and other information in common. Again, we could say that the two particular data frames belong to one general class named data frame. Knowing that a given object belongs to a certain class, we can guess some of its characteristics or about the information it contains.

Furthermore, objects of a certain class do not only share characteristics, they also share behaviour. You can accelerate your car, drive around a bend, or turn on the wind shield wiper. Comparably, you can summarise the content of a data frame, plot the variables, or merge it with another data frame. In fact, there are certain things you can do with proper cars and there are certain other things that you can do with data frames. In R we say that classes possess so-called methods, which define what you can do with exemplars belonging to the class. Put simply, the methods of a class operate on the information that exemplars of the class contain.

Let us create a matrix and a data frame at the R console and discover how classes and their methods work in R. The functions **is()**, **class()**, and **methods()** help us with this.

Ch3_ClassAndMethods.R

```
## Discovering the class and the methods of an object
```

```
d_mx <- cbind(1:5,
              c(21, 22, 20, 21, 21),
              c(25, 33, 26, 23, 27),
              c(7, 9, 15, 8, 9))
d_mx
class(d_mx)

d_df <- as.data.frame(d_mx)
d_df
class(d_df)
is(d_df)

methods(class = "matrix")
summary(d_mx)

methods(class = "data.frame")
stack(d_df)

methods("summary")
methods("stack")
```

Output

```
> d_mx
     [,1] [,2] [,3] [,4]
[1,]    1   21   25    7
[2,]    2   22   33    9
[3,]    3   20   26   15
[4,]    4   21   23    8
[5,]    5   21   27    9
> class(d_mx)
[1] "matrix"
> d_df <- as.data.frame(d_mx)
> d_df
  V1 V2 V3 V4
1  1 21 25  7
2  2 22 33  9
3  3 20 26 15
4  4 21 23  8
5  5 21 27  9
> class(d_df)
[1] "data.frame"
> is(d_df)
[1] "data.frame" "list"        "oldClass"   "vector"
> methods(class = "matrix")
 [1] anyDuplicated as.data.frame as.raster     boxplot       coerce
 [6] determinant   duplicated    edit          head          initialize
[11] isSymmetric   Math          Math2         Ops           relist
```

```
[16] subset          summary         tail            unique
see '?methods' for accessing help and source code
> summary(d_mx)
       V1              V2              V3              V4
 Min.   :1     Min.   :20     Min.   :23.0    Min.   : 7.0
 1st Qu.:2     1st Qu.:21     1st Qu.:25.0    1st Qu.: 8.0
 Median :3     Median :21     Median :26.0    Median : 9.0
 Mean   :3     Mean   :21     Mean   :26.8    Mean   : 9.6
 3rd Qu.:4     3rd Qu.:21     3rd Qu.:27.0    3rd Qu.: 9.0
 Max.   :5     Max.   :22     Max.   :33.0    Max.   :15.0
> methods(class = "data.frame")
 [1] $<-               [              [[            [[<-         [<-
 [6] aggregate         anyDuplicated  as.data.frame as.list      as.matrix
[11] by                cbind          coerce        dim          dimnames
[16] dimnames<-        droplevels     duplicated    edit         format
[21] formula           head           initialize    is.na        Math
[26] merge             na.exclude     na.omit       Ops          plot
[31] print             prompt         rbind         row.names    row.names<-
[36] rowsum            show           slotsFromS3   split        split<-
[41] stack             str            subset        summary      Summary
[46] t                 tail           transform     type.convert unique
[51] unstack           within
see '?methods' for accessing help and source code
> stack(d_df)
   values ind
1       1  V1
2       2  V1
3       3  V1
4       4  V1
5       5  V1
6      21  V2
7      22  V2
8      20  V2
9      21  V2
10     21  V2
11     25  V3
12     33  V3
13     26  V3
14     23  V3
15     27  V3
16      7  V4
17      9  V4
18     15  V4
19      8  V4
20      9  V4
> methods("summary")
```

```
 [1] summary.aov                    summary.aovlist*
 [3] summary.aspell*                summary.check_packages_in_dir*
 [5] summary.connection             summary.data.frame
 [7] summary.Date                   summary.default
 [9] summary.ecdf*                  summary.factor
[11] summary.glm                    summary.infl*
[13] summary.lm                     summary.loess*
[15] summary.manova                 summary.matrix
[17] summary.mlm*                   summary.nls*
[19] summary.packageStatus*         summary.POSIXct
[21] summary.POSIXlt                summary.ppr*
[23] summary.prcomp*                summary.princomp*
[25] summary.proc_time              summary.srcfile
[27] summary.srcref                 summary.stepfun
[29] summary.stl*                   summary.table
[31] summary.tukeysmooth*           summary.warnings
see '?methods' for accessing help and source code
> methods("stack")
[1] stack.data.frame* stack.default*
see '?methods' for accessing help and source code
```

Analysis

First, we create a matrix d_mx with some numbers using **cbind()**. The function **class()** shows that d_mx is indeed an instance of the class matrix. Then, we convert the d_mx into a data frame d_df using **as.data.frame()** and confirm the conversion with class(d_df). We find that the exact class name is data.frame. If we use **is()** with d_df, we find that d_df is not only a data frame but also a list. So, we can expect that functions for lists also apply to our data frame. After confirming the classes of our two objects, we inquire which methods apply to them. The **methods()** function with its class parameter helps with this. We use **methods()** for matrix first and obtain a list of more than ten functions, which can be used with matrices. We then select **summary()** for a test. The function **methods()** applied to data.frame reveals another list of functions, which apply to data frames. Now, we select **stack()** for a test. The **methods()** function also works the other way round: we can not only inquire which methods apply to a class, but which classes can be handled by a particular generic method. Three very popular generic methods are **summary()**, **print()**, and **plot()**. They apply to a long list of classes. Our example shows the many classes handled by **summary()**. In contrast, as can be seen, **stack()** mainly handles data frames.

Why is this tinkering with classes and methods important? Although using R is possible without consideration of this, it helps to better explore unknown territory in R. Several packages supply their functionality by defining one or more classes and associated methods. For example, **mice** for multiple imputations and subsequent modelling defines its own classes and methods to impute, inspect, and model data. Another example is the **zipfR** package for lexical statistics, which also comes with its own data structure and methods. Learning the functionality of an R package often means, first, to inquire the classes the package provides and, second, the available function to apply to these classes.

3.4 Factors - A special type of variable

One variable type needs special attention: factors. The concept of a factor in R corresponds with what researchers call independent variable or experimental factor. Specialised functions prepare factors according to the actual experimental design; for example, to figure in trend analyses or statistical contrasts.

3.4.1 Working with factor variables in R   R LANGUAGE!

Some variables in a data set usually serve as experimental factors, independent variables, or predictors. The variables are usually of nominal scale level and have a small set of categories. Examples are gender, education, marital status, or chronic disease. We use these factors and their interactions in analyses of variance or regression to explain variance in a dependent variable. To use factors in R the corresponding variables need explicit declaration as factors with `factor()`; however, sometimes they are created automatically as a factor in data import. For example, `read.spss()` treats string variables as factors by default, assigning a factor level to every different string contained in the variable. Modelling functions then automatically use information stored in the factor variables to calculate appropriate statistics, like contrasts or trend analyses.

You have to master factors, because *the factor specification often makes a difference in the output of your calculations.* So, the next console session introduces factors and their attributes. A contrived experiment with simulated data regarding the perceived threat by different animals as rated by subjects helps us with this. That is, we take a fly, a rabbit, a crocodile, and a shark, and then take five seconds to conjure up a theory of how threatening these beasts would be perceived by us humans if we happen to blunder into their habitat: we would perceive flies and rabbits as much less threatening compared with crocodiles and sharks. The following console code generates data with animal as a factor variable corresponding to the independent variable in the design, and perceived threat as the dependent variable. We manipulate the factor Animal several times and, finally, use a linear model to evaluate our hypothesis.

Ch3_FactorVariables1.R

```
## Factor variables 1 - Initialise a factor and set a reference category
# Create data frame
d  <-  data.frame(Animal  =  factor(rep(c("Shark","Rabbit","Crocodile","Fly"),
each = 10)),
              Threat = rnorm(40, rep(c(8, 2, 7, 1), each = 10)))

# Enquire factor information
d$Animal
is.ordered(d$Animal)
contrasts(d$Animal)

# Set a reference category
d$Animal <- relevel(d$Animal, "Fly")
d$Animal
contrasts(d$Animal)
```

Output

```
> # Enquire factor information
> d$Animal
 [1] Shark      Shark      Shark      Shark      Shark      Shark      Shark      Shark
 [9] Shark      Shark      Rabbit     Rabbit     Rabbit     Rabbit     Rabbit     Rabbit
[17] Rabbit     Rabbit     Rabbit     Rabbit     Crocodile  Crocodile  Crocodile
Crocodile
[25] Crocodile Crocodile Crocodile Crocodile Crocodile Crocodile Fly        Fly
[33] Fly        Fly        Fly        Fly        Fly        Fly        Fly        Fly
Levels: Crocodile Fly Rabbit Shark
> is.ordered(d$Animal)
[1] FALSE
> contrasts(d$Animal)
          Fly Rabbit Shark
Crocodile   0      0     0
Fly         1      0     0
Rabbit      0      1     0
Shark       0      0     1
>
> # Set a reference category
> d$Animal <- relevel(d$Animal, "Fly")
> d$Animal
 [1] Shark      Shark      Shark      Shark      Shark      Shark      Shark      Shark
 [9] Shark      Shark      Rabbit     Rabbit     Rabbit     Rabbit     Rabbit     Rabbit
[17] Rabbit     Rabbit     Rabbit     Rabbit     Crocodile  Crocodile  Crocodile
Crocodile
[25] Crocodile Crocodile Crocodile Crocodile Crocodile Crocodile Fly        Fly
[33] Fly        Fly        Fly        Fly        Fly        Fly        Fly        Fly
Levels: Fly Crocodile Rabbit Shark
> contrasts(d$Animal)
          Crocodile Rabbit Shark
Fly               0      0     0
Crocodile         1      0     0
Rabbit            0      1     0
Shark             0      0     1
>
```

Analysis

The session begins with the creation of random data of perceived threat in the four conditions of the factor Animal. Then, calling d$Animal returns the data contained in the factor and also the four factor levels. Note that **rep()** was used to create the factor and not **factor()**; nevertheless, it still supplied the data vector of strings automatically as a factor. Next, **is.ordered()** and **contrasts()** return further information about factor attributes, in fact, that there is no order of factor levels, so statistical contrasts would use the first factor level as shown in levels as a reference category. As can be seen, the four levels appear in alphabetical order and the matrix returned by **contrasts()** shows treatment contrasts which would use the crocodile category as intercept in linear modelling. We may designate another one of the categories as the first category; perhaps the least threatening animal could go here, which is supposedly the fly. The **relevel()** function puts fly in place. Checking the contrasts again reveals that now fly is the reference category. Reference categories have many applications in research. Often, one or more experimental conditions are compared with a control condition as a reference category. If the experimental factor is declared as a factor in R, **relevel()** can set the control condition as the reference category.

3.4.2 Ordered factors in R

Let us carry on with the analysis of the foregoing script. An order of factor levels is relevant in trend analysis and in any other analysis where factor levels somehow reflect the idea of a more or less of something, for example, dose of medication or size of musical interval. If factor levels are created from integers, R by default guesses the order of the factor levels according to the size of the integers. In other situations the order must be manually assigned. The functions **is.ordered()** and **as.ordered()** help here.

Ch3_FactorVariables2.R

```
## Factor variables 2 - Ordered factors
source("Ch3_FactorVariables1.R")

# Introduce an order of factor levels
d$Animal <- as.ordered(d$Animal)
d$Animal
contrasts(d$Animal)

# Reorder the factor levels
levels(d$Animal) <- list("Fly" = "Fly", "Rabbit" = "Rabbit", "Crocodile" =
"Crocodile", "Shark" = "Shark")
#reorder(d$Animal, rep(c(4, 2, 3, 1), each = 10)) # Does the same but does not
look nice.
```

Output

```
> # Introduce an order of factor levels
> d$Animal <- as.ordered(d$Animal)
> d$Animal
 [1] Shark     Shark     Shark     Shark     Shark     Shark     Shark     Shark
Shark     Shark     Rabbit
[12] Rabbit    Rabbit    Rabbit    Rabbit    Rabbit    Rabbit    Rabbit    Rabbit
Rabbit    Crocodile Crocodile
[23] Crocodile Crocodile Crocodile Crocodile Crocodile Crocodile Crocodile
Crocodile Fly       Fly       Fly
[34] Fly       Fly       Fly       Fly       Fly       Fly       Fly
```

```
Levels: Fly < Crocodile < Rabbit < Shark
> contrasts(d$Animal)
                .L    .Q          .C
[1,] -0.6708204   0.5 -0.2236068
[2,] -0.2236068  -0.5  0.6708204
[3,]  0.2236068  -0.5 -0.6708204
[4,]  0.6708204   0.5  0.2236068
>
> # Reorder the factor levels
> levels(d$Animal) <- list("Fly" = "Fly", "Rabbit" = "Rabbit", "Crocodile" =
"Crocodile", "Shark" = "Shark")
```

Analysis

The script starts with **source()**, which simply runs the script of the previous section to have the data available. Then, **as.ordered()** declares the factor Animal as an ordered factor. In the example, **as.ordered()** yields the order Fly < Crocodile < Rabbit < Shark. The order relation is indicated here with the < symbol. However, the given order seems to be meaningless in our example. So, to impose the order you want, you first have to use **levels()** to bring the factor levels in the correct order. In the example, a trend of perceived threat could be calculated with the different categories of animal. As can be seen, I prefer to use the **levels()** function instead of the **reorder()** function to reorder the levels. To me, it is more intuitive to be able to read the factor levels in the desired order, rather than to provide a more abstract numerical argument to **reorder()**, which indicates for each entry in the data vector of the factor the new position of the respective factor level. This does not seem to be intuitive to use.

3.4.3 Grouping and dropping of factor levels in R

Moreover, sometimes the effects of all factor levels do not matter in an analysis, but the effect of a group of factor levels against another group of factor levels. In this case a new factor variable can group factor levels under new names. Then, this recoded factor variable serves in subsequent analyses. The function **levels()** can be used for the grouping. In fact, it is not so much a grouping but simply the assignment of a common label for previously distinguishable factor levels.

Ch3_FactorVariables3.R

```
## Factor variables 3 - Grouping and dropping of factor levels
source("Ch3_FactorVariables1.R")
source("Ch3_FactorVariables2.R")

# Group factor levels
d$PredatorPrey <- d$Animal
levels(d$PredatorPrey) <- list("Prey" = "Fly", "Prey" = "Rabbit", "Predator" =
"Crocodile", "Predator" = "Shark")
```

```
# Factor levels in data subsets
d_subset <- d[d$Animal != "Fly",]
d_subset$Animal
d_subset$Animal <- droplevels(d_subset$Animal)
```

Output

```
> # Group factor levels
> d$PredatorPrey <- d$Animal
> levels(d$PredatorPrey) <- list("Prey" = "Fly", "Prey" = "Rabbit", "Predator" =
"Crocodile", "Predator" = "Shark")
>
> # Factor levels in data subsets
> d_subset <- d[d$Animal != "Fly",]
> d_subset$Animal
 [1] Shark     Shark     Shark     Shark     Shark     Shark     Shark     Shark
Shark     Shark     Rabbit
[12] Rabbit    Rabbit    Rabbit    Rabbit    Rabbit    Rabbit    Rabbit    Rabbit
Rabbit    Crocodile Crocodile
[23] Crocodile Crocodile Crocodile Crocodile Crocodile Crocodile Crocodile
Crocodile
Levels: Fly < Rabbit < Crocodile < Shark
```

Analysis

In the example we may hypothesise that the four types of animals belong to the two more inclusive categories of prey and predator, and that perceived threat is higher for animals classified as predator. So, to group fly and rabbit as prey and crocodile and shark as predator, we first create a new factor variable in the data frame as a simple copy of the animal factor. Then, **levels()** obtains a list containing the grouping of animals and applies it to the new factor. This way factors with many factor levels can be simplified.

Grouping of factor levels may be applied with unbalanced data sets, where some factor levels occur rarely; for example, some rare diseases in a clinical data set. In this case, combining factor levels may restrict the interpretation of effects to certain combinations, but may yield better and simpler designs for analysis. Another option is to assign a factor level to every combination of the levels of two or more factors, respectively; that is, to give each level of the interaction of two factors a certain name. For this purpose the function **interaction()** could precede **levels()** to create the levels of the interaction first and to rename them afterwards.

A technical difficulty arises in data subsets based on a selection of factor levels. *If a data subset is created which excludes all instances of a certain factor level, then the factor level is still retained in the factor variable of the data subset.* For example, if all subjects in the fly condition are removed from the data frame, then the animal variable of the subset data frame still contains the factor level Fly. In this situation the **droplevels()** function helps. Applied to a factor, it only retains factor levels that occur in the data frame. In the example, this means that if all subjects in the Fly condition are removed, then the level Fly would be dropped from the factor variable.

3.4.4 The contrast attribute of factors in R

Statistical contrasts are a highly recommended way to compare the means of a dependent variable between experimental conditions. The contrast attribute of factors allows us to set up the comparisons between conditions and then run a regression analysis. To learn the theory of contrasts, see Hays (1994) or Kirk (2012).

Ch3_FactorVariables4.R

```
## Factor variables 4 - The contrast attribute of factors
source("Ch3_FactorVariables1.R")
source("Ch3_FactorVariables2.R")
source("Ch3_FactorVariables3.R")

# Calculate treatment contrast of factor animal regarding perceived threat
contrasts(d$Animal)
contrasts(d$Animal) <- contr.treatment(nlevels(d$Animal))
contrasts(d$Animal)

m1 <- lm(Threat ~ Animal, d)
summary(m1)
cbind(model.frame(m1), model.matrix(m1))

# Calculate special contrasts of factor animal regarding perceived threat
contrasts(d$Animal, how.many = 2) <- cbind(FlyVsRabbit = c(-1, 1, 0, 0),
                                 PreyVsPredator = c(-.5, -.5, .5, .5))
contrasts(d$Animal)

m2 <- lm(Threat ~ Animal, d)
summary(m2)
cbind(model.frame(m2), model.matrix(m2))

str(d)
contrasts(d$Animal) <- NULL
contrasts(d$Animal)
```

Output

```
> # Calculate treatment contrast of factor animal regarding perceived threat
> contrasts(d$Animal)
            .L      .Q        .C
[1,] -0.6708204   0.5  -0.2236068
[2,] -0.2236068  -0.5   0.6708204
[3,]  0.2236068  -0.5  -0.6708204
[4,]  0.6708204   0.5   0.2236068
> contrasts(d$Animal) <- contr.treatment(nlevels(d$Animal))
> contrasts(d$Animal)
        2 3 4
Fly     0 0 0
```

```
Rabbit     1 0 0
Crocodile  0 1 0
Shark      0 0 1
>
> m1 <- lm(Threat ~ Animal, d)
> summary(m1)

Call:
lm(formula = Threat ~ Animal, data = d)

Residuals:
     Min       1Q   Median       3Q      Max
-2.78773 -0.58599  0.09512  0.71491  2.81914

Coefficients:
            Estimate Std. Error t value Pr(>|t|)
(Intercept)   0.8348     0.3547   2.353   0.0242 *
Animal2       1.1960     0.5017   2.384   0.0225 *
Animal3       6.1167     0.5017  12.193 2.42e-14 ***
Animal4       7.2408     0.5017  14.433  < 2e-16 ***
---
Signif. codes:  0 '***' 0.001 '**' 0.01 '*' 0.05 '.' 0.1 ' ' 1

Residual standard error: 1.122 on 36 degrees of freedom
Multiple R-squared:  0.8943,    Adjusted R-squared:  0.8855
F-statistic: 101.5 on 3 and 36 DF,  p-value: < 2.2e-16

> cbind(model.frame(m1), model.matrix(m1))
        Threat    Animal (Intercept) Animal2 Animal3 Animal4
1    8.1576284     Shark           1       0       0       1
2    7.7244039     Shark           1       0       0       1
3    8.0359844     Shark           1       0       0       1
[...]
38  -1.9529563       Fly           1       0       0       0
39   1.2769995       Fly           1       0       0       0
40   0.6434194       Fly           1       0       0       0
>
> # Calculate special contrasts of factor animal regarding perceived threat
> contrasts(d$Animal, how.many = 2) <- cbind(FlyVsRabbit = c(-1, 1, 0, 0),
+                          PreyVsPredator = c(-.5, -.5, .5, .5))
> contrasts(d$Animal)
          FlyVsRabbit PreyVsPredator
Fly               -1           -0.5
Rabbit             1           -0.5
Crocodile          0            0.5
Shark              0            0.5
>
> m2 <- lm(Threat ~ Animal, d)
> summary(m2)
```

```
Call:
lm(formula = Threat ~ Animal, data = d)

Residuals:
    Min      1Q  Median      3Q     Max
-2.7877 -0.6513 -0.0178  0.6875  3.3812

Coefficients:
                    Estimate Std. Error t value Pr(>|t|)
(Intercept)           4.4732     0.1868  23.952   <2e-16 ***
AnimalFlyVsRabbit     0.5980     0.2641   2.264   0.0295 *
AnimalPreyVsPredator  6.0808     0.3735  16.280   <2e-16 ***
---
Signif. codes:  0 '***' 0.001 '**' 0.01 '*' 0.05 '.' 0.1 ' ' 1

Residual standard error: 1.181 on 37 degrees of freedom
Multiple R-squared:  0.8795,    Adjusted R-squared:  0.873
F-statistic: 135.1 on 2 and 37 DF,  p-value: < 2.2e-16

> cbind(model.frame(m2), model.matrix(m2))
        Threat    Animal (Intercept) AnimalFlyVsRabbit AnimalPreyVsPredator
1    8.1576284     Shark           1                 0                  0.5
2    7.7244039     Shark           1                 0                  0.5
3    8.0359844     Shark           1                 0                  0.5
[...]
38  -1.9529563       Fly           1                -1                 -0.5
39   1.2769995       Fly           1                -1                 -0.5
40   0.6434194       Fly           1                -1                 -0.5
>
> str(d)
'data.frame':   40 obs. of  3 variables:
 $ Animal     : Ord.factor w/ 4 levels "Fly"<"Rabbit"<..: 4 4 4 4 4 4 4 4 4 4 ...
  ..- attr(*, "contrasts")= num [1:4, 1:2] -1 1 0 0 -0.5 -0.5 0.5 0.5
  .. ..- attr(*, "dimnames")=List of 2
  .. .. ..$ : chr  "Fly" "Rabbit" "Crocodile" "Shark"
  .. .. ..$ : chr  "FlyVsRabbit" "PreyVsPredator"
 $ Threat     : num  8.16 7.72 8.04 8.79 6.21 ...
 $ PredatorPrey: Ord.factor w/ 2 levels "Prey"<"Predator": 2 2 2 2 2 2 2 2 2 2 ...
> contrasts(d$Animal) <- NULL
> contrasts(d$Animal)
             .L   .Q          .C
[1,] -0.6708204  0.5 -0.2236068
[2,] -0.2236068 -0.5  0.6708204
[3,]  0.2236068 -0.5 -0.6708204
[4,]  0.6708204  0.5  0.2236068
```

Analysis

Animal is an ordered factor and the default contrasts for such factors correspond with trend analyses. That is, the contrast coefficients allow for the test of a linear, quadratic, and cubic trend. With **`contr.treatment()`**, we change the contrast coefficients and set three independent contrasts for treatment levels. Calculating a linear regression with **`lm()`** and with the contrast coefficients as dummy predictors, we obtain estimations and tests of the contrasts. The functions **`model.frame()`** and **`model.matrix()`** reveal how the regression with dummy predictors worked.

Next, we use **`contrasts()`** to assign our own contrast coefficients to the factor. The first contrast compares the perceived threat between Fly and Rabbit without using the other to factor levels. The second contrast uses pairs of factor levels for the comparison. For both contrasts, we also calculate the linear regression and inspect the design matrix with **`model.frame()`** and **`model.matrix()`**. Finally, we reset our contrasts by assigning NULL to the contrasts of Animal.

3.4.5 The use of factors in the statistical analysis

PLANNING ISSUE!

The correct place in your analysis for the preparation of factors is not unambiguous. To set some string variables as factors and to assign value labels can be done right after data import. If other attributes of factors, like contrasts, relate to the planned analysis, it is important not to separate the declaration too far from the analysis. For example, contrasts for a factor may change between analyses. In this case, a contrast may be assigned in the analysis script right before the modelling function and be reset with `contrasts(FactorVariable) <- NULL` afterwards. Moreover, factor levels may be grouped with **`levels()`** only for a specific analysis and be distinguished again later. *It is dangerous if a contrast or reference category is not reset after the analysis.* Then, in other contexts, incorrect settings may affect calculations without notification.

3.5 Structuring the complete analysis I

This section steps back from the programming code and looks at the complete analysis. How do we order the scripts to form the whole analysis? And how do we structure each single script so that it fits in with the analysis?

3.5.1 Structuring the complete statistical analysis

PLANNING ISSUE!

Put all steps of the statistical analysis in code scripts. *These steps may be data import, data cleaning, quality checks, descriptive and inferential statistics, and others. And each step goes in a different code script. Let each script do one thing.* This strategy enables big and comprehensive analyses without clutter and chaos. Each step in the analysis has its place, is retrievable and modifiable even months later. This would be impossible if all steps of an analysis populate a 500+ lines script. Some steps may be more important than others, some may need revision while others are clearly assembled. If the whole analysis resides in one script, clutter grows and after some time the important would be indistinguishable from the unimportant, redundant, misleading, and wrong code. So, *it is best to split the analysis in subtasks*. For example, if an analysis comprises different versions of a linear model, all models should obtain their own script. Then, arrange the scripts so that you always execute a whole script and not parts of a long script line by line.

These are some common steps of a statistical analysis of which each one should have its own script:

- *Import the data* to obtain a data frame on which all further steps build. Perhaps the setting of some variable characteristics like variable labels also figures well here. Furthermore, irrelevant variables may be deleted right after data import.
- *Check the data quality* to arrive at a trustworthy data set, from which all further analyses can start. If this step finds a good end, lots of validity issues about the data set are settled and the analysis can proceed to the more interesting steps.
- *Transform variables and calculate the derived measures* to obtain the target variables for the main analysis. Usually, a vast number of raw variables transform into a handful of measures to test the experimental hypotheses. These measures need meaningful names. This allows us to think in concepts again as in the preparation of the study background.
- *Check distributional assumptions* to judge which inferential statistical method will be adequate for tests of hypotheses. Our decisions here tell us which chapter of the statistics book to consult for an adequate statistical test.
- *Calculate descriptive statistics* and give the reader enough information to completely recalculate the analyses without having to provide the complete data set. With good descriptives, readers can go in this and that direction, follow us or not follow us with the analysis, but be confident to have complete transparency of the data at stake.
- *Visualise the results* with all of R's capabilities to create statistical displays. Because of the many parameter settings and function calls to make a graphic beautiful, always program only one graphic in a script.
- *Test hypotheses* with a modelling function for inferential statistics. Sometimes, a script can create a set of statistical models and display the results for comparison between models.
- *Calculate further exploratory analyses* to take different perspectives of the data and to go beyond the planned analysis. Do not mix your primary analysis to answer the research question with these further calculations.

Doing statistics in the real world is not a unidirectional process where all the planning comes first and all the coding later. You will constantly be revisiting your existing code to look for improvements and enhancements to improve the analysis.

R FUNCTION! ## 3.5.2 source()

Use **source()** to structure the complete analysis in different script files. The function takes the name of another R script as its argument and simply runs the script. This tool enables the splitting of big analyses into several small scripts, each one accomplishing only one specific task. Examples of such tasks are data import, data transformation, calculation of statistics, tabular display of statistics, or visualisation of statistics in a figure. To demonstrate how **source()** works, we need two or three scripts which hook into each other.

Ch3_source_DataCreation.R

```
## Data creation
if(!exists("d"))
{
   d <- data.frame(X1 = c(3, 4, 4, 3, 5),
                   X2 = c(2, 3, 3, 4, 3),
                   X3 = c(4, 3, 4, 5, 3))
}
```

Ch3_source_DataTransformation.R

```
## Data transformation
source("Ch3_source_DataCreation.R")

d$SumX <- rowSums(d)
```

Ch3_source.R

```
## Using source()
source("Ch3_source_DataCreation.R")
source("Ch3_source_DataTransformation.R")

d
mean(d$SumX)
```

Output

```
> d
  X1 X2 X3 SumX
1  3  2  4    9
2  4  3  3   10
3  4  3  4   11
4  3  4  5   12
5  5  3  3   11
> mean(d$SumX)
[1] 10.6
```

Analysis

In a very small scale, the scripts show the structure of a realistic statistical analysis: The first script creates, the second transforms, and the third analyses the data. The data transformation needs the data frame d, so, before doing **rowSums()**, the script uses **source()** to obtain the data. Similarly, before the analysis can calculate the mean of the row sums, the data frame and the row sums must be available. So, the last script uses source with the two other scripts to get access to the data. Notice that the data transformation script calls **source()** and the analysis script does it, too. Including the same file twice is no problem here because the inclusion guard if(!exists("d")){} in the data creation script prevents double creation of d, as will be introduced in more detail next.

3.5.3 Inclusion guards

Run no unnecessary repetitions of code. If you do several analyses in a sequence and each analysis sources the same data preparation script, then the data preparation is executed several times. For example, a script with scale summations could be called again and again by several other scripts where, in fact, one call would suffice. So, it would be helpful to check whether an object or variable already exists before we create it. This is achieved by inclusion guards. They test for the existence of an object. If the object does indeed exist, the code fragment for its creation is skipped from execution. This is an inclusion guard:

Ch3_InclusionGuard.R

```
## Inclusion guard
if(!exists("Object"))
{
        Object <- … # Code for object creation
}
if(!exists("Variable", where = DataFrame))
{
        DataFrame$Variable <- … # Code for variable creation
}
```

Analysis

You can use this at several places in a script, wherever objects or variables are created. Or you can use it once to enclose all code of a script in braces. Usually, good practise is to create one meaningful object in one script of the same name. Then an inclusion guard can enclose the whole script. Inclusion guards can not only be used for objects in the main workspace, but also for variables within a data frame, which is shown by the second example.

At first glance, inclusion guards look complicated and dispensable; however, they help you to write more efficient and faster analyses. Sometimes, creation of an object takes a long while, for example, the import of a large SPSS data set. So, if the data set is available already in the R workspace, each further call of the corresponding script for data import can be skipped.

3.5.4 The structure and content of R scripts

Each statistical analysis comprises several files. These are data sets, scripts, results, and figures. Usually one script imports the data frame. Then, several other scripts do things with the data. Finally, some scripts write results to a file or create a figure. Here are some best practises about how to structure each script:

- Do one thing. Each script should accomplish a well defined task and nothing more.
- A script should output only one object or result (e.g., a table or a figure).
- Write scripts so that you can always run the whole script and not only parts of it.
- Hook the scripts into each other with **source()**, if they run in a meaningful order. Draw the structure of the analysis to see how all scripts hook into each other. You can also draw as structure of the objects that the scripts create.

- Use an inclusion guard to skip execution of the script if the task it accomplishes is already done. For example, you must not read in the data frame 20 times in one analysis session.
- Some say that a whole script should fit on a computer screen.

If you adhere to some simple rules for your analysis, you will not get stuck by the many files that you create in the course of one analysis. Because each script does only one thing, it will be easier to understand if you get back to it after some time.

3.5.5 UPS – Structure of the analysis EXAMPLE!

From the beginning of our analysis of the unpleasant sounds data we should think about how to structure the complete analysis. The analysis will consist of several scripts – one for each particular task. The tasks are data import, data inspection, descriptive statistics, inferential statistics, and graphical display. Let us draw the structure of this analysis.

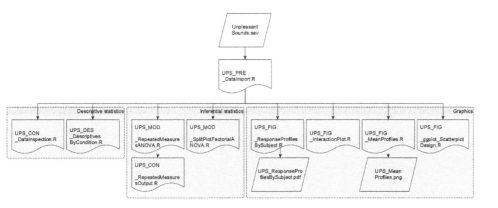

Figure 3.1 Structure of the analysis of the unpleasant sounds example. Each box is an R script or other file.

Figure 3.1 shows that the analysis starts with the data set in IBM SPSS Statistics software ('SPSS') format and a data import script. All other scripts use the data import script to obtain the required data frame. We also indicate some file outputs of graphical displays in pdf and png format. Such a quick drawing of the structure of the analysis helps to keep track of the files involved. We use the `source()` function to connect the scripts.

━━━━━━━━ What you have learned so far ━━━━━━━━

1   To aggregate variables to sum or average scores
2   To handle factor variables
3   To reshape data frames between wide and long format
4   To set variable labels and other attributes
5   To divide a statistical analysis in several scripts, each doing only one thing

---

▬▬▬▬ Exercises ▬▬▬▬

Explore functions

1   Create a matrix A with `matrix(1:16, 4)`. Try **rowSums()**, **colSums()**, **rowMeans()**, and **colMeans()** with this matrix. Then turn one value in the matrix into a missing value with `A[2, 3] <- NA`. Now, apply the functions again and observe the difference.

2   For a music evaluation test, create a factor variable with the two levels "`high pitch`", "`middle pitch`" and "`low pitch`". Each level is to be repeated 20 times. Use **factor()** and **rep()** for this. Set "`low pitch`" as the reference category. Some applications further require breaking up factors into numbers. So, try **as.numeric()** with the factor variable.

3   For the **MusicPreferencesSchool.csv** data, set a `variable.labels` attribute with variables in your preferred language. Use **attr()** for this. Further rename the factor levels in this data frame and use terms from your preferred language. Use **factor()**.

Solve tasks

1   Use **MusicPreferencesSchool.csv.** The data should reveal different preference ratings for musical examples for students in different schools, whereby each subject gave two ratings. Create a new Variable `MeanPref` in the data set containing the mean of the two ratings for each subject: sum the ratings and divide by two.

2   Use **MusicData.csv.** For each subject calculate the average score across all musical examples and store the average scores in a new variable. Use **rowSums()** or **rowMeans()** to complete the task. For one subject there is a missing data point in the ratings. Switch the `na.rm` parameter of the functions so that this person nevertheless obtains an average score.

3   Restructure your analyses of the **MusicPreferencesSchool.csv** analysis: use a script, which only imports the data. Then use **source()** in an analysis script to run the data import script. Make this structure of scripts the basis for all subsequent analyses of this example.

4   Import the data set **MusiciansComplaints.csv.** The data set has a bad structure because the data of half of the subjects stands next to the data of the other half. Use matrix indexing and **rbind()** to place the data from subject 51 to 100 below the data from subject 1 to 50. After this operation the data set is ready for the analysis.

Read code

1   Read the following factor declarations. Describe in your own words what these declarations do. Then create the factors at the R console and check whether your description was correct.

```
# Factor 1
f1 <- factor(rep(c("red", "yellow", "blue", "green"), times = 10))
f1 <- relevel(f1, "blue")

# Factor 2
f2 <- gl(4, 10, labels = c("Bee", "Wasp", "Dog", "Cat"))
levels(f2) <- list("Insect" = "Bee",
```

```
                    "Insect" = "Wasp",
                    "Mammal" = "Dog",
                    "Mammal" = "Cat")

# Factor 3
f3 <- factor(rep(c("Level 1", "Level 2", "Level 3", "Level 4", "Level 5"),
                each = 5),
             ordered = TRUE)
contrasts(f3) <- contr.poly(nlevels(f3))

# Factor 4
f4 <- factor(rep(c("Control",
                   "Intervention 1",
                   "Intervention 2"), each = 20))
contrasts(f4, how.many = 3) <- cbind(Int1VsContr = c(-1, 1, 0),
                                     Int2VsContr = c(-1, 0, 1),
                                     IntVsContr = c(-1, .5, .5))

# Factor 5
A <- factor(rep(c("a1", "a2", "a3"), each = 10))
B <- factor(rep(c("b1", "b2"), times = 15))
f5 <- interaction(A, B)
contrasts(f5, how.many = 1) <- cbind(a1.b1 = c(1, -.2, -.2, -.2, -.2, -.2))
```

Apply in the real world

1   Think about your own project or about a project you work for. Draw a data model for
    the project and present it in the next team meeting where appropriate. Make printouts
    of the model and suggest your colleagues draw associations between variables or to
    highlight parts of the model for a certain analysis. Revise the model after the session
    according to the suggestions you obtained from the team and then send the revision
    out to all team members for their reference. Contemplate how the design can be made
    clearer for everybody with the help of the data model.
2   Do the required data aggregations in your own data frame to prepare for the analysis.
    For example, create sum or average scores or set up appropriate levels for factor
    variables.

Become a statistics programmer

1   Create a template structure for all your scripts. This structure should consist of an
    adequate header comment showing, for example, creation date, title, author, and purpose
    of a script. Furthermore, the template should consist of a section to invoke libraries and
    other scripts. Finally, all code should be surrounded by an inclusion guard.

WBD   EXAMPLE!

1   Use the data of the WBD example about web-based cognitive bias modification. Make a
    plan to structure the analysis. Draw which steps are necessary for the analysis.

(Continued)

2   For the example `WBD`, create a set of scripts for data import, data transformation, and data analysis. Hook the scripts into each other with **`source()`**.

3   For the variables BDI, anhedonia, and prospective imagery test, reshape the data from wide to long format. To do so, invoke the script from the main text in your analysis or write your own script using **`reshape()`**.

EXAMPLE! **ITC**

1   After data import of `stanovich_data_used.sav`, inspect the attributes of the data set. Add variable labels for the variables age, gender, and sat. Use **`attr()`** for this.

2   Find out in the variable code book how the card selection variables `P1P`, `P1NotP`, `P1Q`, and `P1NotQ` have been aggregated to `correct_responses_p1`. The aggregation could have been different. For example, calculate the count of correct responses of the four card selection variables. Ensure that this count has the range of 0 to 4 correct card selections. Use **`as.numeric()`** to transform the factor variables in integers, first. Then, use **`rowSums()`** to aggregate the four variables.

EXAMPLE! **UPS**

1   Go back to our sketch of the analysis structure in the main text. Set up a folder for the analysis and create a set of R files with file names displayed in this sketch. Use the **`source()`** function and appropriate file names to connect all scripts.

EXAMPLE! **SCB**

1   Import the conflict reconciliator data at the console. Then save the data frame in the R file format .Rdata. All your declarations about the data frame are preserved in this file format.

2   Add another variable.labels attribute with labels from your preferred language. Use **`attr()`** for this. Optionally, also change the labels for factor levels in this data set to labels from your preferred language. This makes presentation of the results easier if English is not the presentation language.

3   The desired statistical analyses are several univariate two-factor analyses of variance to assess the effectiveness of the intervention program and the interaction between intervention program and type of school regarding the frequency of conflicts in school-yards. To enable a univariate analysis, the variables Y in the data frame (count of conflicts in a given month) have to be combined appropriately. So, compute the following new variables:

4   For the months 1 and 2 as well as the months 3 and 4 calculate the mean of the measurements (e.g., `(Y1C1 + Y1C2) / 2`). Calculate for all three types of conflicts, respectively. In total, six new variables should be created and added to the data frame. Document all calculations in an R script.

5   Calculate pretest-posttest difference values for the six new variables. Subtract the posttest mean (months 3 and 4) from the pretest mean (months 1 and 2). In total, three new variables should be created. Regarding these variables, a positive effect of the intervention should lead to positive group means. Now, you have created the dependent measures of pretest-posttest differences in conflict count.

4

DESCRIPTIVE AND EXPLORATORY DATA ANALYSIS

========================= Chapter overview =========================

This chapter introduces functions for descriptive and exploratory statistics in univariate, bivariate, and multivariate contexts. We practise the description of categorical and quantitative variables. We write the results in HTML reports to be shared with the research team. The programming technique of vectorisation further facilitates the calculation of statistics. Finally, we will visit more options to structure scripts and the whole analysis. Programming features: quick results report, vectorised operations, lists, search path.

4.1 Data exploration and data quality

We start this chapter with some reflections about the importance of exploratory data analysis. We also gather examples of questions for data exploration that you and your colleagues might answer with the data.

STATISTICAL
ISSUE!

4.1.1 The need for data exploration

We may be committed to the idea of accepting any data that comes in from our previously devised experimental plan. Because we specified a hypothesis and specified the experimental plan and methods – the rules of data collection – we assume that everything in the experiment happened according to this rule. After data collection, we would directly proceed to inferential statistics (Chapters 6 and 7) without concern about data quality. We feel that any descriptive knowledge of the data is ultimately suspicious of data snooping, that is, stating the hypothesis after the fact. However, *research in real life never perfectly corresponds with an experimental plan and we should always expect deviations from truly valid data collection.* Measurement devices do not allow for perfect replication due to technical problems: subjects do not show up at their appointed experimental sessions, some questionnaire item characteristics may depend on time of day of questionnaire completion, some experimental group worked in a fresh air laboratory, the next group worked in a not so fresh laboratory. Conversely, our inferential statistical analyses later are only valid in an ideal set-up. How do we get out of this dilemma? *We have to inspect the data for experimental errors and for influences other than the hypothesised ones. We have to look out for associations between variables that we could not preconceive, but which could have an impact on our analysis.*

Let us assume that we study the impact of different social situations relating to social identity on evaluations of musical examples. Without much thought, but as we always do, we let subjects note their gender on the questionnaire. Later in the analysis, again without much thought, we check whether the musical evaluations depend on gender. If not, we may forget the factor for subsequent analyses, though it may still interact with other factors. But if we find an association, we may leave it in the analysis even at later steps to enhance the precision of the design. This exploratory approach is surely invalid if we conceive of our experiment as a perfect system where we take replications to decide about population parameters. However, if we ignore the imperfect aspects of our experiments in practise, we may end up with highly biased conclusions and accept data as valid which is, in fact, full of data collection errors and further uncontrolled dependencies between variables.

Exploratory data analysis was born with the idea by Tukey (1997) that a researcher has to assume multiple viewpoints on his data set, which could reveal hidden tendencies or associations. The data should be scrutinised from different angles, with summary statistics and graphical displays. This approach at first glance contradicts the approach of hypothesis testing, that is, a researcher states his hypotheses before data collection and then collects only the amount of data directly sufficient to run a statistical test, then decides between the null-hypothesis and the alternative hypothesis, and is then finished with the data. In this approach the data is only needed for a single decision. On the contrary, exploratory data analysis tells the researcher to immerse themself in the data and to get a feeling of what is important. The hypothesis tester would cynically call this data snooping. However, let us follow a strategy which takes the best of both worlds: the strictness and commitment of rigorous hypotheses testing and the wise pragmatism of exploratory data analysis in a world with dirty data. Behrens' (1997) approach to exploratory data analysis compares both perspectives and describes various options of graphical displays, which highlight specific aspects of the data like outliers, extreme values, and skewed distributions. He suggests that model development is an iterative process of tentative model specification and the analysis of residuals. In sum, *exploratory data analysis has two aims: (1) to reveal errors and (2) to reveal hidden effects in the data*. The first is a prerequisite for further analysis; the second can be an end in itself.

4.1.2 Questions to ask about data quality   TEAMWORK!

Brainstorm questions about the data which make you more confident about data quality. Among others, such questions could focus on the allocation of subjects between experimental conditions, the even distribution of demographic variables across experimental conditions, or the correlations between independent variables which were not subject to randomisation. Furthermore, the distributions of the dependent variables across conditions or possible outliers might interest you. Is it reasonable to assume that measurements of dependent variables were sampled from a normal distribution? Is departure of measurements from the normal distribution problematic because of a small sample size? For your collection of steps I suggest creating an empty R script and writing all ideas as one-line comments in the script using # at the beginning of each line. Perhaps you can arrange them in a logical sequence. Then afterwards you can use this list to insert the actual steps of the analysis at their appropriate location. This script is the draft of the analysis, and it can be refined in several iterations, perhaps after you showed the initial results to your colleagues. As the single steps of this data exploration grow longer, you may split the script into several small scripts, each one doing only one exploratory task.

4.2 Descriptive statistics

Data description is the most basic job in the analysis. We will practise many different functions, which enhance your grasp on the data. The more of these functions you memorise, the faster you can work and the more fun it becomes.

4.2.1 table()   R FUNCTION!

To display counts, `table()` is most easy to use. Simply supply it with one or more variables from your data set and obtain the univariate or multivariate frequency distribution. The function easily goes together with `addmargins()` for marginal distributions in crosstabs, `ftable()` for a readable display of the joined distribution of more than two variables, and `chisq.test()` to test the strength of association between the variables used in a table. In the first script, we create the data for Chapter 4 and store it as an R file Ch4_DataCreation.R. Then we use `source()` to include the data creation script in the descriptive analysis.

Ch4_DataCreation.R

Data creation

```
if(!exists("d"))
{
  set.seed(12345)
  temp <- sample(1:6, 300, replace = TRUE)
  d <- data.frame(A = gl(3, 100, labels = c("a1", "a2", "a3")),
                  B = gl(2, 50, 300, labels = c("b1", "b2")),
                  X = temp,
                  Y = 10 + temp + round(rnorm(300,
                                        rep(c(0, .5, 0, -.2, .3, 1), 2,
                                        each = 50)))))
  rm(temp)
}
```

Ch4_table.R

table()

```
source("Ch4_DataCreation.R")

# Univariate distributions
table(d$A)
table(d$X)
table(d$Y)
table(cut(d$Y, breaks = 4))

# Bivariate distributions
t_AY <- table(d$A, d$Y)
print(t_AY)
addmargins(t_AY)
margin.table(t_AY, margin = 2)
prop.table(t_AY, margin = 1)

# Multivariate distributions
t_ABY <- table(d$A, d$Y, d$B)
print(t_ABY)
ftable(d$A, d$B, d$Y, col.vars = 3)

# Tests for independence
summary(t_AY)
chisq.test(d$A, d$Y)
mantelhaen.test(t_ABY)
```

Output

```
> # Univariate distributions
> table(d$A)
```

```
 a1  a2  a3
100 100 100
> table(d$X)


 1  2  3  4  5  6
44 66 49 49 49 43
> table(d$Y)


 8   9 10 11 12 13 14 15 16 17 18 19 20
 1   2  7 30 50 54 52 40 39 16  7  1  1
> table(cut(d$Y, breaks = 4))


(7.99,11]   (11,14]    (14,17]    (17,20]
       40       156         95          9
>
> # Bivariate distributions
> t_AY <- table(d$A, d$Y)
> print(t_AY)


      8  9 10 11 12 13 14 15 16 17 18 19 20
  a1  0  1  4  8 20 20 17 10 13  5  1  1  0
  a2  1  1  3 13 14 14 17 18 13  4  2  0  0
  a3  0  0  0  9 16 20 18 12 13  7  4  0  1
> addmargins(t_AY)


       8   9  10  11  12  13  14  15  16  17  18  19  20 Sum
  a1   0   1   4   8  20  20  17  10  13   5   1   1   0 100
  a2   1   1   3  13  14  14  17  18  13   4   2   0   0 100
  a3   0   0   0   9  16  20  18  12  13   7   4   0   1 100
  Sum  1   2   7  30  50  54  52  40  39  16   7   1   1 300
> margin.table(t_AY, margin = 2)


 8  9 10 11 12 13 14 15 16 17 18 19 20
 1  2  7 30 50 54 52 40 39 16  7  1  1
> prop.table(t_AY, margin = 1)


        8    9   10   11   12   13   14   15   16   17   18   19   20
  a1 0.00 0.01 0.04 0.08 0.20 0.20 0.17 0.10 0.13 0.05 0.01 0.01 0.00
  a2 0.01 0.01 0.03 0.13 0.14 0.14 0.17 0.18 0.13 0.04 0.02 0.00 0.00
  a3 0.00 0.00 0.00 0.09 0.16 0.20 0.18 0.12 0.13 0.07 0.04 0.00 0.01
>
> # Multivariate distributions
> t_ABY <- table(d$A, d$Y, d$B)
> print(t_ABY)
, ,  = b1
```

```
       8  9 10 11 12 13 14 15 16 17 18 19 20
  a1   0  0  4  7  9  5 11  4  5  3  1  1  0
  a2   1  1  0  7  5  6 10 10  7  2  1  0  0
  a3   0  0  0  5  9  6 12  7  5  3  3  0  0


, ,  = b2


       8  9 10 11 12 13 14 15 16 17 18 19 20
  a1   0  1  0  1 11 15  6  6  8  2  0  0  0
  a2   0  0  3  6  9  8  7  8  6  2  1  0  0
  a3   0  0  0  4  7 14  6  5  8  4  1  0  1

> ftable(d$A, d$B, d$Y, col.vars = 3)
          8  9 10 11 12 13 14 15 16 17 18 19 20

a1 b1    0  0  4  7  9  5 11  4  5  3  1  1  0
   b2    0  1  0  1 11 15  6  6  8  2  0  0  0
a2 b1    1  1  0  7  5  6 10 10  7  2  1  0  0
   b2    0  0  3  6  9  8  7  8  6  2  1  0  0
a3 b1    0  0  0  5  9  6 12  7  5  3  3  0  0
   b2    0  0  0  4  7 14  6  5  8  4  1  0  1
>
> # Tests for independence
> summary(t_AY)
Number of cases in table: 300
Number of factors: 2
Test for independence of all factors:
        Chisq = 20.081, df = 24, p-value = 0.6922
        Chi-squared approximation may be incorrect
> chisq.test(d$A, d$Y)


        Pearson's Chi-squared test

data:  d$A and d$Y
X-squared = 20.081, df = 24, p-value = 0.6922


Warning message:
In chisq.test(d$A, d$Y) : Chi-squared approximation may be incorrect
> mantelhaen.test(t_ABY)


        Cochran-Mantel-Haenszel test

data:  t_ABY
Cochran-Mantel-Haenszel M^2 = 20.116, df = 24, p-value = 0.6901
```

Analysis

The first script creates a simple data frame from a balanced two-factor design with factors A and B, a dependent variable Y, and a covariate X. The use of the temp variable establishes the linear relationship between X and Y. The script for data creation should reside in the same directory as the script for the analysis. This separation of data creation and analysis allows us to use the data creation script again in other analyses. We can simply use **source()** with it if we need the data.

The analysis script then proceeds from univariate over bivariate and multivariate analyses to some tests for independence in the context of contingency tables. The **table()** function is central in all analyses. Supplied with a single variable, it returns the univariate frequency distribution of the variable categories. If there are too many categories, it combines with **cut()** to display the frequencies of a reduced number of categories. In the bivariate application, the function takes two variables, either qualitative or quantitative. We store the table in a first step in the object t_AY and then use **addmargins()** with it to add the marginal distributions for rows and columns. If only the marginal distribution is of interest, **margin.table()** suffices. Proceeding now from counts to relative frequencies, **prop.table()** provides the conditional frequencies of the Y categories, given the levels of A. Going to the multivariate distributions, the question arises of how to present higher-dimensional data in the two dimensions of rows and columns. If we run **table()** with the three variables A, B, and Y, it creates contingency tables of two variables for all levels of the third variable, in this case A x Y distributions for the two levels of B. Compact displays of higher-dimensional data are aggregated in a so-called flat table produced by **ftable()**. It virtually allows for frequency displays for any number of variables, some assigned and nested in the rows, others in the columns. Notice how easily this function lets you inspect the distribution of a dependent variable for all levels in the experimental design. Finally, we test for independence between variables with a chi^2 test, which reveals whether we can reject the assumption of conditional independence between two variables. The functions **summary()** and **chisq.test()** give almost the same information about this. A generalisation of this test for independence is supplied with the Cochran–Mantel–Haenszel statistic to test for conditional independence between two variables for each level of a third variable.

Use **table()** often in the initial descriptive analysis of your data. The function easily combines with **barplot()** to give a graphical representation of the count data and with **htmlTable()** from the **htmlTable** package to create a beautiful output. Another interesting table function for you to explore is **xtabs()**, which uses the formula notation (see Chapter 6) to specify contingency tables. The formula approach is even enhanced with the **tables** package.

4.2.2 ITC – Table of correct and incorrect responses   EXAMPLE!

The Wason card selection task performance includes the counting of correct and incorrect responses. Frequency tables can display these counts.

ITC_DES_TablesCorrectIncorrect.R

```
## ITC - Frequency tables of correct and incorrect responses
if(!exists('TablesCorrectIncorrect'))
```

```
{
  source("./Scripts/ITC_PRE_LoadData.R")

  TablesCorrectIncorrect <- with(ITC$stvu,
                                 {
                                   ftable(correct_responses_p1,
                                          correct_responses_p2,
                                          correct_responses_p3)
                                 })
}
```

Output

```
> TablesCorrectIncorrect
                                              correct_responses_p3 incorrect correct
correct_responses_p1 correct_responses_p2
incorrect            incorrect                                            57       3
                     correct                                               3      14
correct              incorrect                                            10       5
                     correct                                               6      79
```

Analysis

The script creates one object as its output: TablesCorrectIncorrect. It uses an inclusion guard to prevent repeated execution of the script in an analysis consisting of several scripts. To use **with()** enables direct access to the variables of the data frame without indexing them. The script then uses the powerful **ftable()** function to force a multivariable analysis in a two-dimensional table, that is, in a flat table.

After running the script, call the object TablesCorrectIncorrect at the console. It shows the counts of all eight combinations of correct and incorrect responses yielded by the three-card selection problems. The count of 79 for all problems correct, shows that many subjects correctly solved all the card selections. In contrast, there were also many subjects who gave incorrect responses to all three selection tasks. The results pattern suggests that if a person knows the principle to solve the problems, he or she can safely apply it to multiple problems. Furthermore, the high frequency of all responses incorrect, suggests that many subjects followed a biased strategy to solve the problems.

R LANGUAGE! ## 4.2.3 The output of statistical results

It is good to have a big collection of functions available for the inspection of data frames and descriptive statistics at the console. Table 4.1 consists of several functions sorted by topic, which are especially handy to use. Some functions like **mean()**, **var()**, **sd()**, **median()**, and **summary()** supply standard univariate statistics. Bivariate analyses usually start with **cov()** or **cor()**, which produce covariance and correlation matrices. To analyse the dimensionality of quantitative variables, **princomp()** and **factanal()** produce objects consisting of principal component and exploratory factor analyses. The last two functions produce objects of classes `princomp` and `factanal`, which consist of more than single descriptive statistics; they consist of all information necessary to validate the dimensionality of questionnaire items or other correlated variables.

In R, control of decimal places and rounding are an issue since the default number of decimal places used for descriptive statistics often clutters the results display. The **print()** function consists of the parameter `digits` to control for the number of decimal places in the results display. This preserves all decimal places in the internal representation of a given object in R. In contrast, using **round()** or its associates **trunc()**, **floor()**, or **ceiling()** cuts off decimal places at the specified point and throws information away. So, in most cases you do not want to round values that you use in subsequent calculations. Beware which solution to use in any given context. However, for quick results it often works to nest the function for descriptives inside a call of **print()** with the digits parameter set to the number of desired decimal places. Try things like `print(mean(VariableX), digits = 2)` or `print(cor(VariableX, VariableY), digits = 3)`.

Table 4.1 Functions for data exploration at the console.

Information		
Data frame		
str() data structure	attributes() data attributes	attr() manipulate attributes
names() variable names	View() data viewer	rowSums() row sums
colSums() column sums	nrow() number of cases	ncol() number of variables
Variables and objects		
is() class information	levels() factor categories	duplicated() duplicate cases
is.na() missing cases	length() element count	sort() sort elements
Workspace and directory		
dir() available files	getwd() working directory	
Quick helpers		
c() concatenate elements	as.data.frame() object conversion	as.matrix() object conversion
unlist() break-up list	? function help	?? keyword help
rbind() aggregate cases	cbind() aggregate columns	merge() merge data frames
subset() subset data frame	paste() string vectors	grep() string matching
Statistics		
Descriptives		
summary() summary statistics	table() frequency table	addmargins() marginal distribution
prop.table() relative frequencies	ftable() flat table	margin.table() marginal distribution
min() minimum	max() maximum	range() range
mean() mean	sd() standard deviation	cov() covariance matrix
cor() correlation matrix		

(Continued)

Table 4.1 (Continued)

Models		
t.test() t-test	aov() analysis of variance	lm() regression analysis
predict() predicted values	princomp() principal component analysis	factanal() factor analysis

Quick helpers		
apply() vectorised operation – matrix	tapply() vectorised operation – univariate design	by() vectorised operation – multivariate design
lapply() vectorised operation – list	print() format output	round() rounding
function() user defined function	cut() categorise variable	capture.output() results file output

Graphics		

High-level plotting		
plot() generic plotting	hist() histogram	barplot() bar plot
boxplot() box plot	matplot() visualise matrix	qqnorm() quantile-quantile diagram

Quick helpers		
par() graphics parameters	layout() multiple graphic layout	plot.new() empty diagram
plot.window() plot coordinates	legend() plot legend	pdf() pdf output
png() png output		

R LANGUAGE! ## 4.2.4 Extend the functionality with add-on packages

Although the basic installation of R is very powerful to create any type of analysis, R also lives by its numerous add-on packages. For your analysis, do some search in the internet on Task Views to find out which packages you need. Install the packages with one of the menu options in the R program and then include them in the analysis with **library()** or **require()**. For example, from the **psych** package a call of library(psych) makes all its functions available. Another option to include certain single functions from add-on packages uses the double colon operator with package::function(). This includes the package only for the single function call and also tells the reader of the script where a possibly exotic function belongs. Let us try some packages and see what output they produce with little effort.

Ch4_DataExplorationAddOnPackages.R

```
## Add-on packages for data exploration
source("Ch4_DataCreation.R")

library(gdata)
library(memisc)
library(descr)
library(sjstats)
library(Hmisc)
library(psych)
library(car)
```

```
attach(d)

# gdata
ll()
nobs(Y)

# memisc
Descriptives(Y)
Table(Y)
codebook(d)

# descr
freq(Y)
CrossTable(A, X)
compmeans(Y, A)

# sjstats
prop(d, Y > 17)
props(d, A == "a1" & Y > 12)
tab <- table_values(table(A, Y))
tab$cell
tab$expected
means_by_group(d, Y, A)

# Hmisc
contents(d)
bystats(Y, A, B, fun = mean)
rcorr(X, Y)

# psych
describeBy(Y, list(A, B), mat = TRUE)
mardia(Y)
corr.test(X, Y)

# car
leveneTest(Y, A)
vif(lm(Y ~ A + B + X))

detach(d)
```

Output

```
> # gdata
> ll()
      Class KB
d data.frame  8

[…]
```

```
> # memisc

[…]

> codebook(d)
==============

    A

-------------

    Storage mode: integer
    Factor with 3 levels

    Levels and labels      N Valid

    1 'a1'               100  33.3
    2 'a2'               100  33.3
    3 'a3'               100  33.3

==============

    B

-------------

    Storage mode: integer
    Factor with 2 levels

    Levels and labels      N Valid

    1 'b1'               150  50.0
    2 'b2'               150  50.0

==============

    X

-------------

    Storage mode: integer

         Min:   1.000
         Max:   6.000
        Mean:   3.407
    Std.Dev.:   1.658
```

```
   Skewness:  0.116
   Kurtosis: -1.233

==============

   Y

------------

   Storage mode: double

        Min:  8.000
        Max: 20.000
       Mean: 13.723
   Std.Dev.:  2.010
   Skewness:  0.167
   Kurtosis: -0.307

>
> # descr

[…]

> CrossTable(A, X)
   Cell Contents
|-----------------------|
|                     N |
| Chi-square contribution |
|           N / Row Total |
|           N / Col Total |
|         N / Table Total |
|-----------------------|

===============================================================
          X
A          1        2        3        4        5        6   Total
---------------------------------------------------------------
a1        17       22       19       18       11       13     100
        0.371    0.000    0.435    0.170    1.741    0.124
        0.170    0.220    0.190    0.180    0.110    0.130   0.333
        0.386    0.333    0.388    0.367    0.224    0.302
        0.057    0.073    0.063    0.060    0.037    0.043
---------------------------------------------------------------
a2        13       19       15       17       20       16     100
        0.189    0.409    0.109    0.027    0.823    0.194
        0.130    0.190    0.150    0.170    0.200    0.160   0.333
```

```
        0.295    0.288    0.306    0.347    0.408    0.372
        0.043    0.063    0.050    0.057    0.067    0.053
------------------------------------------------------------------
a3         14       25       15       14       18       14      100
        0.030    0.409    0.109    0.333    0.170    0.008
        0.140    0.250    0.150    0.140    0.180    0.140    0.333
        0.318    0.379    0.306    0.286    0.367    0.326
        0.047    0.083    0.050    0.047    0.060    0.047
------------------------------------------------------------------
Total      44       66       49       49       49       43      300
        0.147    0.220    0.163    0.163    0.163    0.143
==================================================================
```

[...]

```
> # sjstats
> prop(d, Y > 17)
[1] 0.03
> props(d, A == "a1" & Y > 12)
[1] 0.2233
```

[...]

```
> means_by_group(d, Y, A)

# Grouped Means for Y by A

Category | Mean |  N  |  SD  |  SE  |    p
-------------------------------------------
a1       | 13.55 | 100 | 1.98 | 0.20 | 0.437
a2       | 13.60 | 100 | 2.05 | 0.21 | 0.453
a3       | 14.02 | 100 | 1.99 | 0.20 | 0.215
Total    | 13.72 | 300 | 2.01 | 0.12 |

Anova: R2=0.011; adj.R2=0.004; F=1.651; p=0.194
>
> # Hmisc
```

[...]

```
> rcorr(X, Y)
     x    y
x 1.00 0.86
y 0.86 1.00

n= 300
```

```
P
   x  y
x     0
y  0
>
> # psych
> describeBy(Y, list(A, B), mat = TRUE)
        item  group1  group2  vars    n     mean           sd  median  trimmed
mad min max range         skew     kurtosis        se
X11       1          a1      b1      1   50  13.44  2.232963    13.5   13.325
2.2239  10  19      9  0.37945341 -0.6215803 0.3157886
X12       2          a2      b1      1   50  13.72  2.109647    14.0   13.800
1.4826   8  18     10 -0.44245452 -0.1945942 0.2983492
X13       3          a3      b1      1   50  13.98  1.963857    14.0   13.900
2.2239  11  18      7  0.31248486 -0.7652701 0.2777313
X14       4          a1      b2      1   50  13.66  1.709637    13.0   13.625
1.4826   9  17      8  0.06823902 -0.3869375 0.2417791
X15       5          a2      b2      1   50  13.48  2.012613    13.0   13.450
2.9652  10  18      8  0.13900148 -0.9031716 0.2846265
X16       6          a3      b2      1   50  14.06  2.034498    13.5   13.950
2.2239  11  20      9  0.57588629 -0.2025389 0.2877215
> mardia(Y)
Call: mardia(x = Y)

Mardia tests of multivariate skew and kurtosis
Use describe(x) to get the univariate tests
n.obs = 300    num.vars =  1
b1p =  0.03    skew =  1.38  with probability =  0.24
 small sample skew =  1.41  with probability =  0.23
b2p =  2.68    kurtosis =  -1.15  with probability =  0.25> corr.test(X, Y)
Call:corr.test(x = X, y = Y)
Correlation matrix
[1] 0.86
Sample Size
[1] 300
Probability values  adjusted for multiple tests.
[1] 0

 To see confidence intervals of the correlations, print with the short=FALSE
option
>
> # car
> leveneTest(Y, A)
Levene's Test for Homogeneity of Variance (center = median)
       Df F value Pr(>F)
group   2  0.1592 0.8529
      297
```

```
> vif(lm(Y ~ A + B + X))
      GVIF Df GVIF^(1/(2*Df))
A 1.008435  2         1.002102
B 1.001323  1         1.000661
X 1.009758  1         1.004867
```

Analysis

Our example uses the packages **gdata** (Warnes et al., 2017), **memisc** (Elff, 2021), **descr** (Aquino, 2021), **sjstats** (Lüdecke, 2021), **Hmisc** (Harrell Jr, 2021a), **psych** (Revelle, 2021), and **car** (Fox & Weisberg, 2019). If you have not already done so, install them with the menu option' install packages' in the R program. All packages come with several functions to manage and explore the data. We use only a fraction of the available functions. Use library(help = "package name") to obtain a list of all functions in any one package. At the console we select two or three functions from each package and inspect their output. As can be seen in the output, **ll()** in **gdata** improves on the **ls()** function to display objects in the workspace. In **memisc** the **codebook()** function gathers variable information about all variables in the data set and returns it in a format that we often find as a supplement to real empirical data sets. In **descr** the **CrossTable()** function includes more information by default than the **table()**. In **sjstats** **prop()**, and **props()** are easy-to-use short cuts to obtain the proportions of subsets of the data relative to the whole data set. The **means by group()** function in this package shows descriptive statistics of a dependent variable for the different levels of an independent variable. However, the **describeBy()** function in the **psych** package returns even more information about this. Furthermore, in **psych** the **mardia()** function helps us to assess the shape of the distributions of our quantitative variables. Finally, **leveneTest()** and **vif()** in the **car** package help us to assess homogeneity of variance and independence of factors or predictors and, thus, help us with the prerequisites of analysis of variance and regression analysis.

Try these functions with your own data, too. However, I do not want to advocate using these particular functions but to stress that there are hundreds or thousands of good helpers out there in the add-on packages, and that you could explore how they work and what they return at the console, like we just did. Different packages implement similar functions with different specifications and parameters. Some functions with similar output are easier to use than others. Spend some time to find your favourite ones. Then use them in your analysis scripts. Task views of the R project at https://cran.r-project.org/web/views/ lists some general statistical topics and describes associated packages.

EXAMPLE! ## 4.2.5 WBD - Descriptive statistics

We now practise descriptive statistics with the example about web-based cognitive bias modification. Because the main analysis of the study is about the comparison between intervention and a control group regarding depression severity over time, the descriptive statistics between the two groups for the depression scores are important for us. From the **psych** package, **describeBy()** helps us with this.

WBD_DES_DescriptiveStatistics.R

```
## WBD - Depression statistics by condition
if(!exists("DepressionStatisticsByConditions"))
{
  source("Scripts/WBD_PRE_DataImport.R")
  source("Scripts/WBD_PRE_ReshapeToLongFormat.R")
  library(psych)

  DepressionStatisticsByConditions <- with(WBD$CPS_LongFormat,
                                {
describeBy(cbind(BDI, ANH, PITPV),
                                                      group =
list(Measurement, Condition),
                                                      mat =
TRUE)
                                })
}
```

Output

```
> print(DepressionStatisticsByConditions, digits = 2)
            item              group1   group2 vars  n mean      sd median
trimmed   mad  min  max range   skew kurtosis   se
BDI1          1             Baseline  Imagery    1 76 30.0    8.63   29.5
29.9  8.15 13.0 50.0  37.0  0.178   -0.602 0.990
BDI2          2 Post-treatment  Imagery    1 72 23.2   12.26   22.0
22.6 11.86  1.0 62.0  61.0  0.585    0.288 1.444
BDI3          3             1-month  Imagery    1 71 20.5   14.52   18.0
19.1 13.34  0.0 62.0  62.0  0.808    0.449 1.724
BDI4          4             3-month  Imagery    1 66 19.3   12.40   18.0
18.7 15.57  0.0 50.0  50.0  0.361   -0.678 1.527
BDI5          5             6-month  Imagery    1 69 17.2   13.08   17.0
16.0 13.34  0.0 61.0  61.0  0.869    0.713 1.575
BDI6          6             Baseline  Control    1 74 31.1   10.17   30.0
31.0 10.38  6.0 52.0  46.0  0.127   -0.564 1.182
BDI7          7 Post-treatment  Control    1 69 23.6   11.94   22.0
22.6 13.34  7.0 54.0  47.0  0.620   -0.408 1.437
BDI8          8             1-month  Control    1 69 21.5   13.48   19.0
20.7 14.83  0.0 54.0  54.0  0.548   -0.660 1.622
BDI9          9             3-month  Control    1 63 22.2   15.18   18.0
21.7 19.27  0.0 52.0  52.0  0.287   -1.183 1.913
BDI10        10             6-month  Control    1 64 18.8   12.71   16.0
17.9 13.34  0.0 50.0  50.0  0.521   -0.703 1.589
ANH1         11             Baseline  Imagery    2 76  3.2    1.36    3.0
3.1  1.48  0.0  6.0   6.0  0.324   -0.504 0.156
```

```
ANH2            12  Post-treatment  Imagery    2  72   2.3   1.40        2.0
2.3  1.48  0.0  6.0    6.0   0.488    -0.271 0.165
ANH3            13               1-month  Imagery    2  71   2.1   1.56        2.0
2.0  1.48  0.0  6.0    6.0   0.396    -0.549 0.185
ANH4            14               3-month  Imagery    2  66   1.9   1.62        2.0
1.7  1.48  0.0  6.0    6.0   0.648    -0.258 0.199
ANH5            15               6-month  Imagery    2  69   1.5   1.44        1.0
1.3  1.48  0.0  6.0    6.0   0.914     0.449 0.173
ANH6            16              Baseline  Control    2  73   3.5   1.35        3.0
3.4  1.48  0.0  6.0    6.0   0.055    -0.553 0.159
ANH7            17  Post-treatment  Control    2  69   2.8   1.52        3.0
2.7  1.48  0.0  6.0    6.0   0.240    -0.644 0.183
ANH8            18               1-month  Control    2  69   2.4   1.83        2.0
2.3  1.48  0.0  6.0    6.0   0.548    -0.696 0.221
ANH9            19               3-month  Control    2  64   2.7   1.86        2.0
 2.6  2.97  0.0  6.0    6.0   0.152    -1.013 0.232
ANH10           20               6-month  Control    2  64   2.2   1.69        2.0
2.0  1.48  0.0  6.0    6.0   0.339    -0.817 0.212
PITPV1          21              Baseline  Imagery    3  76   2.8   0.88        2.9
2.8  0.89  1.0  5.0    4.0  -0.086    -0.541 0.101
PITPV2          22  Post-treatment  Imagery    3  69   3.2   0.81        3.2
3.2  0.89  1.6  5.0    3.4   0.193    -0.665 0.098
PITPV3          23               1-month  Imagery    3  67   3.1   0.84        3.1
3.1  0.89  1.5  4.9    3.4   0.020    -0.788 0.102
PITPV4          24               3-month  Imagery    3  66   3.1   0.86        2.9
3.1  0.89  1.3  5.0    3.7   0.039    -0.615 0.106
PITPV5          25               6-month  Imagery    3  66   3.1   0.85        3.1
3.1  1.04  1.6  5.0    3.4   0.185    -0.936 0.105
PITPV6          26              Baseline  Control    3  74   2.9   0.81        3.0
2.9  0.82  1.0  5.0    4.0   0.067    -0.029 0.094
PITPV7          27  Post-treatment  Control    3  69   3.0   0.83        3.1
3.0  0.74  1.0  4.9    3.9  -0.184    -0.464 0.100
PITPV8          28               1-month  Control    3  67   2.9   0.86        3.0
2.9  0.89  1.0  4.7    3.7  -0.161    -0.763 0.105
PITPV9          29               3-month  Control    3  62   2.9   0.78        2.8
2.9  0.96  1.0  4.3    3.3  -0.113    -0.821 0.099
PITPV10         30               6-month  Control    3  62   2.8   0.80        3.0
2.9  1.04  1.0  4.3    3.3  -0.270    -0.862 0.102
```

Analysis

The script first includes our script from Chapter 2 to load the data. It also uses the data in long format. The first glance over the results usually checks whether the baseline characteristics of the dependent variables are comparable between conditions. Furthermore, we can already assess how the group means develop over time.

4.2.6 Handling of missing data in R

In research, missing data occur. Whether it be that a measurement device breaks while collecting data or a participant oversees some items in a questionnaire, it is helpful to know the options to inspect missings in the data set. In R, missings are coded as NA, which means that the data entry is *not available*. The following script shows a few quick options to find and display missings in a data frame. It rests on the function **is.na()**.

Ch4_FindingMissingData.R

```
## Finding missing data
# Data creation
d <- data.frame(ID = paste0("id", 1:10),
                X = c(3, 2, NA, 3, 5, 5, NA, 2, 4, 3),
                Y = c(NA, NA, 5, 3, 4, 2, NA, 4, 2, 5),
                row.names = "ID")

# Find missings
View(d)
anyNA(d)
is.na(d)
rowSums(is.na(d))
colSums(is.na(d))
which(is.na(d))
is.na(d$X)
View(d[is.na(d$X),])
View(d[!complete.cases(d),])

# Handle missings in calculations
mean(d$X)
mean(d$X, na.rm = TRUE)
rowSums(d)
rowSums(d, na.rm = TRUE)
table(d$X)
table(d$X, useNA = "always")
```

Output

```
> # Find missings
> View(d)
> anyNA(d)
[1] TRUE
> is.na(d)
        X     Y
id1  FALSE  TRUE
id2  FALSE  TRUE
```

	X	Y
id1	3	*NA*
id2	2	*NA*
id3	*NA*	5
id4	3	3
id5	5	4
id6	5	2
id7	*NA*	*NA*
id8	2	4
id9	4	2
id10	3	5

Figure 4.1 Data set with missings.

Source: RStudio and Shiny are trademarks of RStudio, PBC

```
id3    TRUE FALSE
id4   FALSE FALSE
id5   FALSE FALSE
id6   FALSE FALSE
id7    TRUE  TRUE
id8   FALSE FALSE
id9   FALSE FALSE
id10 FALSE FALSE
> rowSums(is.na(d))
 id1  id2  id3  id4  id5  id6  id7  id8  id9 id10
   1    1    1    0    0    0    2    0    0    0
> colSums(is.na(d))
X Y
2 3
> which(is.na(d))
[1]  3  7 11 12 17
> is.na(d$X)
 [1] FALSE FALSE  TRUE FALSE FALSE FALSE  TRUE FALSE FALSE FALSE
> View(d[is.na(d$X),])
> View(d[!complete.cases(d),])
> # Handle missings in calculations
> mean(d$X)
[1] NA
> mean(d$X, na.rm = TRUE)
[1] 3.375
```

```
> rowSums(d)
 id1  id2  id3  id4  id5  id6  id7  id8  id9 id10
  NA   NA   NA    6    9    7   NA    6    6    8
> rowSums(d, na.rm = TRUE)
 id1  id2  id3  id4  id5  id6  id7  id8  id9 id10
   3    2    5    6    9    7    0    6    6    8
> table(d$X)

2 3 4 5
2 3 1 2
> table(d$X, useNA = "always")

   2    3    4    5 <NA>
   2    3    1    2    2
```

Analysis

To get a feeling for the distribution of missings in my data, I often call several descriptive functions and combinations of functions to show different aspects of missingness. The script demonstrates this with a small data frame containing missings, coded with NA. After initial inspection of the data with **View()**, **anyNA()** tells us that there are missings somewhere in the data and **is.na()** tests for every entry in the data whether it is available or not. The functions **rowSums()** and **colSums()** provide information as to how many missings we have in rows and columns. Note that in R we sometimes apply a summation function to count TRUE values. The function team **which()** and **is.na()** return the index numbers in the data frame, where the missings are. The function **is.na()** applied to d$X lets us focus on the variable X in particular. With View(d[is.na(d$X),]) we obtain an overview of the cases which have a missing in the particular variable X. Notice in this code that we test for missings in X only but then obtain all data columns. Another practical function to inspect complete or incomplete cases is **complete.cases()**. As can be seen, the ! can be used to switch between complete and incomplete cases. Besides inspection of missings, they can be handled in several statistical functions with appropriate parameter settings. The script includes three examples, which consist of a parameter to handle missings. The **mean()** function returns NA if applied to a vector consisting of missings. We need to switch its na.rm parameter to ignore missings and calculate the mean with the available data.

This section showed how to find missings in a data frame using **is.na()**. The next important step would be to explain why the data is missing. The last decades saw the increasing importance of explaining missing data and of the benefits of data imputation to amend missingness. Starting with the seminal work by Donald Rubin (Rubin, 1976, 1987), the literature distinguishes different types of missingness, which differently hamper statistical conclusion validity. Sophisticated methods to inspect missing data and to impute data are implemented in the R packages **mice** (Van Buuren & Groothuis-Oudshoorn, 2011), **mi** (Su, Gelman, Hill & Yajima, 2011), and **Hmisc** (Harrell Jr, 2021a). The R Task View about missing data https://cran.r-project.org/web/views/MissingData.html offers even more tools for missing data analysis and data imputation.

4.2.7 Results report for the next team meeting

How nice would it be to have the exploratory data analysis in a good-looking report to present your team or supervisor. R can do this with the **knitr** package. *One has to create a so-called Rhtml file, which is plain text made of HTML code and R code in one document. In this document, between the HTML code are chunks of R code.* To create the report, install the package **knitr**. The document then has to be processed by **knitr**, which executes the R code and creates an HTML file with the results. A web browser can display this file. The following is an example of the use of **knitr**. R code in Rhtml is delimited by `<!--begin.rcode` and `end.rcode-->`. Store this Rhtml script as a file with the ending Rhtml.

Ch4_ReportTemplate.Rhtml

```
<!--##Source R code in an Rhtml file-->
<html>

<head>
<title>Source R code in an Rhtml file</title>
</head>

<body>

<p>A minimal report with knitr.</p>

<!--begin.rcode
set.seed(12345)
y <- rnorm(10)
y
sum(y)
end.rcode-->

</body>
</html>
```
Ch4_ReportTemplate_ReportCreation.R

```
## Create the report from the Rhtml file
library(knitr)
knit("Ch4_ReportTemplate.Rhtml")
```

Output

Ch4_ReportTemplate.html

```
A minimal report with knitr.

set.seed(12345)
y <- rnorm(10)
y
```

```
## [1]   0.5855288   0.7094660  -0.1093033  -0.4534972   0.6058875  -1.8179560
## [7]   0.6300986  -0.2761841  -0.2841597  -0.9193220
```

```
sum(y)
```

```
## [1] -1.329441
```

Analysis

The Rhtml file is a combination of R code and HTML. The main structure of the file is HTML including the `<html>`, `<head>`, and `<body>` elements. Please look at https://htmlreference. io/ to learn more about HTML. The R code between `<!--begin.rcode` and `end.rcode-->` is simply the creation and display of ten random numbers. Save this script in the current working directory as an Rhtml file. Then use the **knit()** function from the **knitr** package to turn the Rhtml file into an HTML file consisting of the results. The output of **knit()** is the results file with the ending HTML. There are several options to modify appearance of the R results in the final HTML file, for example, whether the R code should be displayed in the resulting HTML file or not. All options can be found in Xie (2015) and Xie (2021a). All the rest of the Rhtml file is normal HTML.

4.3 Powerful helper functions and function teams for data exploration

The advice in this section makes your data exploration more efficient. We will practise how to analyse several variables at once and how to assemble the results.

4.3.1 tapply()   R FUNCTION!

The function **tapply()** *conveniently returns descriptive statistics within groups defined by an experimental factor.* In our simulated data for this chapter, the group means of Y for the three factor levels of A and for the combinations of factors A and B are returned by **tapply()**.

Ch4_tapply.R

```
## Statistics for levels of factor variables with tapply()
source("Ch4_DataCreation.R")

tapply(d$Y, d$A, mean)
with(d, tapply(Y, list(A, B), mean))
```

Output

```
> tapply(d$Y, d$A, mean)
   a1    a2    a3
13.55 13.60 14.02
> with(d, tapply(Y, list(A, B), mean))
```

```
      b1    b2
a1 13.44 13.66
a2 13.72 13.48
a3 13.98 14.06
```

Analysis

The first parameter of **tapply()** requests a data vector, the second calls for a factor variable and the third for the desired statistical function, in this case the mean. Note that the function argument goes without parentheses. The function mean can also be replaced by **sd** or by any user-defined function, however complex, which can calculate on a data vector split for different groups. This makes **tapply()** a very powerful function, which can even apply self-written functions to grouped data. The following function returns a sum of the Y values within the factor levels of A: tapply(X = dY, dA, function(X){sum(X)}). The dependent variable Y is declared as the requested X variable in **tapply()** and is passed to the self-written function argument later in the call to **tapply()**. Between round braces { and } is given what the function does, namely the summation of the vector entries. Note that the function only sums for one vector; that this is achieved for all factor levels of A is managed by **tapply()**.

R LANGUAGE! ## 4.3.2 Vectorised operations in R

The function **by()** is one of the most powerful members of the so-called apply function family. *It allows for the application of a function to several data columns within the levels of one or more grouping variables. It generalises* **tapply()** *from one to more than one data columns.* The following code example shows a standard application of **by()**. Usually you have more than one dependent measures and place these as a data frame as the first argument and you want to calculate summary statistics within the levels of a given factor, which is placed in the second position.

Ch4_by.R

```
## Using by() to calculate for several columns split by group.
# Data creation
Data <- data.frame(
    A = gl(n = 2, k = 4, labels = c("a1","a2")),
    y1 = c(2, 3, 4, 5, 6, 7, 8, 9),
    y2 = c(1, 1, 2, 3, 4, 2, 1, 6))

# Apply function
Statistics <- with(Data,
                {
                    by(cbind(y1,y2), A, function(X)
                    {
                        list(Mean = colMeans(X),
                            Sum = colSums(X))
                    })
                })
```

```
Statistics$a1$Mean["y1"]
StatisticsUnlisted <- unlist(Statistics)
names(StatisticsUnlisted)
StatisticsUnlisted[grep("Mean", names(StatisticsUnlisted))]
```

Output

```
> Statistics$a1$Mean["y1"]
 y1
3.5
> StatisticsUnlisted <- unlist(Statistics)
> names(StatisticsUnlisted)
[1] "a1.Mean.y1" "a1.Mean.y2" "a1.Sum.y1"  "a1.Sum.y2"   "a2.Mean.y1" "a2.
Mean.y2"
[7] "a2.Sum.y1"  "a2.Sum.y2"
> StatisticsUnlisted[grep("Mean", names(StatisticsUnlisted))]
a1.Mean.y1 a1.Mean.y2 a2.Mean.y1 a2.Mean.y2
      3.50       1.75       7.50       3.25
```

Analysis

In the script, first, a data frame is created with one experimental factor A and two dependent measures y1 and y2. *Our aim with this data set is to calculate data means and data summation of the two dependent variables within the levels of factor A, perhaps in order to calculate the intermediate steps of an analysis of variance. Next, we use a powerful combination of four functions, with(), by(), cbind(), and function() to do the job.* The function with() gives access to the variables of Data, saving us from $ constructions. We will learn more about its benefits later. Next, by() defines the central job that for the two dependent variables y1 and y2, a user-defined function should be applied within the levels of A. Next, cbind() collects the two variables to one object, because by() needs one object as the first argument. Finally, function() contains the collection of functions that will be applied to the data vectors specified with by(). This was the description of what the functions do; however, let us translate the code into human language: we access the data frame Data and select two dependent variables and one grouping variable. Within the groups, we do some calculations on the dependent variables. Unfortunately, by() creates a somewhat complex list output with named elements and sub elements. You can access all content of this list by recursively using the indexing operators $ and []. However, this is sometimes unpractical and the brutal unlist() is easier to use to get all content as a vector of named elements. You could further combine this with grep(), for example, to extract all the elements containing a mean value.

In contrast to vectorised operations, there is the more traditional approach of a loop using for() to apply the same set of actions to all elements of an object. This technique is also available in other programming languages like C++ and sometimes we also need it in R. However, if a certain operation can be executed either as a loop or vectorised, the vectorised version is preferred in R because it is usually faster (Matloff, 2011).

EXAMPLE! ## 4.3.3 SCB - Description of several variables

The student reconciliators data consists of several measurements of bullying or violence on schoolyards over the project duration of four months. These measurements yield the Y variables in the data frame. Because these variables can all be treated equally, vectorised function calls to apply one statistic to all of them in one line of code is feasible.

SCB_DES_DescriptivesAllViolenceMeasurements.R

```
## SCB - Descriptive statistics for all measurements of violence
if(!exists("DescrViolenceVar"))
{
  source("./Script/SCB_PRE_DataImport.R")
  source("./Script/SCB_PRE_VariableLabels.R")
  library(psych)

  DescrViolenceVar <- with(SCB,
                {
                  DependentVariables <- c("Y1C1", "Y2C1", "Y3C1",
                                          "Y1C2", "Y2C2", "Y3C2",
                                          "Y1C3", "Y2C3", "Y3C3",
                                          "Y1C4", "Y2C4", "Y3C4")

                  # Vectorised function application
                  MeansViolence <- apply(SCB[DependentVariables],
                                        MARGIN = 2,
                                        FUN = mean)
                  ConditionalSummariesViolence <- by(SCB[DependentVariables],
                                                     SCB$A,
                                                     summary)
                  list(MeansViolence = MeansViolence,
                       ConditionalSummariesViolence =
ConditionalSummariesViolence)
                })
}
```

Output

```
> DescrViolenceVar
$MeansViolence
      Y1C1      Y2C1      Y3C1      Y1C2      Y2C2      Y3C2      Y1C3      Y2C3
  19.80000 20.06111 20.17222 19.90556 20.01111 19.79444 20.21667 20.07222
      Y3C3      Y1C4      Y2C4      Y3C4
  19.96111 19.75000 19.76667 19.93333

$ConditionalSummariesViolence
SCB$A: Conflict reconciliators
```

```
        Y1C1                Y2C1                Y3C1                Y1C2
Min.   :16.0       Min.   :16.00      Min.   :14.00      Min.   :14.00
1st Qu.:18.0       1st Qu.:19.00      1st Qu.:18.00      1st Qu.:18.00
Median :20.0       Median :20.00      Median :20.00      Median :20.00
Mean   :19.7       Mean   :20.02      Mean   :20.08      Mean   :19.82
3rd Qu.:21.0       3rd Qu.:21.00      3rd Qu.:22.00      3rd Qu.:21.00
Max.   :25.0       Max.   :24.00      Max.   :25.00      Max.   :24.00
        Y2C2                Y3C2                Y1C3                Y2C3
Min.   :15.00      Min.   :15.00      Min.   :15.00      Min.   :16.00
1st Qu.:18.75      1st Qu.:18.00      1st Qu.:18.00      1st Qu.:19.00
Median :20.00      Median :20.00      Median :20.00      Median :20.00
Mean   :19.63      Mean   :19.57      Mean   :19.58      Mean   :19.73
3rd Qu.:21.00      3rd Qu.:21.00      3rd Qu.:21.00      3rd Qu.:21.00
Max.   :24.00      Max.   :25.00      Max.   :24.00      Max.   :24.00
        Y3C3                Y1C4                Y2C4                Y3C4
Min.   :15.00      Min.   :14.00      Min.   :14.0       Min.   :13.00
1st Qu.:18.00      1st Qu.:18.00      1st Qu.:18.0       1st Qu.:18.00
Median :19.00      Median :19.00      Median :19.0       Median :19.50
Mean   :19.27      Mean   :19.12      Mean   :19.2       Mean   :19.28
3rd Qu.:20.00      3rd Qu.:20.00      3rd Qu.:20.0       3rd Qu.:21.00
Max.   :24.00      Max.   :24.00      Max.   :22.0       Max.   :23.00
------------------------------------------------------------
SCB$A: Enhanced teacher attendance
        Y1C1                Y2C1                Y3C1                Y1C2
Min.   :13.00      Min.   :14.00      Min.   :15.00      Min.   :14.00
1st Qu.:18.75      1st Qu.:18.00      1st Qu.:18.00      1st Qu.:18.00
Median :19.00      Median :20.00      Median :19.50      Median :20.00
Mean   :19.63      Mean   :19.73      Mean   :19.57      Mean   :19.42
3rd Qu.:21.00      3rd Qu.:21.00      3rd Qu.:21.00      3rd Qu.:21.00
Max.   :24.00      Max.   :27.00      Max.   :24.00      Max.   :24.00
        Y2C2                Y3C2                Y1C3                Y2C3
Min.   :14.00      Min.   :14.00      Min.   :15.00      Min.   :15.00
1st Qu.:18.00      1st Qu.:18.00      1st Qu.:18.00      1st Qu.:18.00
Median :20.00      Median :19.00      Median :20.00      Median :19.50
Mean   :19.75      Mean   :19.48      Mean   :19.42      Mean   :19.52
3rd Qu.:22.00      3rd Qu.:21.00      3rd Qu.:21.00      3rd Qu.:21.00
Max.   :24.00      Max.   :25.00      Max.   :24.00      Max.   :24.00
        Y3C3                Y1C4                Y2C4                Y3C4
Min.   :15.00      Min.   :15.00      Min.   :14.00      Min.   :15.00
1st Qu.:18.00      1st Qu.:18.00      1st Qu.:18.00      1st Qu.:18.00
Median :20.00      Median :19.00      Median :19.00      Median :19.50
Mean   :19.63      Mean   :19.22      Mean   :19.25      Mean   :19.47
3rd Qu.:21.00      3rd Qu.:21.00      3rd Qu.:20.00      3rd Qu.:21.00
Max.   :26.00      Max.   :24.00      Max.   :25.00      Max.   :25.00
------------------------------------------------------------
SCB$A: No intervention
```

```
        Y1C1                Y2C1                Y3C1                Y1C2
 Min.   :16.00      Min.   :14.00      Min.   :16.00      Min.   :15.00
 1st Qu.:18.00      1st Qu.:18.00      1st Qu.:19.00      1st Qu.:19.00
 Median :20.00      Median :21.00      Median :21.00      Median :20.00
 Mean   :20.07      Mean   :20.43      Mean   :20.87      Mean   :20.48
 3rd Qu.:21.25      3rd Qu.:22.00      3rd Qu.:23.00      3rd Qu.:22.00
 Max.   :27.00      Max.   :27.00      Max.   :26.00      Max.   :26.00
        Y2C2                Y3C2                Y1C3                Y2C3
 Min.   :15.00      Min.   :16.00      Min.   :16.00      Min.   :15.00
 1st Qu.:19.75      1st Qu.:19.00      1st Qu.:20.00      1st Qu.:19.75
 Median :21.00      Median :20.00      Median :22.00      Median :21.00
 Mean   :20.65      Mean   :20.33      Mean   :21.65      Mean   :20.97
 3rd Qu.:22.00      3rd Qu.:22.00      3rd Qu.:23.00      3rd Qu.:22.00
 Max.   :26.00      Max.   :25.00      Max.   :27.00      Max.   :26.00
        Y3C3                Y1C4                Y2C4                Y3C4
 Min.   :15.00      Min.   :14.00      Min.   :16.00      Min.   :16.00
 1st Qu.:19.00      1st Qu.:19.00      1st Qu.:19.75      1st Qu.:20.00
 Median :21.00      Median :21.00      Median :21.00      Median :21.00
 Mean   :20.98      Mean   :20.92      Mean   :20.85      Mean   :21.05
 3rd Qu.:22.25      3rd Qu.:22.25      3rd Qu.:23.00      3rd Qu.:22.00
 Max.   :26.00      Max.   :27.00      Max.   :27.00      Max.   :27.00
```

SCB_DES_CovarianceCorrelation.R

```
## SCB - Covariance and correlation
if(!exists('CovarianceCorrelation'))
{
  source("./Script/SCB_PRE_DataImport.R")
  source("./Script/SCB_PRE_VariableLabels.R")
  library(psych)
  library(Hmisc)

  DependentVariables <- names(SCB)[grep("Y", names(SCB))]
  CovarianceCorrelation <- list(Covariances = cov(SCB[DependentVariables]),
                                Correlations = cor(SCB[DependentVariables]),
                                CorrTest1 = cor.test(SCB$Y1C1, SCB$Y1C2),
                                CorrTest2 = corr.test(SCB[DependentVariables]))
  rm(DependentVariables)
}
```

Output

```
> print(CovarianceCorrelation, digits = 2)
$Covariances
      Y1C1 Y2C1 Y3C1 Y1C2 Y2C2 Y3C2 Y1C3 Y2C3 Y3C3 Y1C4 Y2C4 Y3C4
Y1C1   5.0  2.3  2.1  1.5  1.8  1.8  1.8  1.5  2.1  2.2  2.1  1.9
Y2C1   2.3  5.2  2.3  2.3  2.1  2.2  2.2  2.1  2.7  2.0  2.1  2.2
```

```
Y3C1  2.1  2.3  5.4  2.3  2.1  2.0  2.4  1.6  2.4  2.3  2.1  2.6
Y1C2  1.5  2.3  2.3  4.6  2.2  1.9  1.7  2.2  2.3  2.0  1.6  1.8
Y2C2  1.8  2.1  2.1  2.2  5.1  2.2  2.0  2.1  2.5  1.9  2.0  2.5
Y3C2  1.8  2.2  2.0  1.9  2.2  4.9  2.2  1.4  1.9  2.1  1.7  1.9
Y1C3  1.8  2.2  2.4  1.7  2.0  2.2  5.9  2.3  2.6  2.8  2.7  3.0
Y2C3  1.5  2.1  1.6  2.2  2.1  1.4  2.3  5.3  1.9  2.6  2.0  2.2
Y3C3  2.1  2.7  2.4  2.3  2.5  1.9  2.6  1.9  5.5  2.1  2.6  2.4
Y1C4  2.2  2.0  2.3  2.0  1.9  2.1  2.8  2.6  2.1  5.9  2.9  2.6
Y2C4  2.1  2.1  2.1  1.6  2.0  1.7  2.7  2.0  2.6  2.9  4.9  2.4
Y3C4  1.9  2.2  2.6  1.8  2.5  1.9  3.0  2.2  2.4  2.6  2.4  5.8
```

```
$Correlations
      Y1C1 Y2C1 Y3C1 Y1C2 Y2C2 Y3C2 Y1C3 Y2C3 Y3C3 Y1C4 Y2C4 Y3C4
Y1C1 1.00 0.44 0.41 0.31 0.36 0.36 0.32 0.29 0.40 0.40 0.43 0.35
Y2C1 0.44 1.00 0.43 0.47 0.41 0.43 0.41 0.41 0.50 0.37 0.41 0.41
Y3C1 0.41 0.43 1.00 0.47 0.39 0.40 0.43 0.29 0.44 0.41 0.41 0.47
Y1C2 0.31 0.47 0.47 1.00 0.45 0.41 0.33 0.45 0.47 0.38 0.33 0.34
Y2C2 0.36 0.41 0.39 0.45 1.00 0.45 0.37 0.40 0.47 0.35 0.41 0.47
Y3C2 0.36 0.43 0.40 0.41 0.45 1.00 0.41 0.28 0.36 0.39 0.36 0.36
Y1C3 0.32 0.41 0.43 0.33 0.37 0.41 1.00 0.41 0.46 0.48 0.51 0.50
Y2C3 0.29 0.41 0.29 0.45 0.40 0.28 0.41 1.00 0.34 0.46 0.40 0.40
Y3C3 0.40 0.50 0.44 0.47 0.47 0.36 0.46 0.34 1.00 0.37 0.51 0.42
Y1C4 0.40 0.37 0.41 0.38 0.35 0.39 0.48 0.46 0.37 1.00 0.54 0.44
Y2C4 0.43 0.41 0.41 0.33 0.41 0.36 0.51 0.40 0.51 0.54 1.00 0.44
Y3C4 0.35 0.41 0.47 0.34 0.47 0.36 0.50 0.40 0.42 0.44 0.44 1.00
```

```
$CorrTest1


        Pearson's product-moment correlation

data:  SCB$Y1C1 and SCB$Y1C2
t = 4, df = 178, p-value = 2e-05
alternative hypothesis: true correlation is not equal to 0
95 percent confidence interval:
 0.17 0.44
sample estimates:
 cor
0.31
```

```
$CorrTest2
Call:corr.test(x = SCB[DependentVariables])
Correlation matrix
      Y1C1 Y2C1 Y3C1 Y1C2 Y2C2 Y3C2 Y1C3 Y2C3 Y3C3 Y1C4 Y2C4 Y3C4
Y1C1 1.00 0.44 0.41 0.31 0.36 0.36 0.32 0.29 0.40 0.40 0.43 0.35
Y2C1 0.44 1.00 0.43 0.47 0.41 0.43 0.41 0.41 0.50 0.37 0.41 0.41
```

```
Y3C1 0.41 0.43 1.00 0.47 0.39 0.40 0.43 0.29 0.44 0.41 0.41 0.47
Y1C2 0.31 0.47 0.47 1.00 0.45 0.41 0.33 0.45 0.47 0.38 0.33 0.34
Y2C2 0.36 0.41 0.39 0.45 1.00 0.45 0.37 0.40 0.47 0.35 0.41 0.47
Y3C2 0.36 0.43 0.40 0.41 0.45 1.00 0.41 0.28 0.36 0.39 0.36 0.36
Y1C3 0.32 0.41 0.43 0.33 0.37 0.41 1.00 0.41 0.46 0.48 0.51 0.50
Y2C3 0.29 0.41 0.29 0.45 0.40 0.28 0.41 1.00 0.34 0.46 0.40 0.40
Y3C3 0.40 0.50 0.44 0.47 0.47 0.36 0.46 0.34 1.00 0.37 0.51 0.42
Y1C4 0.40 0.37 0.41 0.38 0.35 0.39 0.48 0.46 0.37 1.00 0.54 0.44
Y2C4 0.43 0.41 0.41 0.33 0.41 0.36 0.51 0.40 0.51 0.54 1.00 0.44
Y3C4 0.35 0.41 0.47 0.34 0.47 0.36 0.50 0.40 0.42 0.44 0.44 1.00
Sample Size
[1] 180
Probability values (Entries above the diagonal are adjusted for multiple
tests.)
     Y1C1 Y2C1 Y3C1 Y1C2 Y2C2 Y3C2 Y1C3 Y2C3 Y3C3 Y1C4 Y2C4 Y3C4
Y1C1   0    0    0    0    0    0    0    0    0    0    0    0
Y2C1   0    0    0    0    0    0    0    0    0    0    0    0
Y3C1   0    0    0    0    0    0    0    0    0    0    0    0
Y1C2   0    0    0    0    0    0    0    0    0    0    0    0
Y2C2   0    0    0    0    0    0    0    0    0    0    0    0
Y3C2   0    0    0    0    0    0    0    0    0    0    0    0
Y1C3   0    0    0    0    0    0    0    0    0    0    0    0
Y2C3   0    0    0    0    0    0    0    0    0    0    0    0
Y3C3   0    0    0    0    0    0    0    0    0    0    0    0
Y1C4   0    0    0    0    0    0    0    0    0    0    0    0
Y2C4   0    0    0    0    0    0    0    0    0    0    0    0
Y3C4   0    0    0    0    0    0    0    0    0    0    0    0

To see confidence intervals of the correlations, print with the short=FALSE option
```

Analysis

The two scripts for descriptive statistics and correlations produce much output at the console. However, all results are stored in two objects, DescrViolenceVar and CovarianceCorrelation. These results give us a first impression of whether the intervention program could have an effect on the number of incidents relating to bullying or violence.

R FUNCTION! ### 4.3.4 list()

There are objects which do not fit in the rectangle form of a table, but which themselves contain complex information. Such objects usually are lists. Lists are very general objects in R and can contain almost everything. A list is like a shopping bag, which can hold various related or unrelated things without any constraint of the types or number of objects stored inside it; in fact, my shopping bag does not care if I buy bread or cheese, it simply keeps my groceries together. Some operations may need something out of the list and simply grasp the required thing using the correct indexing. Lists can hold anything and they accompany your workflow in a statistical analysis. They allow for efficient programming. For example, if you want to pass on several objects in one go, simply put the objects in a list and pass on the list. The function `list()` instantiates a list, and the content of the list goes between parentheses. List elements can obtain names, and indexing list elements can later use these names. An example list, which holds all the results of a statistical analysis, is the following.

Ch4_list.R

```
## Using lists to collect objects
source("Ch4_DataCreation.R")

# Collect results of an analysis in a list
myResults <- list(Descriptives = by(data = data.frame(d$X, d$Y),
                                     INDICES = list(d$A, d$B),
                                     FUN = summary),
                  Correlations = cor(d$X, d$Y),
                  ANOVA = aov(Y ~ A * B, data = d))


# Access the elements of the results list
myResults$Descriptives
myResults$Correlations
summary(myResults$ANOVA)
names(myResults$ANOVA)
myResults$ANOVA$coefficients
```

Output

```
> # Access the elements of the results list
> myResults$Descriptives
: a1
: b1
       d.X                d.Y
 Min.    :1.00    Min.    :10.00
 1st Qu.:2.00    1st Qu.:12.00
 Median :3.00    Median :13.50
 Mean    :3.36    Mean    :13.44
 3rd Qu.:5.00    3rd Qu.:15.00
 Max.    :6.00    Max.    :19.00
---------------------------------------------------------------
: a2
: b1
       d.X                d.Y
 Min.    :1.0    Min.    : 8.00
 1st Qu.:2.0    1st Qu.:12.00
 Median :4.0    Median :14.00
```

```
Mean   :3.6   Mean    :13.72
3rd Qu.:5.0   3rd Qu.:15.00
Max.   :6.0   Max.    :18.00
----------------------------------------------------------------
: a3
: b1
      d.X             d.Y
 Min.   :1.00   Min.    :11.00
 1st Qu.:2.00   1st Qu.:12.00
 Median :3.00   Median :14.00
 Mean   :3.44   Mean    :13.98
 3rd Qu.:5.00   3rd Qu.:15.00
 Max.   :6.00   Max.    :18.00
----------------------------------------------------------------
: a1
: b2
      d.X             d.Y
 Min.   :1.0    Min.    : 9.00
 1st Qu.:2.0    1st Qu.:12.25
 Median :3.0    Median :13.00
 Mean   :3.1    Mean    :13.66
 3rd Qu.:4.0    3rd Qu.:15.00
 Max.   :6.0    Max.    :17.00
----------------------------------------------------------------
: a2
: b2
      d.X             d.Y
 Min.   :1.0    Min.    :10.00
 1st Qu.:2.0    1st Qu.:12.00
 Median :3.0    Median :13.00
 Mean   :3.6    Mean    :13.48
 3rd Qu.:5.0    3rd Qu.:15.00
 Max.   :6.0    Max.    :18.00
----------------------------------------------------------------
: a3
: b2
      d.X             d.Y
 Min.   :1.00   Min.    :11.00
 1st Qu.:2.00   1st Qu.:13.00
 Median :3.00   Median :13.50
 Mean   :3.34   Mean    :14.06
 3rd Qu.:5.00   3rd Qu.:16.00
 Max.   :6.00   Max.    :20.00
> myResults$Correlations
[1] 0.8551011
> summary(myResults$ANOVA)
```

```
          Df Sum Sq Mean Sq F value Pr(>F)
A          2   13.3   6.663   1.638  0.196
B          1    0.0   0.030   0.007  0.932
A:B        2    2.8   1.390   0.342  0.711
Residuals 294 1195.9  4.068
> names(myResults$ANOVA)
 [1] "coefficients"  "residuals"    "effects"      "rank"
 [5] "fitted.values" "assign"       "qr"           "df.residual"
 [9] "contrasts"     "xlevels"      "call"         "terms"
[13] "model"
> myResults$ANOVA$coefficients
(Intercept)        Aa2         Aa3         Bb2     Aa2:Bb2     Aa3:Bb2
      13.44       0.28        0.54        0.22       -0.46       -0.14
```

Analysis

The script includes the script for data creation from the beginning of this chapter. Then it uses **list()** to create the list myResults. The function initialises three list elements, Descriptives, Correlations, and ANOVA, and assigns some content to these elements. The remainder of the script shows access to the list elements. It simply works with the **$**-operator like in data frames. In fact, a data frame is merely a list, where each of the list components has the same length. Note how functions can also take list elements as arguments, as **summary()** and **names()** demonstrate here. If a list element is itself a list, like ANOVA in the example, the same way of indexing works. However, beside **$** there is another way of list indexing using the double square brackets [[]]. Simply put in the name of the element to be indexed. In the example it could be myResults[["ANOVA"]]. Lists populate all parts of the statistical analysis whenever a meaningful bunch of objects is needed which is not a data frame or matrix. So, master their creation and the access to their contents.

4.3.5 Function teams that work well together

R LANGUAGE!

There are some functions which often work together. For instance, on the help page of **seq()** scroll down almost to the bottom. The section "See also" provides similar functions, helpful for some similar programming problem. In Table 4.2 I give an overview of some common tasks and useful function teams to handle them.

Table 4.2 Examples of function teams.

Purpose	Team
Create a vector according to a rule	seq(), rep()
Enquire and clean up objects at the console	ls(), names(), rm(), CTRL + L
Enquire attributes of an object	str(), attributes(), attr()
Enquire the type of a given object	class(), methods(), is()
Concatenate and search strings according to pattern	paste(), grep()

(Continued)

Table 4.2   (Continued)

Purpose	Team
Create a loop	for(), length()
Manage the search path	search(), attach(), detach()
Manage the working directory	getwd(), setwd(), dir(), load(), save()
Create a data frame for an experiment with independent and dependent variables	data.frame(), rnorm(), factor(), interaction(), expand.grid()
Create a matrix	matrix(), vector(), cbind(), rbind(), diag(), dimnames(), t()
Cross tabulation	table(), ftable(), addmargins(), chisq.test(), prop.table(), margin.table(), xtabs()
Write your own function and apply it to data	with(), by(), tapply(), function()
Create a multiple graph layout	layout(), matrix(), layout.show()
Create a diagram from scratch	plot.new(), par(), plot.window()
Calculate an analysis of variance and inspect the results	aov(), formula(), lm(), summary(), model.tables(), Error(), model.matrix()
Include add-on packages and learn about their functionality	library(), vignette(), ??, ?, demo()
Write results to a text file	capture.output(), cat(), print()

R LANGUAGE!   ## 4.3.6 The search path in R

We already talked about the global environment holding the R objects and the working directory containing files for the analysis. We now talk about the location where R finds all the functions used in an analysis. This is important because some packages consist of functions with the same name, which can cause unwanted ambiguity of which version of the function to use in a function call. For example, the packages **gplots** (Warnes et al., 2020) and **plotrix** (Lemon, 2006) both have a function **plotCI()** to display confidence intervals. If both packages are in use in an R session, how does R decide which version of the function to use? *To use a certain function, R goes through the so-called search path to match the function call with the functions it has available.* We can inspect the path with **search()**. This prints out a numbered list of locations where functions or objects reside. As can be seen, the global environment is the first entry in this list, meaning that R looks up anything in the global environment first and then moves on to the other entries in the list. Two other entries are the names of packages that directly tell us about their purpose: **stats** and **graphics**. Obviously, **stats** contains many statistics functions and **graphics** contains functions for graphical display. In the case of a function that is available in two packages, R uses the version it finds first in the search path starting from the global environment and travelling through all the entries of the search path.

In the case of **plotCI()**, if **gplots** has been put on the search path with **library()** more recently than **plotrix**, it comes earlier in the search path and its version of **plotCI()** would be used. If you want to use the function implemented in the later path entry, you can remove the earlier entry from the path with **detach()** like in detach(package:PackageName). You can also directly address the function in the later entry with the double colon operator **::**.

So, if **plotrix** comes behind **gplots** in the path, a call of **plotrix::plotCI()** would use the **plotCI()** function of **plotrix**. In the case of an R object, the function call attach (object) places Object at the second position behind the global environment, so that all elements contained in the object are now directly available. For objects, detach(Object) would remove the object from the search path.

So, if you load a certain package with **library()** or make the elements of an R object available with **attach()**, you basically modify the search path and tell R more locations where to look up objects and functions. In sum, remember to know the entries in the search path and do not overload the search path to avoid the danger of conflicts about which object or function R is going to use in your function calls.

4.4 A graphical user interface to explore data – The RCommander

R programming is cool, but sometimes to point-click some results for the next team meeting is also OK. R incorporates the point-click solution with the Rcommander, a graphical user interface where you specify the statistical analysis.

If you want quick results and do not want to program, the **RCommander** (Fox, 2005) is your solution. It offers a rich set of functions for data exploration and modelling. It is a point and click interface and contains menu options and several dialogues to specify the calculations. It automatically creates a protocol of the used functions. However, the protocol you get in the **RCommander** will contain different functions than those which you would program yourself to get the statistics. There are more than 30 other packages related to the **Rcommander**.

Install the **Rcommander** *and start it with* library(Rcmdr). A graphical user interface opens which is structured in five parts: a menu, buttons for selecting data frames and models, an R script or markdown field, an output field, and a messages field (Figure 4.2). With the menu options you can specify statistical calculations and run them; the corresponding script is stored in the script field. You can also write your own code in the script and run chunks of

Figure 4.2 Graphical user interface of the R Commander.

code if you select some lines and hit Ctrl+R or the run button. The output of the calculations together with the currently executed code are written in the output field. If you use graphics, however, they are placed in the R window which is still present in the background. With the **Rcommander** you can run a complete statistical analysis from start to finish. You can load and analyse data and store the specifications in a script which can be saved and further edited even with functions which are not available in the **Rcommander** itself.

4.5 Structuring the complete analysis II

Earlier, we talked about how to structure the complete analysis. Now, we practise the <u>with()</u> function as a highly efficient tool to structure your scripts.

PROGRAMMING
ISSUE!

4.5.1 Using a main script to structure the analysis

The statistical analysis usually involves the programming of many scripts, which go in a certain logical order from data acquisition to inferential statistics. So, it may be a good idea to reflect this order in a master script or main script, which uses <u>source()</u> with all the other scripts in the correct order. In particular, calculations that build on previous calculations have to be placed in the appropriate order in the main script. Then, the complete analysis could be reproduced by simply running the main script. However, I suggest making each script also self-contained in the sense that each script sources all the scripts that must run before it. Redundant inclusions of scripts are no problem if we adhere to the inclusion guards. That is, each intermediate object in the analysis is produced only once. Try it out with your analyses to see whether you prefer to use a main script or not.

R FUNCTION! ## 4.5.2 with()

The <u>with()</u> *function allows us to keep steps in the analysis together, which belong together. It declares an object as the main focus of interest and then takes lines of code inside braces {}, which operate on this object. As output,* <u>with()</u> *returns the last line of code inside {}. The following script demonstrates the use of* <u>with()</u> *to calculate means in a data frame.*

Ch4_with.R

```
## Using with() to structure the analysis
set.seed(12345)
d <- data.frame(A = rep(c("a1", "a2", "a3"), 10),
                y = rnorm(30, c(5, 6, 7)))

Means <- with(d,
              {
                GrandMean <- mean(y)
                GroupMeans <- tapply(y, A, mean)
                list(GrandMean = GrandMean, GroupMeans =  GroupMeans)
              })
Means
```

Output

```
> Means
$GrandMean
[1] 6.078807

$GroupMeans
       a1       a2       a3
5.262950 6.262840 6.710631
```

Analysis

In this code example, another helpful element of structured programming in R is used, the function `with()`. It takes a data frame as the first argument and a block of function calls as the second argument, which are applied to the data frame. Inside `with()`, the variables of the data frame are directly accessible and do not need to be addressed with the $-operator. In this case, a few function calls are written between the braces { and }. The last line inside the {} collects the results of earlier calculations in a list, which is returned as the output of `with()`. The braces are only necessary if more than one line of code is executed inside `with()`. For one single line, a construction like `with(d, mean(y))` would have been sufficient.

Using `with()` leads to well written scripts in several respects:

- The function `with()` replaces the need for `attach()` and `detach()` to make variables available for calculations. In some scripts `attach()` and `detach()` can be far apart, so there is the danger to forget the `detach()` for a data frame after completion of the calculations. In this case, the variable names of the data frame remain directly accessible.
- Using `with()` also makes the script more accessible. Everything between braces can be indented in a block and can be grasped as a meaningful unit immediately. You automatically grasp the function calls that belong together even if you read the script after half a year.
- The function `with()` also keeps the workspace clean, because all objects created after { disappear after }. Within the block marked by the braces, R creates a separate environment for the function calls. To return something at the console, we therefore had to collect the results of the calculations in a list.

The function `within()` is a relative of `with()` made for data transformations. Let us quickly check at the console that `with()` only makes the data in the data frame available for operations, whereas `within()` modifies a copy of the data frame.

Ch4_within.R

```
## within()
source("Ch4_DataCreation.R")

head(d)
d1 <- with(d, Y <- Y+10)
head(d1)
d2 <- within(d, Y <- Y+10)
```

```
head(d2)
d3 <- within(d, Z <- Y+20)
head(d3)
d4 <- within(d, X <- NULL)
head(d4)
with(d, d$YY <<- Y + 50)
head(d)
```

Output

```
> head(d)
   A  B X  Y
1 a1 b1 6 17
2 a1 b1 3 12
3 a1 b1 2 12
4 a1 b1 4 16
5 a1 b1 2 11
6 a1 b1 5 15
> d1 <- with(d, Y <- Y+10)
> head(d1)
[1] 27 22 22 26 21 25
> d2 <- within(d, Y <- Y+10)
> head(d2)
   A  B X  Y
1 a1 b1 6 27
2 a1 b1 3 22
3 a1 b1 2 22
4 a1 b1 4 26
5 a1 b1 2 21
6 a1 b1 5 25
> d3 <- within(d, Z <- Y+20)
> head(d3)
   A  B X  Y  Z
1 a1 b1 6 17 37
2 a1 b1 3 12 32
3 a1 b1 2 12 32
4 a1 b1 4 16 36
5 a1 b1 2 11 31
6 a1 b1 5 15 35
> d4 <- within(d, X <- NULL)
> head(d4)
   A  B  Y
1 a1 b1 17
2 a1 b1 12
3 a1 b1 12
4 a1 b1 16
```

```
5 a1 b1 11
6 a1 b1 15
> with(d, d$YY <<- Y + 50)
> head(d)
   A  B X  Y YY
1 a1 b1 6 17 67
2 a1 b1 3 12 62
3 a1 b1 2 12 62
4 a1 b1 4 16 66
5 a1 b1 2 11 61
6 a1 b1 5 15 65
```

Analysis

To illustrate `within()` at the console, we use the data creation script from the chapter beginning, which creates the data frame d with two factors and two quantitative variables. The `head()` function to display only the top rows of a data frame keeps the output short. For comparison, we use `with()` first and calculate a variable. The result d1 is merely the Y variable plus ten. In contrast, the first use of `within()` returns a modification of the whole d data frame, where the values of Y are increased by ten. The modified data frame is stored in d2. The third call creates the new variable z within the data frame and then stores it in d3. The fourth call shows how to delete a variable from the data set by assigning NULL to the variable. Finally, we use the superassignment operator <<- (Matloff, 2011) to modify the data frame inside `with()`. Because `with()` creates its own environment, objects in the global environment are not modified by `with()`. However, the superassignment operator allows for such modifications.

In sum, use `with()` wherever you can to structure your analysis and keep related operations together. Use `within()` for more complex modifications inside a data set beyond the addition of single variables.

4.5.3 UPS – Descriptive statistics by condition

EXAMPLE!

The example unpleasant sounds measured user ratings of pleasantness for aversive sounds in several conditions of sound manipulation. Let us inspect the descriptive statistics. We use `with()` to structure the script.

UPS_DES_DescriptivesByCondition.R

```
## UPS - Descriptive statistics by sound condition
if(!exists('DescriptivesBySound'))
{
  source("Scripts/UPS_PRE_DataImport.R")
  library(psych)

  DescriptivesBySound <- with(UPS,
                    {
```

```
                                    describeBy(rating,
                                        group = list(sound,
                                                    variation),
                                        digits = 2,
                                        mat = TRUE)
                            })
}
```

Output

```
> DescriptivesBySound
        item          group1              group2 vars   n mean    sd median
trimmed  mad min max range  skew kurtosis   se
X11           1      Styrofoam  Without  pitch    1  96  4.84  1.05      5
4.97 1.48   1   6     5 -1.10      1.46 0.11
X12           2  Tschiritsch  Without  pitch    1  96  5.01  1.17      5
5.19 1.48   1   6     5 -1.18      1.18 0.12
X13           3       Vomiting  Without  pitch    1  96  4.98  0.89      5
5.05 1.48   1   6     5 -1.10      2.63 0.09
X14           4  Fingernails  Without  pitch    1  96  4.35  0.96      4
4.37 1.48   2   6     4 -0.32     -0.09 0.10
X15           5      Styrofoam    Only  pitch    1  96  3.36  0.90      3
3.41 1.48   1   5     4 -0.51      0.00 0.09
X16           6  Tschiritsch    Only  pitch    1  96  4.03  1.27      4
4.09 1.48   1   6     5 -0.39     -0.27 0.13
X17           7       Vomiting    Only  pitch    1  96  3.80  1.06      4
3.82 1.48   1   6     5 -0.02     -0.35 0.11
X18           8  Fingernails    Only  pitch    1  96  3.56  1.03      4
3.56 1.48   1   6     5  0.03     -0.54 0.11
X19           9      Styrofoam     Band-stop    1  96  4.56  1.02      5
4.63 1.48   1   6     5 -0.69      0.59 0.10
X110        10  Tschiritsch     Band-stop    1  96  4.45  1.01      4
4.46 1.48   1   6     5 -0.19      0.01 0.10
X111        11       Vomiting     Band-stop    1  96  4.16  1.15      4
4.19 1.48   1   6     5 -0.22     -0.40 0.12
X112        12  Fingernails     Band-stop    1  96  4.10  1.02      4
4.17 1.48   1   6     5 -0.50      0.06 0.10
X113        13      Styrofoam     Band-pass    1  96  4.58  0.97      5
4.64 1.48   2   6     4 -0.54      0.00 0.10
X114        14  Tschiritsch     Band-pass    1  96  4.78  1.00      5
4.88 1.48   1   6     5 -0.88      1.16 0.10
X115        15       Vomiting     Band-pass    1  96  4.72  1.07      5
4.81 1.48   1   6     5 -0.70      0.26 0.11
X116        16  Fingernails     Band-pass    1  96  4.66  1.12      5
4.74 1.48   2   6     4 -0.45     -0.58 0.11
X117        17      Styrofoam      Low-pass    1  96  4.12  0.95      4
4.14 1.48   1   6     5 -0.32      0.43 0.10
```

```
X118      18  Tschiritsch            Low-pass      1  96  4.92  0.95              5
5.00 1.48    3    6      3 -0.42    -0.85 0.10
X119      19         Vomiting        Low-pass      1  96  4.45  1.16              5
4.53 1.48    1    6      5 -0.56    -0.20 0.12
X120      20  Fingernails            Low-pass      1  96  3.78  0.93              4
3.77 1.48    2    6      4  0.21    -0.44 0.10
X121      21       Styrofoam         High-pass     1  96  4.99  0.96              5
5.12 1.48    2    6      4 -0.84     0.13 0.10
X122      22  Tschiritsch            High-pass     1  96  3.56  1.13              4
3.55 1.48    1    6      5  0.08     0.08 0.12
X123      23         Vomiting        High-pass     1  96  4.84  1.13              5
4.97 1.48    1    6      5 -1.00     1.00 0.12
X124      24  Fingernails            High-pass     1  96  4.57  0.97              5
4.64 1.48    1    6      5 -0.72     1.16 0.10
X125      25       Styrofoam         Original      1  96  3.67  0.91              4
3.65 1.48    1    6      5  0.12     0.40 0.09
X126      26  Tschiritsch            Original      1  96  4.85  1.10              5
4.97 1.48    1    6      5 -0.81     0.34 0.11
X127      27         Vomiting        Original      1  96  3.53  1.02              4
3.55 1.48    1    6      5 -0.17     0.07 0.10
X128      28  Fingernails            Original      1  96  3.90  0.99              4
3.96 1.48    1    6      5 -0.37     0.02 0.10
```

Analysis

The script returns the descriptive statistics for the pleasantness rating for several combinations of sound quality and sound variation. It uses the **describeBy()** function of the **psych** package. The object DescriptivesBySound contains the statistics and you can enter it at the console to see its content. All statistics yield from the same sample size, so there was apparently no condition leading to missing data. Furthermore, all means lie in the upper half of the rating scale; however, almost always the complete rating scale was used by the subjects.

The script returns its numbers stored in an object, which is ready for use in figures and reports. The script thereby has a good structure: it uses an inclusion guard to prevent double execution. It uses **with()** to structure the calculation and to hide preparatory variable manipulations. In fact, sound_labels and variation_labels are auxiliary variables just for the purpose of the present calculation. The variables consist of the explicit labels of the sound conditions, for example, Styrofoam or Low-pass. Without the two variable transformations using **factor()**, the results display would merely include numbered conditions.

━━━━━━━━━━ **What you have learned so far** ━━━━━━━━━━

1   R functions for exploratory analysis
2   To use vectorised operations to gain efficiency
3   R functions to structure the analysis and make it more readable for others

(Continued)

4   To use lists to aggregate results
5   To create a quick HTML report of the analysis
6   The R commander as a graphical user interface for analysis

━━━ Exercises ━━━

Explore functions

1   Compare between different covariance and correlation functions. Try `cov()`, `cor()`, `psych::corr.test()`, `psych::pairs.panels()`, `psych::r.test()`, or `Hmisc::rcorr()` with the `MusicPreferencesSchool.csv` data set. Use the variables PreferenceSong1, PreferenceSong2, and Creativity for the analyses. Which functions would you prefer for your own analyses?

2   Use the data set `MusiciansComplaints.csv`. Create a cross table of frequencies for the categorical variables Complaint and Instrument. Use `table()`. Further explore the frequencies with `addmargins()` and `prop.table()`.

3   Restructure your analyses of the previous exercises in this section with `with()`. Note how `with()` makes variable access easier.

Solve tasks

1   Create a vector of 100 random numbers with theoretical mean of 100 and standard deviation 10. Classify the values into ten categories and calculate a table of frequencies for the classes. Use `rnorm()`, `cut()`, and `table()`.

Read code

1   The following code fragments use several good structuring elements - `with()`, `tapply()`, `list()`, and `function()`. Read the code and describe in your own words what it does. What are the particular roles of the four helper functions? Run the code in R. Do not forget to put the data set `MusicData.csv` in the working directory.

```
d <- read.csv("MusicData.csv")

# Statistics by groups
tapply(d$Ex1, d$Gender, mean)
tapply(d$Ex1, d$Gender, sd)

# Easy variable access
with(d, tapply(Ex2, Gender, mean))

# More than one statistic as output
with(d,
    {
        M <- tapply(Ex1, Gender, mean)
        SD <- tapply(Ex1, Gender, sd)
```

```
      list(Mean = M, `Standard deviation` = SD)
  })


# Arbitrary function by groups
with(d,
     {
       tapply(Ex1, Gender, function(X)
       {
         M <- mean(X, na.rm = TRUE)
         SD <- sd(X, na.rm = TRUE)
         N <- length(X)
         list(Mean = M,
                 `Standard deviation` = SD,
                 `Sample size` = N)
       })
     })
```

Analyse data

1   Use **MusicData.csv**. Calculate a cross tab for Gender and Age. Use **table()**. Do we have a two-factor balanced design? Add marginal distributions to the cross tabs.

2   Use **MusicData.csv**. Calculate the covariance and correlation matrices for the music evaluations. Calculate a correlation test with **corr.test()** from the **psych** package and with **rcorr()** from **Hmisc**.

3   Use **MusicData.csv**. Calculate descriptive statistics (mean, median, variance, standard deviation, and quartile information) for the evaluations of some musical examples. Recalculate the statistics for both levels of Gender.

4   Use **MusicPreferencesSchool.csv**. Are students of different school types different in their average musical preferences? To answer the question, calculate some descriptive statistics with **tapply()**. Repeat the calculations with **psych::describe.by()**.

5   Use **MusicPreferencesSchool.csv**. Take more than one dependent variable into account and use **by()** to calculate descriptive statistics for several variables divided by factor levels.

6   Use **MusicPreferencesSchool.csv**. Calculate a one-way ANOVA for the influence of school type on musical preferences.

7   Use **MusicPreferencesSchool.csv**. Divide the variable Creativity in three categories of low, medium, and high creatives. Create a new variable CreatCat with **cut()** and select appropriate new names for the categories with levels.

8   Use **MusicPreferencesSchool.csv**. Create tables with absolute and relative frequencies of the new categories of creativity.

9   Use **MusicPreferencesSchool.csv**. Is creativity associated with musicality? Inspect cross tabs between the categories of creativity and musicianship. Do creatives more often play a musical instrument? Use the function **table()** and also add marginal distributions.

10 Use **MusicPreferencesSchool.csv**. Return the group means of the two preference ratings for all combinations of the factors School and Musician. Use the function **by()**.

(Continued)

11 Use `MusicPreferencesSchool.csv`. Focus now on the preference ratings for song 1. Create histograms for all combinations of `School` and `Musician` and arrange them in one display by using `layout()`. Set uniform y-axes and category widths by using the parameters `ylim` and `breaks`. If possible, create the six histograms with one call of `tapply()`.

12 Use `MusicPreferencesSchool.csv`. Calculate a two-factor ANOVA for the preference ratings for song 1. Use the factors `School` and `Musician`. Write the ANOVA table and effect estimators (`model.tables()`) to a results file with `capture.output()`. Display the group means with `interaction.plot()`.

Apply in the real world

1 Calculate descriptive statistics for your own data. Program frequency tables with `table()` and `ftable()` for categorical variables. Calculate means and variances for quantitative variables. Use `tapply()` to calculate statistics for different experimental groups.

2 Think about a project you work for which suffered from missing data. Draw a data model of the project and discuss with your team, which variables are candidates for data imputation and which variables could serve as sources of information for the imputation.

Become a statistics programmer

1 Take a data set that many colleagues in your team know and schedule a live exploratory analysis of the data. In the meeting everybody can throw in questions about the data and all are asked to come up with useful functions to answer the questions. Use the console for this session.

EXAMPLE! **WBD**

1 Run the script `WBD_DES_DescriptiveStatistics.R` from the main text. Now use the template `Ch4_ReportTemplate.Rhtml` from Chapter 4 to store the descriptive statistics in an HTML report.

2 In the foregoing task, `psych::describeBy()` produces several descriptive statistics. From the results object, select only mean, standard deviation, and sample size. Use `[]` and an index vector for this.

3 Calculate covariance and correlation matrices for the outcomes depression, anhedonia, and prospective imagery test. Use `cov()` and `cor()` with the data set in wide format.

4 Use the data set in wide format to recalculate the descriptive statistics mean and standard deviation from the supplementary file to Blackwell and colleagues (2015). Use `tapply()`. Can you recalculate the exact numbers?

EXAMPLE! **ITC**

1 Do a console session with the Wason task data about the independence of thinking bias and cognitive capability. Use many different functions to get a feeling for the data. Which variables are available? What are the different scale levels of the variables? What are the frequency distributions of some key variables?

2 Check descriptively whether the SAT score is independent of age and gender in this sample. For this, calculate mean SAT scores between gender groups and between different age groups.

3 Find the few missing values in the data set. Use `is.na()` and `complete.cases()` for this. How are the missing values distributed among the cases and the variables?

UPS

EXAMPLE!

1 Use `ftable()` to create a flat table of frequencies for all categories of the pleasantness ratings. Put the categories of group, sound, and variation in the rows of this table and rating in the columns. Closely observe the frequency distributions in all rows. Can you find bottom or ceiling effects in these distributions.

2 Store the flat table of the foregoing task in an HTML report. Use the `knitr` package for this.

SCB

EXAMPLE!

1 Check if the experimental design is balanced; that is, all experimental groups should contain the same number of schools. Use `table()`.

2 Check whether the two covariates are distributed evenly across factor levels. Use `cut()` to classify the covariates first; then, use `table()` and `barplot(table())` to visually compare the distributions of the covariates.

3 Calculated descriptive statistics for the dependent variable of pretest-posttest differences for all three types of conflict. Use `tapply()` to calculate the statistics for the different experimental groups. Also try `psych::describeBy()` to explore the data.

4 Further inquire about other meaningful exploratory statistics functions with the help pages of R. Use `??` to search for key terms.

5

GRAPHICAL DATA ANALYSIS

━━━━━━━━━━━━━━━━━━━━━━━━━━ Chapter overview ━━━━━━━

This chapter complements Chapter 4 as an integral part of exploratory analysis and describes
R's graphical facilities to get the best overview of the data. For example, we implement
response profiles of all subjects for the detection of response sets, and also displays of multiple
histograms for the evaluation of item characteristics. These are quick and convenient options
to get the maximum immersion into the data. Programming features: high-level and low-level
graphics, simultaneous output of multiple graphics with uniform layout.

In the next section we will create graphics with one function call. Simply try all functions and select your favourite and apply it to your own data. Afterwards, we will focus on the parameters of the function calls to manipulate the display. Graphics are not only fun to create and to look at; they play an important role in the checking of assumptions for statistical modelling and the display of statistical effects (Loftus, 1996; Masson & Loftus, 2003). For example, normality checks can be supported with histograms or Q-Q diagrams. Error bars and scatter plots give an idea about the homogeneity of variances (i.e. homoscedasticity). Moreover, profiles of the raw data let us easily spot extreme values and also response tendencies of the subjects.

5.1 Quick graphical data exploration

Some R functions generate full diagrams with only one function call. Such functions are high-level graphics functions. This helps you to quickly get results for the next team meeting. We will practise bar diagrams, scatter plots, histograms, and functions of add-on packages.

R LANGUAGE! ### 5.1.1 High-level graphics in R

Table 5.1 shows a collection of high-level graphic functions. *High-level means that complete graphics can be produced with one function call. In using a high-level function, many things happen internally in R to produce the display; for example, the set-up of the graphics window, creation of the axes and annotations, and the creation of the data display.* In contrast, low-level graphic functions create single elements in a graphics display; for example, points, lines, polygons, or text. R also consists of different graphic systems which follow their own rules of graphics creation. The functions of the different systems cannot easily be combined; however, they all use R programming. We will introduce different systems in Chapter 9.

Let us start with the different approaches *of how we create graphics in R*:

- create a display only with one high-level function,
- use a high-level function first and add further elements with low-level functions,
- create a display from scratch with low-level functions only,
- put more than one high-level graphics output on top of each other,
- program graphics in a different graphical system,
- combine (with some effort) different graphical systems in one display,
- divide the graphical device to include several graphics in one display,
- use the generic function `plot()` with an R object, to visualise information pertaining to the object, for example, to use `plot()` with an analysis of variance object.

Table 5.1 Useful high-level graphics functions.

Function	Display
plot()	Scatter plot of one or two variables and generic plot function for diverse R objects (e.g., aov, lm)
hist()	Frequency distribution of categorised quantitative variables
boxplot()	Quartile information and outliers of empirical distributions

lattice::histogram()	Frequency distribution of categorised quantitative variables using grid graphics
plotCI()	Confidence intervals and error bars
gplots::plotCI()	Another confidence and error bars function
lattice::xyplot()	Scatter plot using grid graphics
coplot()	Scatter plot conditioned by factors
scatterplot3d::scatterplot3d()	Three-dimensional scatter plots
stripchart()	Unidimensional scatter plots or scatter plots by categories
pairs()	Matrix of scatter plots
psych::pairs.panels()	Matrix of scatter plots, histograms, and correlation coefficients in one display
spineplot()	Bar diagrams of conditional frequencies and marginal distributions
barplot()	Bar plots
plotrix::barp()	Bar plots
contour()	Display of two axes and one height dimension
persp()	Three-dimensional plotting of surfaces
vcd::binreg_plot()	Plots for logistic regression
vcd::assoc()	Visualisation of contingency tables
vcd::cd_plot()	Distribution of a categorical variable for different levels of a numerical variable
vcd::strucplot()	Visualisation of contingency tables
matplot()	Plot columns of one matrix against columns of another
qqnorm()	QQ-plot of data quantiles against theoretical quantiles
stem()	Console output of a stem-leaf diagram

One of the basics is a histogram of the univariate distribution of a dependent variable Y. *The histogram depicts the counts of measurements in a data set.* It shows bars at different sections of the Y scale, which correspond with how many measurements fall in the respective subdivision of the scale. This is usually informative for the unconditional distribution of Y and also for the conditional distribution given the different categories of an independent variable A. Let us inspect an example with random data.

Ch5_hist.R

```
## Using hist() to display the empirical distribution of a quantitative variable
set.seed(12345)
Y <- round(rnorm(100, 50, 5))
hist(Y)
table(Y)
range(Y)
hist(Y, col = "grey", ylim = c(0, 40), breaks = seq(from = 37.5,
                                                    to = 62.5,
                                                    by = 5))
```

Output

```
> table(Y)
Y
38 41 42 43 44 45 46 47 48 49 50 51 52 53 54 55 56 57 58 59 60 61 62
 2  2  3  3  1  5  5  7  7  5  3  4  4 15  9  5  2  3  3  5  2  3  2
> range(Y)
[1] 38 62
```

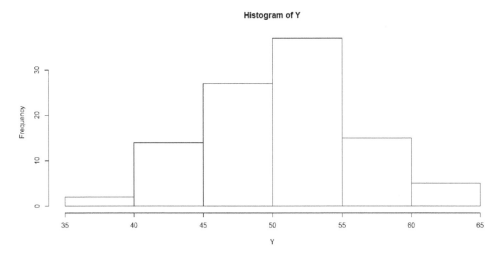

Figure 5.1 Basic histogram created with `hist()`.

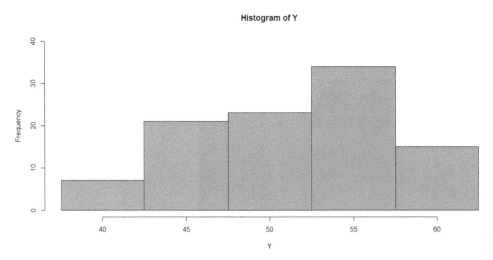

Figure 5.2 Histogram created with `hist()` with parameter modifications.

Analysis

The script generates random data first and stores it in the variable Y. The high-level function **hist()** displays the distribution of Y within its observed range (Figure 5.1). Note that **hist()** automatically aggregates Y into fewer categories than actually observed in Y. We then obtain a table of the distribution of Y with **table()**, which produces the counts for all data values. The next call of **hist()** then adds parameters to the call to manipulate the display. One of these parameters is breaks, which allows for an arbitrary categorisation along the range of Y (Figure 5.2). In this case, breaks obtains a vector created by **seq()**, which defines the category boundaries for the Y values. The information we obtained from **table()** and **range()** help us to decide on good category boundaries.

Other R packages offer even more histogram functions. A call of ??histogram gives an overview of functions. Near the top of the resulting list, the function graphics::hist is given, which we just used. Perhaps you like the function lattice::histogram(y) more, or sfsmisc::hist.bxp(y). The function **histogram()** of the package **lattice** (Sarkar, 2008) allows for histograms of a dependent variable y for all levels of a factor A with a function call lattice::histogram(~y|A).

5.1.2 plot()

R FUNCTION!

The main function to create a graphical display is **plot()**. Throwing in a vector displays the vector entries against their index number. Two vectors yield a scatter plot of the two-dimensional distribution, which often gives an idea about the correlation. Important to know is that **plot()** works with objects of many different types; so, in exploring unfamiliar objects it is always a good idea to use **plot()** with them and to observe the result. Let us try it.

Ch5_plot.R

```
## Using the generic plot() function for graphical display
# Data creation
set.seed(12345)
d <- data.frame(A = factor(rep(c("a1", "a2"), 50)),
                X = rep(1:5, each = 20))
d$Y <- rep(c(5, 7), 50) + .2 * d$X + rnorm(100)

# Data display
plot(d$X)
plot(d$A)
plot(d$A, d$Y)
plot(d$X, d$Y)
plot(aov(Y ~ A, data = d))
```

Output

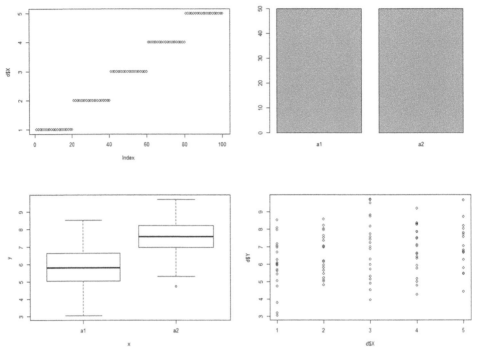

Figure 5.3 Four uses of the generic `plot()` function.

Analysis

In the script we create a small data frame consisting of an experimental factor A, a covariate X, and a dependent variable Y. Then, five simple calls of `plot()` with one or two variables as arguments show how `plot()` adapts to the type of variable it obtains. The covariate X appears as a scatter plot using the index numbers on the X-axis. In contrast, A appears as a bar chart showing frequencies. The next two calls show bivariate distributions. Y displayed against the factor levels of A yields box plots, whereas Y displayed against the covariate X yields a scatter plot. Please explore further what happens if the result of the **aov()** function – the analysis of variance – is directly handed over to `plot()`. The examples show that `plot()` is very flexible. If you encounter a new and unfamiliar object, use `plot()` with it. It will often reveal an informative display.

R LANGUAGE! ## 5.1.3 Function parameters, arguments, and defaults in R

Confusion sometimes arises about the order of the parameters in function calls. So, let us take **seq()** as an example and practise how parameters work. What is the result of seq(1, 20, .01) or of seq(1, -20, -.01)? The help page for **seq()**, which ?seq opens, says that the function creates numbers according to a rule. The section usage on the help page shows how it works: **seq()** generates a sequence of number from a starting point to an end point by a certain increment. For practise the explicit use of parameters in front of the arguments is always a good idea. So, seq(from = 1, to = 20, by = .01) is much more readable. If a function call omits the parameter names, then the order of arguments is fixed and must comply with the order of parameters on the help page. However, the order of arguments can differ from the function definition if the parameter name is used together with the argument. Furthermore, sometimes you want to skip a parameter in the function call, but want to pass an argument to a parameter further down in the parameter list. In this case you have to use the parameter name. In **seq()**, for example, you may skip the to parameter, but then have to use the parameter names of by and length.out.

Moreover, many functions define default arguments for most of their parameters. If the user does not explicitly hand an argument to a parameter, R uses the default argument. An available default follows the parameter name in the form "= argument" in the parameter list on the help page. Defaults allow for the use of the function with only few necessary specifications. If a parameter occasionally requires an argument other than the default argument, the function call must call the parameter explicitly and hand over the argument. Defaults make the work with R convenient, so use them whenever possible. For example, figure out which parameters and defaults come into play with a call of rnorm(10). Often you can find NULL as a default argument. The R object NULL says that a thing does not exist (Matloff, 2011), so the parameter can go unspecified. Furthermore, R often uses the arguments TRUE and FALSE for binary decisions, whereby one of the two usually acts as a default. The two arguments often reside in the parameter lists of functions. If you do not feel like typing too much, simply use T instead of TRUE and F instead of FALSE.

5.1.4 High-level graphics with add-on packages   R LANGUAGE!

Some R packages contain several graphics functions which add on to the base graphics. Some of them are compatible with base graphics, which means that they share many parameters with base graphics and that they can be enhanced with low-level graphics functions. However, other packages are incompatible. In this section, we describe the general features of useful graphics packages and some prominent functions.

The **plotrix** *package (Lemon, 2006) contains many high-level functions of several categories.* Functions like **barp()**, **gap.boxplot()**, **plotCI()**, or **histStack()** produce standard diagrams for data exploration. There are also many playful functions, for example, **violin plot()**, **pyramid.plot()**, **pie3D()**, **ladderplot()**, **fan.plot()**, **gantt.chart()**, or **battleship.plot()**. With **plotrix** installed, library(help = "plotrix") lists the available functions. Furthermore, **plotrix** adds helpful low-level functions like **staxlab()**, **textbox()**, or **draw.circle()**. Most interesting are two features of the package: (1) all functions come with examples that you can simply copy-paste to the R console to see how they work, and (2) unlike other packages, this package only depends on other packages of the core distribution of R. The first feature makes it very user-friendly; the second one suggests that the package will indeed work now and in the future because it does not rely on packages which are likely to be discontinued. The package is maintained by Jim Lemon and has several other contributors; it comes in a high version number, which indicates that it has undergone considerable development and several improvements in the past. In sum, it is pure fun to explore all the possibilities with this package.

The `gplots` package, maintained by Gregory R. Warnes et al. (2020), also supplies high-level and low-level functions. It contains **`barplot2()`**, **`boxplot2()`**, or **`plotCI()`** for graphical data exploration. More special functions are **`venn()`** or **`balloonplot()`**. It also contains low-level functions like **`angleAxis()`** to display an axis with rotated labels or **`sinkplot()`** to display text output of R in the graphics window. The package also comes in a high version number, does not depend on other libraries, and has several contributors. In sum, the package contains helpful enhancements of standard diagrams and further display options. It can safely be used.

The `lattice` package by Deepayan Sarkar (2008) consists of several high-level graphics functions as alternatives to the standard high-level graphics. It has functions like **`barchart()`**, **`bwplot()`** to display box plots, **`histogram()`**, and **`xyplot()`** to display scatter plots. Several functions allow for a conditional display of the diagrams split by the levels of an experimental factor, using the formula notation in R. For example, **`cloud()`** accepts formulas like cloud(Z ~ X * Y | A) to display a three-dimensional scatter plot of z against X and Y, split by the levels of factor A. Lattice is a highly established package, which has been available for more than ten years. It not only implements graphical functions, but also an aesthetics of graphics described by Cleveland (1994). However, its functions are implemented in the grid graphics system (Murrell, 2002, 2016) and it is incompatible with base graphics.

The **vcd** package by Meyer, Zeileis, and Hornik (2020) specialises in the display of categorical data. Like lattice it is implemented in the grid graphics system. It offers the **`strucplot()`** or **`assoc()`** function for the visualisation of contingency tables, **`binreg plot()`** for the display of conditional frequency distributions of binary responses in a logistic regression. It further offers several data sets for demonstration of the functions and for teaching purposes. All functions come with example code on the help pages which can simply be copied to the console. The package offers several playful options for all the functions included.

There are other packages which create graphics relating to different topics. For example, the **igraph** package creates networks of nodes and edges for network analyses (Csardi & Nepusz, 2006), the **mice** package uses diagnostic diagrams for the analysis of missing data patterns and multiple imputation performance (Van Buuren & Groothuis-Oudshoorn, 2011), the **Hmisc** package contains several alternatives to standard diagrams (Harrell Jr, 2021a), and the **sjPlot** package contains diagram functions for linear models (Lüdecke, 2020). Finally, **ggplot2** consists of the powerful functions **`qplot()`** and **`ggplot()`** to create complex graphics with little code as we will practise in Chapter 9 (Wickham, 2016).

EXAMPLE! ## 5.1.5 WBD – Frequency distributions and means of outcome variables

The clinical trial about a web-based intervention for cognitive bias modification requires the inspection of frequency distributions and group means. The following script provides an impression of the data. Histograms for the five BDI variables show the univariate distributions across conditions.

WBD_FIG_BDI_Histogram_by_ConditionMeasurement.R

```
## WBD - BDI Histograms by condition and measurement
source("./Scripts/WBD_PRE_DataImport.R")
source("./Scripts/WBD_PRE_ReshapeToLongFormat.R")
```

```
library(lattice)

with(WBD$CPS_LongFormat,
    {
            histogram(~ BDI | Condition + Measurement,
                    main = "WBD - BDI distribution by condition")
    })
```

Output

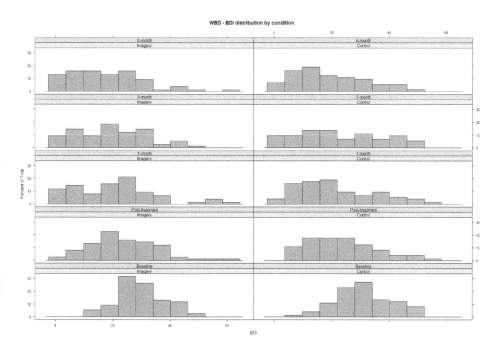

Figure 5.4 Frequency distribution of BDI variable split by experimental condition.

Analysis

The function **lattice::histogram()** is an alternative to **hist()** and creates histograms for outcome variables split by the conditions in the design (Figure 5.4). In this case, it creates the histogram of the depression distribution at the five time-points between the two experimental conditions. For the baseline measurement the distribution peaks between scores of 25 to 30 at the border between moderate and severe depression. So, most participants fall in the middle of the possible scores. *With further measurements the distribution moves more and more to the left, indicating that depression severity declines on average as time passes.* This development may be due to a regression effect towards the (yet unknown) mean (Kirk, 2012).

The next diagram shows group means for the five measurements. The function **interaction. plot()** produces the diagram using the data in long format. It puts the time points on the x-axis and draws a line of means for each of the two groups.

WBD_FIG_BDI_Mean_by_ConditionMeasurement.R

```
## WBD – Mean BDI values by condition and measurement
source("./Scripts/WBD_PRE_DataImport.R")
source("./Scripts/WBD_PRE_ReshapeToLongFormat.R")

with(WBD$CPS_LongFormat,
    {
        interaction.plot(x.factor = Measurement,
                        trace.factor = Condition,
                        response = BDI,
                        fun = function(X) mean(X, na.rm = TRUE),
                        ylab = "BDI",
                        las = 1)
    })
```

Output

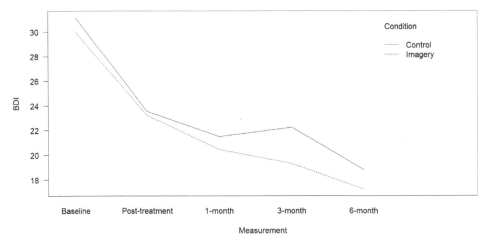

Figure 5.5 Means and error bars of depression severity development over time. The two lines allow for group comparisons.

Analysis

The function **interaction.plot()** creates a display of the group means of the two experimental conditions across the five measurements. The y-axis shows the BDI scores. The lower the means fall in the diagram, the lower the depression severity. We see an almost constant decline of the group means across measurements. Moreover, on average, depression severity remains in the region of mild to moderate depression and the lines of the two groups develop almost parallel. However, significant mean differences cannot be inferred from the diagram. Though the strongest decline in depression severity comes between baseline and post-treatment measurements, this effect may simply be due to a statistical regression effect (Kirk, 2012). This display gives an outlook for the test of the statistical hypothesis. We get an idea about the mean differences between conditions of the post-treatment and baseline differences.

5.1.6 SCB – Scatter plots of associations between quantitative variables

EXAMPLE!

The example of student reconciliators against school bullying consists of several quantitative variables. We have twelve measures of the dependent variable of school bullying and two quantitative covariates. The graphical analysis of their associations feeds into our upcoming decisions for inferential statistics. *The following script generates several scatter plots of variable associations.* Scatter plots are clouds of points, in which each point corresponds to the data values of a single case.

SCB_FIG_Scatterplots.R

```
## SCB - Graphical data exploration - Scatter plots
source("./Script/SCB_PRE_DataImport.R")
source("./Script/SCB_PRE_VariableLabels.R")
source("./Script/SCB_PRE_DataAggregation.R")

# Association between Pre- and Post-measurement for variable Y1
with(SCB, plot(PersonPerson_PRE, PersonPerson_POST))

# Association between baseline and post-measurement for all Y variables
with(SCB, matplot(jitter(cbind(PersonPerson_PRE,
                  PersonGroup_PRE,
                  GroupGroup_PRE)),
             jitter(cbind(PersonPerson_POST,
                  PersonGroup_POST,
                  GroupGroup_POST)),
             pch = 3:5,
             xlab = "Baseline measurement",
             ylab = "Post measurement"))

# Matrix of scatter plots
pairs(SCB[,grep("Y", names(SCB))])
psych::pairs.panels(SCB[,grep("Y", names(SCB))])

# Conditional distribution of the dependent variable
stripchart(SCB$PersonPerson_PREPOST ~ A, SCB, method = "stack")

# Associations between dependent variables and covariates
# conditional on program and type of school
lattice::xyplot(SCB$PersonPerson_PREPOST ~ X1 | A + B, SCB)
lattice::xyplot(SCB$PersonPerson_PREPOST ~ X2 | A + B, SCB)

# Conditional association between dependent variable and covariate
coplot(SCB$PersonPerson_PREPOST ~ X1 | A + B, SCB)

# Association between dependent variables in 3D
rgl::plot3d(SCB[,c("PersonPerson_PREPOST",
                  "PersonGroup_PREPOST",
                  "GroupGroup_PREPOST")])
```

Output

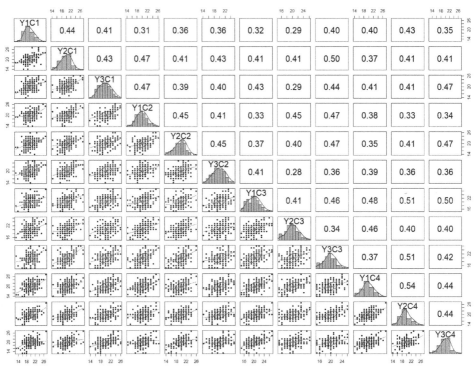

Figure 5.6 Matrix of scatterplots and correlations produced with **pairs.panels()** from the **psych** package.

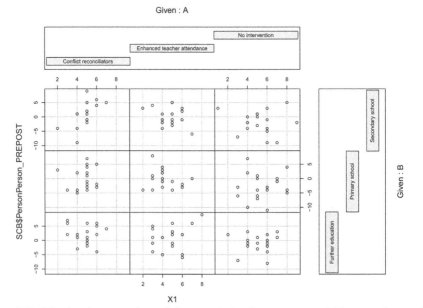

Figure 5.7 Matrix of scatter plots of the association between covariable X1 and one of the dependent variables conditional on the two factors A and B. Produced with **coplot()**.

Analysis

As usual, the script starts with data import using **source()**. The subsequent call to **plot()** displays the association between the post-measurements and pre-measurements of school bullying of person-against-person incidents. Without knowing the exact correlation coefficient, the display shows a cloud of points from the bottom left pointing to the top right of the display. However, the points all seem to fall on a grid, which is due to the limited levels of measurements of each variable. Therefore, due to superposing of points, we cannot see if some points in the display represent more than one case in the data. *The next function call uses* **matplot()** *to show the associations between baseline- and post-measurements for all three* Y *variables in one display. The function displays the data columns of one matrix against the columns of another.* In this case, the baseline measurement of person-against-person incidents is drawn against the post measurement of person-against-person incidents. The same applies for the two other types of bullying. Three additional functions help us with the use of **matplot()**: **with()**, **cbind()**, and **jitter()**. That is, **with()** allows for easy access to the Y variables, two calls of **cbind()** form the two matrices of baseline- and post-measurements out of raw data, and **jitter()** spreads out the displayed symbols in the diagram to help against superposing of data points. As can be seen, **matplot()** helps if there are several pairs of variables which are meaningfully related and which can be displayed in the same scale of measurement. The next display is a matrix of scatter plots for the associations between all Y variables in the data frame. Another implementation of such a matrix of scatter plots in the **psych** package is **psych::pairs.panels()**. The function additionally displays the univariate distributions in the main diagonal and the correlation coefficients between pairs of variables (Figure 5.6).

The **stripchart()** *function provides the univariate distribution of the post-pre difference of* Y1. It looks almost like a histogram, but all data points are still discriminable. The function **lattice::xyplot()** uses the formula notation of R to specify the variables for the scatter plot. The tilde operator ~ specifies the two variables for the diagram, the vertical bar | specifies the variables used for conditioning, in this case the two main independent variables intervention program and type of school. The diagrams of the associations between Y1 and each of the two covariates X1 and X2 show rather flat or round clouds of data point; in fact, this indicates rather low correlations. Since a high correlation between dependent variable and covariate is one prerequisite for the efficient use of analysis of covariance, maybe this method is counter-indicated. The literature suggests a correlation of at least r = .60 for analysis of covariance to become useful. However, the covariate distributions seem to be at least independent of the conditions and type of school, as indicated by the data points spreading nicely in the same span of the x-axis for all distributions. Independence between covariate and independent variable is another prerequisite for analysis of covariance. The function **coplot()** also implements conditional scatter plots (Figure 5.7), and you can decide whether **lattice::xyplot()** or **coplot()** pleases more.

Finally, three-dimensional scatter plots show the association between the three dependent variables. The function **rgl::plot3d()** is only the basic interface to the complex **rgl** library, which offers numerous functions for three-dimensional display. Both diagrams show that the data points rise in a nice cloud with almost the shape of a cigar from negative to positive values suggesting a positive association between the three variables.

In sum, several functions in R produce scatter plots, which can be one-, two-, and three-dimensional. Some functions use a formula interface, others use lists of arguments to pass variables to the functions. Again, play around with the functions to find your favourite.

5.2 Enhancing the graphical display

The high-level functions comprise many parameters to manipulate the appearance of a diagram. Some parameters are shared across functions and it is useful to know them. Furthermore, R offers low-level functions to add elements to an existing display, for example, points, lines, or text.

R LANGUAGE! ### 5.2.1 The graphics device in R

High-level graphics functions are very convenient. They fashion a view of the data with few specifications. This is useful for quick results. *But these graphics functions often can do more; they accept arguments to many parameters for fine control of the output.* A little play with the parameters can considerably enhance a quick and dirty initial graphic and you can sometimes finalise it for publication in a manuscript. Paul Murrell (2016) comprehensively describes the manipulation of base graphics parameters.

In order to learn about the parameters, we first have to acquire the structure of graphics devices in R. *All graphics in R are displayed in graphics devices. Think of such devices as blank sheets of paper on which figures are printed.* However, the devices in R can be the screen, the printer, and graphics files like pdf, png, or tiff. As a default, a high-level graphics function, like the ones we saw earlier, sends its output to a window on the screen.

Many textbooks about R introduce the structure of the graphics device for standard graphics (Figure 5.8) (Murrell, 2016). *The device consists of three rectangular regions: the device region, the figure region, and the plot region.* The regions are embedded within each other; that is, a device consists of one or more figures, and a figure consists of a plot region. The plot region contains the actual data display, for example, a scatter plot. The plot region is surrounded by the figure region consisting of annotation of the x- and y-axis or the main title of the diagram. The figure region, in turn, is surrounded by the device region, which itself can contain annotation. With device and figure region you can create layouts with several diagrams on one page.

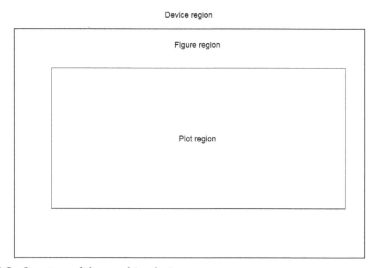

Figure 5.8 Structure of the graphics device.

5.2.2 par()

Device, figure, and plot region enable fine control of their size and appearance. Though high-level graphics functions take care of these settings themselves, the function **par()** is the key to them. The function consists of many parameters (Table 5.2); see also Murrell (2016). In the following script we manipulate several of the parameters to show how they work.

Table 5.2 Commonly used graphical parameters.

Parameter	Argument	Purpose
ann	TRUE/FALSE	Annotate figure or not
bg	Colour specification, e.g., "blue"	Background of the device region
cex	Numerical value	Magnifying or shrinking of graphics annotation relative to the default
col	Colour specification	Default colour for display of data
din	Numerical value of width and height, e.g. c(6, 8)	Size of the device region in inches
fin	Numerical value of width and height, e.g. c(6, 8)	Size of the figure region in inches
lty	String, e.g. "solid", "dashed"	Line type for the diagrams
lwd	Numerical value	Line width
mar	Numerical vector, e.g. c(2, 3, 0, 0)	Size of margins of the plot region measured in number of lines
oma	Numerical vector, e.g. c(2, 3, 0, 0)	Size of margins of the figure region measured in number of lines
pch	Numerical value	Symbol used for display
xaxt	Character, e.g. "n" for suppression of axis	Type of x-axis
yaxt	Character, e.g. "n" for suppression of axis	Type of y-axis

Ch5_par.R

```
## Manipulate graphical parameters with par()
set.seed(12345)

# General settings
layout(matrix(1:2, nrow = 1))
par(oma = c(2,2,2,2), bg = "grey 90", fg = "blue")

# Left diagram
hist(rnorm(100))
box(col = "orange")
box(which = "figure", col = "grey 50")

# Right diagram
par(mar = c(5, 3, 3, 3))
```

```
hist(rnorm(100),
     xlim = c(-10,10),
     xlab = "Random variable",
     xaxt = "n",
     yaxt = "n",
     main = "My histogram",
     col = "red")
box(col = "orange")
box(which = "figure")
par(oma = c(1, 2, 1, .5))
box(which = "figure")
```

Output

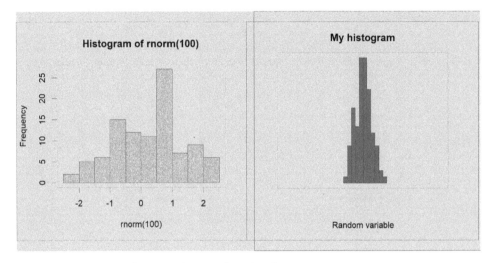

Figure 5.9   Figure with manipulation of graphical parameters.

Analysis

The script first divides the graphical device in two panels and then fills each panel with a diagram. The output looks ugly, but it shows parameter variations, which could make your figures beautiful. This is how it works:

- The function **par()** uses graphical parameters for global settings of the figure.
- The high-level function **hist()** and the low-level function **box()** use parameters to change their appearance.

The functions **layout()** and **matrix()**, together, divide the device region. The subsequent call of **par()** with parameters oma, bg, and fg sets some general properties of the device. The parameter oma is a short form for outer margin and it assigns space around all figure regions. The four integers are numbers of lines in the order bottom, left, top, and right of the figure regions. To use number of lines and not inch or cm as a unit is obviously good, because we usually write text in the margins. For example, we could write the label of a y-axis, which is common to several stacked diagrams into the left margin. The parameters bg and fg of **par()** relate to background and foreground colours: the whole device is given a grey background and the bars of the histogram are blue. This diagram also shows the margins with two calls of **box()**. In this diagram, **hist()** draws some random data first. Then, an orange box high-lights the border between figure and plot region, and a further grey box highlights the one between figure region and device region.

The diagram uses parameters in the histogram call. The parameters xaxt and yaxt set to "n" suppress display of the axes in high-level functions. This is useful in some applications when the default axis should be replaced with a specially designed one using **axis()**. Finally, **par()** with the parameter oma comes in a strange place after the creation of the diagram. Usually, oma is set before all graphics functions. But for demonstration purposes this shows that the borders between device, figure, and plot region are not fixed when the first line has been drawn. In fact, the borders can be changed at any time. Subsequent drawing of elements then respects the new settings. This complies with the so-called painters model of R graphics: what is drawn on the device stays there and cannot be changed any more, even if **par()** changes the settings.

5.2.3 UPS - Diagram of the interaction between sound and variation

EXAMPLE!

The unpleasant sounds analysis builds on a large experimental design with many within-subjects factors. In fact, the interaction between the four levels of sound and seven sound variations amount to 28 combinations of factor levels. To overlook the results of the pleasant-ness ratings, an interaction diagram of the mean ratings may help. This could use the levels of sound variation on the x-axis and use different lines for the levels of sound. We will create the diagram with one function call of **interaction.plot()**. However, **par()** will help us to set up adequate margins for the graphics device.

UPS_FIG_InteractionPlot.R

```
## UPS - Interaction plot
source("Scripts/UPS_PRE_DataImport.R")

with(UPS,
    {
      par(mar = c(5, 6, 1, 1))
      interaction.plot(x.factor = variation,
                       trace.factor = sound,
                       response = rating,
                       col = paste0("grey", c(10, 30, 50, 70)),
                       lwd = 2,
                       lty = 1:4,
                       ylim = c(1, 6),
                       las = 1,
                       xlab = "Variation",
                       ylab = "Mean rating\n(1: pleasant, 6: unpleasant)",
                       trace.label = "Sound")
    })
```

Output

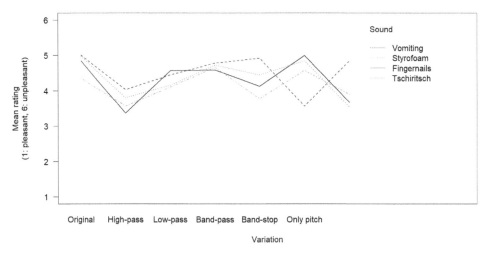

Figure 5.10 UPS interaction plot of average ratings of unpleasantness.

Analysis

All four of the resulting zigzag profiles lie in the upper half of the scale, indicating general feelings of unpleasantness when hearing the sounds. In some parts the profiles run in parallel, in other parts they differ.

R LANGUAGE!

5.2.4 Low-level graphics in R

Low-level graphics enhance diagrams with additional elements. Adding annotation or a legend to the display often makes it more comprehensible for readers. Several so-called low-level functions add to the display of a high-level graphic (Table 5.3). However, low-level functions also easily generate a graphical display from scratch. Let us practise with two displays.

Ch5_LowLevelGraphics.R

```
## Using low-level graphics functions
set.seed(12345)

# General settings
layout(matrix(1:2, nrow = 1))

# Left diagram
plot.new()
plot.window(xlim = c(1,10), ylim = c(0,100))
box()
axis(1)
axis(2)
points(1:10, rnorm(10, 50, 5), pch = 11:20)
```

```
# Right diagram
par(fg = "red", las = 1, mar = c(3, 3, 3, 3))
plot.new()
plot.window(xlim = c(1,10), ylim = c(0,100))
box()
axis(1)
axis(2, cex.axis = 2)
lines(1:10, rnorm(10, 50, 5), lty = "dashed", lwd = 3)
```

Output

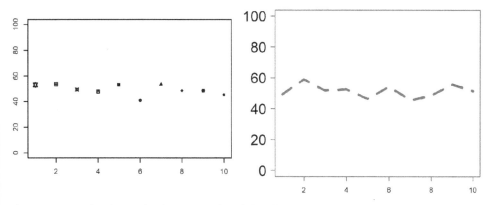

Figure 5.11   Graphical displays created with low-level functions.

Analysis

The left diagram is created from scratch using some low-level graphics functions. The function **plot.new()** starts the diagram. Then **plot.window()** initialises the coordinate system with the parameters xlim and ylim, each taking minimum and maximum of the x- and y-axis, respectively. Note that it is not before the two calls of **axis()** that the x- and y-axis are actually drawn. The axes show the correct limits, even if the specification of limits was done in **plot.window()**. Finally, random data points can be drawn with the low-level function **points()** using the parameter pch to specify the symbol used for each data point. In fact, this example shows a quick way to look at the available symbols in R to represent data points. Simply create a dummy diagram with several different symbols with plot(1:20, pch = 1:20) and then choose one symbol for the diagram you work on.

Table 5.3   Low-level graphics functions.

Function	Description
points()	Draw points at specified locations
lines()	Draw connected lines at specified locations
segments()	Draw non-connected using four coordinates for each line

(Continued)

Table 5.3 (Continued)

arrows()	Segments with arrow heads at end points
box()	Draw a border around the plot, figure, or device region
axis()	Add a coordinate axis to one of the four borders of the plot region
text()	Add text to a diagram (e.g., to annotate data points)
mtext()	Add text to the figure or device region
legend()	Add a legend to the diagram
polygon()	Draw filled contours
rect()	Draw rectangles
abline()	Draw straight lines through the whole plot region (e.g., regression lines)
symbols()	Draw symbols (e.g., circles, stars) at specified locations
curve()	Take a mathematical function and draws the corresponding graph
grid()	Add a grid to a diagram
title()	Add a title and axis labels to a diagram
rug()	Add small lines as a one-dimensional representation of a variable to one axis.

Table 5.4 Common parameters of low-level graphical functions.

Parameter	Description
x, y	x and y location
type	Type of symbol to be displayed
lty	Line type (e.g., dashed, dotted)
lwd	Line width
labels	Text to be displayed

The right diagram introduces more parameters. Instead of **points()** it uses **lines()** to draw the data. The **par()** function switches fg, the foreground colour, to red; it further uses the las parameter to change the orientation of the axis labels. The parameter knows four different settings, numbered from zero to three. Setting it to one yields horizontal annotation. Check with the help page of **par()** what the other three settings mean. The parameter mar sets the margins of the figure region; it also uses lines as a unit and the four numbers are for bottom, left, top, and right margin. The diagram also has two axes at positions one and two, that is, bottom and left. One axis uses the cex.axis parameter to set the relative size of axis annotation. It means character enlargement as a multiple of the default character size. In this case the axis annotation is to be drawn at double size. Furthermore, with the call of **lines()**, the two parameters lty for line type and lwd for line width show two more ways to refine the data display (Table 5.4). Again, the options for lty can be found on the **par()** help page.

In sum, graphical parameters allow for the fine control of the graphical appearance. They can be used with high-level graphics functions and with low-level functions. In Chapter 9, we will use more graphical parameters. Furthermore, the book R Graphics *by Paul Murrell (2016) is highly recommended for further practice.*

5.2.5 legend()

Legends are useful in complex graphical displays to show what certain symbols, lines, or other marks in the display mean. The base graphics functions in R do not automatically add a legend when it is necessary. It has to be added with the **legend()** function after the call of a high-level graphics function. Let us try it with a scatter plot with data points which belong to different groups.

Ch5_legend.R

```
## Using legend() to annotate graphics

# Data creation
set.seed(12345)
d <- data.frame(A = factor(rep(c("a1", "a2"), 50)),
                X = rep(1:5, each = 20))
d$Y <- rep(c(5, 7), 50) + .2 * d$X + rnorm(100)

# Data display
DisplaySymbols <- as.numeric(d$A) + c(14, 15)
plot(d$X, d$Y, pch = DisplaySymbols)
legend(x = 1.2, y = 10.5,
       title = "Factor A",
       legend = levels(d$A),
       pch = DisplaySymbols)
```

Output

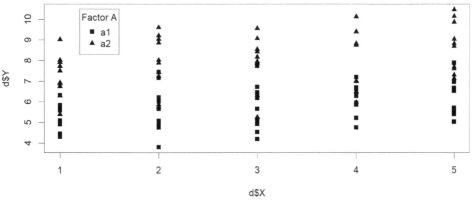

Figure 5.12 Using the **legend()** function to annotate graphics.

Analysis

Because the data points in the display show to which level of factor A each data point belongs, the pch parameter of **plot()** obtains the vector DisplaySymbols consisting of a numeric recoding of the levels of A. The same object is also passed to the pch parameter of **legend()**.

This function further obtains the information of what the symbols denote and the position of the legend in the figure region.

5.2.6 File formats to save graphics in R

Naturally, after programming of a figure you want to store it in a file. Different file formats are available and you should select according to where you want to use the figure – in a paper, in a presentation, or on the internet. Let us compare some of the prominent file formats for graphics: pdf, png, tiff, and jpeg. Each format has its own R function.

Ch5_SavingGraphics.R

```
## Saving of graphics in different file formats
plot(1:25, pch = 1:25)

png(filename = "pngplot.png")
plot(1:25, pch = 1:25)
dev.off()

jpeg(filename = "jpgplot.jpg")
plot(1:25, pch = 1:25)
dev.off()

tiff(filename = "tiffplot.tif")
plot(1:25, pch = 1:25)
dev.off()

pdf(file = "pdfplot.pdf")
plot(1:25, pch = 1:25)
dev.off()
```

Output

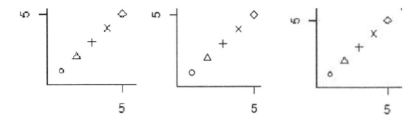

Figure 5.13 Output of png, pdf, and jpg graphics devices.

Analysis

We used all graphics only with the filename parameter and put the same figure in it. Notice that each graphics file closes with **dev.off()**. All functions, whether high-level or low-level, go between the function to start the graphic device in a certain format and **dev.off()**. The pdf format returns the best-looking result, as can be seen in a pdf viewer. Zoom in and out and see that the result is always sharp; print it on paper and it will look good. That's because pdf is a vector-based format, whereas png, tiff, and jpeg are in bitmap format. It is useful to learn more about these formats and their differences, if you like to produce graphics very often (https://en.wikipedia.org/wiki/Image_file_formats). The pdf format also produces a small file size, so issues in sending it by e-mail to your colleagues will not arise. In contrast, the tif format without further specification produces the biggest file size, which is more than 100 times bigger than png or pdf. The jpg format also produces comparably small file sizes; however, open the graphic we just created and zoom in on some displayed number. In graphics containing only a small number of colours and sharp contrasts between coloured areas, jpeg produces ugly graphical artefacts in the vicinity of colour boundaries. The jpeg format is good for storage of coloured photographs and allows for high compression; however, in statistical displays with only few colours, the other three file types are more useful.

5.2.7 ITC – Diagram of categorical variables   EXAMPLE!

The example ITC about the Wason card selection task contains many categorical variables; most of these variables are dichotomous. The variables code whether a subject selected the correct cards from the four cards in the three tasks, respectively. In the data frame stvu, all variables are coded as correct or incorrect. Let us proceed to inspect the relationship between the number of correct card selections and the quantitative independent variable of the SAT score of cognitive ability. We use the spineplot and save it in a png file.

ITC_FIG_SpineplotNumberCorrectBySAT.R

```
## ITC – Spineplot of categorical variables
source("./Scripts/ITC_PRE_LoadData.R")

png(filename = "./Figures/ITC_SpineplotNumberCorrectBySAT.png",
    width = 1800, height = 900, pointsize = 8,
    bg = "white", res = 300)
with(ITC$stvu,
    {
        NumberCorrect <- rowSums(cbind(p1_pnotq_correct == "correct",
                                       p2_pnotq_correct == "correct",
                                       p2_pnotq_correct == "correct"))
        spineplot(factor(NumberCorrect) ~ ITC$stvu$sat,
                  xlab = "SAT score",
                  ylab = "Number of correct card selections")
    })
dev.off()
```

Output

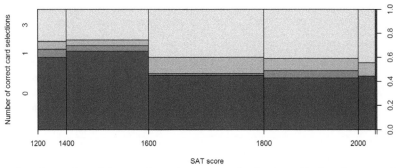

Figure 5.14 Spineplot of the number of correct card selections displayed against SAT score categories.

Analysis

The script first loads the data and then initialises a png graphics file to store the figure. With the `stvu` data frame we then create a new variable of the number of correct responses. The three variables coding the correct responses for the three problems are factors with two levels and the underlying numerical codes of 1 for incorrect and 2 for correct. Because of this coding we cannot simply use `rowSums()` to sum the values of the variables for each subject. Instead, for the three variables, we test each value whether it is correct and then use `rowSums()` to sum the truth values of these tests. This is a manoeuvre which is generally useful for binary responses not coded with 0 and 1.

Then we call `spineplot()` to relate a qualitative dependent variable to a quantitative independent variable. *The diagram shows the conditional distributions for the levels of a qualitative dependent variable within certain intervals of a quantitative independent variable.* It looks a little bit like a stacked bar chart. But it also shows the marginal distribution of the independent variable in different widths of the bars. This graphical exploration suggests that higher numbers of correct card selections are more frequent with higher SAT scores.

5.3 Checking model assumptions with graphics

Besides data exploration, graphics serve in the preparation of statistical modelling; for example, to inspect the deviation of empirical distributions from the normal distribution or to compare variances between experimental conditions.

STATISTICAL
ISSUE!

5.3.1 Graphical options to check model assumptions

Analysis of variance (ANOVA) or linear regression depend on whether some assumptions regarding the data generation apply. For the analysis of variance, Hays (1994, p. 390) lists three assumptions:

- the distributions of errors are assumed normal within each condition,
- the error variances are equal for all conditions, and
- the errors of any pair of observations are independent.

If the assumptions do not hold, bias enters the experimental design. For example, the actual type I error level could be different from the nominal level of $\alpha = 0.05$, leading to more than 5% false rejections of a true null-hypothesis. Moreover, non-independent predictors or experimental conditions can lead to high variability in the estimates of the design impairing valid conclusions. Some statistical tests can check the assumptions: the normality assumption may be checked with a Shapiro–Wilk test with **shapiro.test()**, and homogeneity of variances can be tested with the Levene test with **leveneTest()**. However, the independence of errors usually relies on rigorous experimental conduct, which ensures that the recording of one data case does not affect the recording of other cases.

Furthermore, there are some graphical options to help us to assess whether the assumptions hold. In the next sections, we will use the quantile-quantile diagram (Q-Q diagram) to get an idea about normal distribution. We will use error bars to compare variances and profile diagram within subjects to assess the independence of observations. Notice however, neither statistically nor graphically, is there proof of the validity of the assumptions. Data is never distributed normally, it can only more or less comply with the assumption that it has been sampled from the normal distribution. Some even argue for the robustness of the analysis of variance or the *t*-test against violations of its assumptions (Wilcox, 2012).

5.3.2 WBD – Checking the normality assumption EXAMPLE!

In the example of web-based cognitive bias modification for depression, the five depression measurements figure as the core outcomes of the study. We will now check these measures for normality across the two experimental conditions using the Q-Q diagram.

WBD_FIG_QQplots.R

```
## WBD - Q-Q plots for testing normality
source("./Scripts/WBD_PRE_DataImport.R")
source("./Scripts/WBD_PRE_ReshapeToLongFormat.R")

with(WBD$CPS_LongFormat,
    {
        layout(matrix(1:10, nrow = 5, byrow = TRUE))
        par(mar = c(2,3,2,0), oma = c(3,3,3,0))
         InteractionConditionMeasurement <- interaction(Condition, Measurement,
sep = " ")
        lapply(levels(InteractionConditionMeasurement), function(X)
        {
            qqnorm(BDI[InteractionConditionMeasurement == X],
                main = X,
                ylim = c(0, 60))

        })
```

```
mtext("Theoretical quantiles", 1, line = 1, outer = TRUE)
mtext("BDI Data", 2, line = 1, outer = TRUE)
mtext("WBD - Q-Q plots to check normality assumption", 3, line = 1, outer
= TRUE)
    })
```

Output

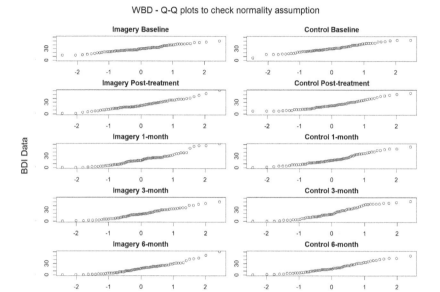

Figure 5.15 Q-Q norm diagram for BDI variables across conditions and measurements.

Analysis

The script produces a layout with several diagrams, one for each combination of condition and measurement of depression. It uses the data in long format and hides all its calculations in an environment created with **with()**. To generate the layout of multiple diagrams we use the **layout()** function, which will be explained in more detail below. In multiple diagram layouts the margins often need some adjustment; so we use **par()** with the mar and oma parameters for this. Then, we reduce the two variables Condition and Measurement to one interaction variable and apply **qqnorm()** for the depression scores in all conditions of this interaction variable. The functions **lapply()** and **function()** help us here. That is, for all levels of InteractionConditionMeasurement we apply the code in the environment created with **function()**. Each level of the interaction variable becomes an argument for the parameter x and is subsequently used to specify which scores of BDI to select, and which main title each diagram obtains. It could have been easier to use tapply(BDI, InteractionConditionMeasurement, qqnorm), but we would then have no opportunity to write the correct main title in each diagram. Finally, **mtext()** allows us to write the common annotation of all diagrams in the margins of the whole display.

The Q-Q diagrams relate the quantiles of the empirical data to the theoretical quantiles of the normal distribution. They are basically scatter plots. However, all points fall on a sort of straight line from the bottom left to the top right. Given the sample size, the theoretical quantiles show where to expect the data points, if the data were truly sampled from the normal distribution. It uses the standard normal distribution as reference, centred around the mean of zero and spread out up to little more than two standard deviations to the left and right. Looking at the data only from the viewpoint of the x-axis, the data points become more dense towards the middle of the diagrams and spread out evenly on both sides, just as what is true about the normal distribution, which is more dense towards the theoretical mean and less dense as one moves farther away from the centre. The points of the empirical distribution are now rank ordered and matched with the points of the equal rank of the theoretical distribution. If the empirical data points also show the same kind of density distribution across the scale range, all points should fall on a straight line from the bottom left to the top right. There are character-istic deviations from this ideal pattern indicating some sort of skewed empirical distribution. For example, if many data points fall towards the lower end of the distribution, the Q-Q diagram bends down; if many data points fall towards the upper end, the diagram bends up.

Regarding the depression data, we can observe that the shape of the diagrams closely resem-ble straight lines, but some diagrams bend down a little bit and some single points break out of the lines towards the upper ends. However, according to this display, we would not hesitate to calculate parametric statistics in the analysis. *Remember that the data will never be normally distributed; they can only evoke the impression that they have been sampled from the normal dis-tribution.* It is not easy to decide whether a parametric statistical test assuming a normal distribution is robust against distributional violations or if a non-parametric test needs to be found to give you valid statistical results (Ceyhan & Goad, 2009; Serlin & Harwell, 2004; Wilcox, 2012). In fact, given a reasonable sample size, say N > 30, the distribution of the means can be expected to approach the normal distribution very closely.

5.3.3 SCB - Graphically inspecting variation and confidence intervals

EXAMPLE!

The student reconciliator study uses three experimental groups. Graphical inspection of vari-ances with error bars is a good option to check whether they are different between groups. We use **plotCI()** and **plotmeans()** from the **gplots** package for this.

SCB_FIG_ErrorBars.R

```
## SCB - Error bars
source("./Script/SCB_PRE_DataImport.R")
source("./Script/SCB_PRE_VariableLabels.R")
source("./Script/SCB_PRE_DataAggregation.R")
library(gplots)

# Prepare statistics before plotting
descr <- with(SCB,
          {
              list(means = tapply(PersonPerson_PRE,
                                A, mean),
```

```
                       sd = tapply(PersonPerson_PRE,
                                    A, sd))
              })
pdf(file = "./Figures/SCB_ErrorBars.pdf", pointsize = 18)
# Plot standard deviations as error bars
plotCI(x = 1:3,
       y = descr$means,
       uiw = descr$sd,
       xlim = c(.5, 3.5),
       ylim = c(0, 50),
       pch = 16,
       gap = 0,
       xlab = "Intervention program",
       ylab = "Person against Person conflict",
       xaxt = "n",
       las = 1)
axis(side = 1,
     at = 1:3,
     labels = names(descr$means))

# Plot 95% confidence intervals for the group means
plotmeans(PersonPerson_PRE ~ A,
          data = SCB,
          p = .95,
          xlab = "Intervention program",
          ylab = "Person against Person conflict",
          las = 1)
dev.off()
```

Output

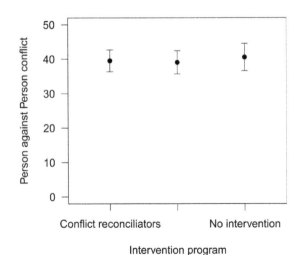

Figure 5.16 Error bars using plotCI() from package gplots.

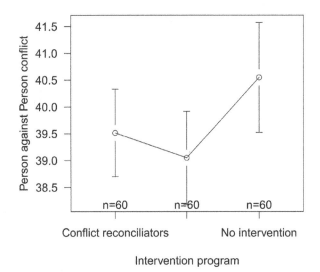

Figure 5.17 Confidence limits using `plotmeans()` from package `gplots`.

Analysis

The package `gplots` consists of two graphics functions for the display of variation in the data. The first one is `plotCI()`. It is not so easy to use and consists of many parameters. In fact, it requires you to calculate the descriptive statistics with another function and then supply it with the statistics. Thus, the script uses two `tapply()` calls to calculate group means and standard deviations. The function also has no good control of the x-axis, so we need to call `axis()` to fix it. The second function is `plotmeans()`. It displays confidence intervals for group means with a specified level of confidence. The function accepts a formula as a first argument to specify the variable association to be displayed, in this case the intervention program as an independent variable and the number of person-against-person conflicts as independent variable. Comparing the two functions, `plotCI()` is more versatile and lets you display anything as error bars, whereas `plotmeans()` is easier to use and calculates the necessary statistics itself. However, the meaning of its error bars is restricted to confidence intervals. The `pdf()` and `dev.off()` functions enclose the calls of `plotCI()` and `plotmeans()`, so both graphics go in one pdf file.

Comparing standard deviations in Figure 5.16, they look similar between conditions. Furthermore, inspecting the confidence limits in Figure 5.17 one would not expect to find significant differences between group means of the pre-measurements before the experimental intervention.

5.4 Arranging multiple graphics in one display

Graphical data analysis often produces many figures. It is often helpful to arrange similar or conceptually linked figures on one page and submit this page for team discussion. We will discuss some easy ways to create multi-figure layouts.

5.4.1 layout()

The function team which produces graphical layouts is **layout()** *and* **matrix()**. *Together, these two functions divide the page in squares and rectangles of which each one can hold a diagram.* Moreover, with **par()** and the mar and oma parameters, very beautiful and flexible layouts are possible. There are other options for the page layout, which use the mfrow parameter of **par()** or the **split.screen()** function; however, I find **layout()** most flexible. It treats the graphical device as a matrix of figure regions numbered according to the entries of the matrix. Try the following code at the console to divide the device region into six figure regions arranged in three rows and two columns. Inspect the display after each line of code.

Ch5_layout.R

```
## Using layout() and matrix()
layout(matrix(1:6, nrow = 3))
par(mar = c(4,3,2,1), oma = c(1,2,3,4))
layout.show(6)

plot.new()
box()
box(which = "figure")
text(.5, .5, "Plot region")
mtext("Figure region", 1:4)
mtext("Device region", 1:4, outer = TRUE)
plot.new()
box()
box(which = "figure")
plot.new()
box()
box(which = "figure")
plot.new()
box()
box(which = "figure")
plot.new()
box()
box(which = "figure")
plot.new()
box()
box(which = "figure")
box(which = "outer")
```

Output

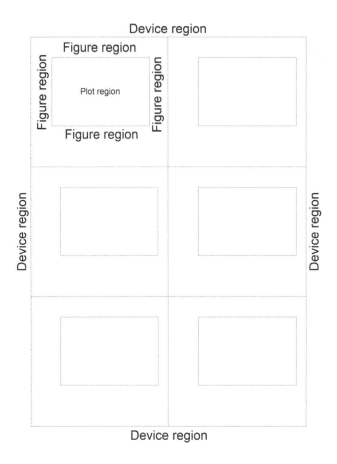

Figure 5.18 Using the functions **layout()** and **matrix()** to create multiple figures on one graphics device.

Analysis

The output of these lines of code is given in Figure 5.18. It shows the division of the graphics device into six figure regions according to the following matrix: $\begin{pmatrix} 1 & 4 \\ 2 & 5 \\ 3 & 6 \end{pmatrix}$.

Boxes, created with **box()**, and text fields, created with **text()** and **mtext()**, indicate the plot, figure, and device regions. Note how function arguments were used to specify which box or where text is to be drawn. The figure regions can be filled in with high-level graphics like, for example, histograms or scatter plots. In this case, we use **plot.new()** as a dummy graphics function with no graphical output; it simply allows us to proceed from one figure region to the next. The figure regions are filled according to their index numbers as given in the matrix. So, in our case, we start at the top left region and go through all regions column by column until we reach the bottom right field. By default, if we now call a seventh high-level graphic, then R would erase the previous diagrams and would start drawing again in the top left corner.

EXAMPLE! ## 5.4.2 ITC – Histograms of SAT distributions by gender

To establish the association between SAT scores and performance in the card selection task, the former should be independent of as many other variables in the data set as possible. One such variable is gender. Let us graphically check whether SAT score and gender can be treated as independent in this data set by using histograms of the SAT scores split by gender.

ITC_FIG_HistogramSATByGender.R

```
## ITC – Histograms of SAT score distributions by gender
source("./Scripts/ITC_PRE_LoadData.R")

with(ITC$stvu,
    {
        SATRange <- range(sat, na.rm = TRUE)
        layout(1:2)
        par(mar = c(4.5, 4, 1, .5))
        lapply(levels(gender),
            function(X)
            {
              hist(sat[gender == X],
                  breaks = seq(SATRange[1]-50,
                               SATRange[2]+50,
                               by = 100),
                  ylim = c(0, 30),
                  col = "grey",
                  main = paste("SAT distribution", X),
                  xlab = "SAT score",
                  xaxt = "n",
                  las = 1)
              axis(1, at = seq(SATRange[1], SATRange[2], 100))
            })
    })
```

Output

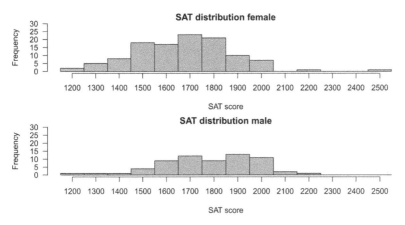

Figure 5.19 Histograms of age distributions by gender.

Analysis

The script uses **layout()** to stack two histograms on top of each other, one histogram for the SAT distribution for female participants the other for male participants. As can be seen, there were not only more female participants, but also the average SAT scores for females and males seem to differ.

To use **hist()** inside the call of **lapply()** allows us to easily create common x- and y-axes between histograms. The second argument to **lapply()** is **function()**, which defines a so-called anonymous function. A function of this type is not stored in an object, but is only for one-time use. It allows us to collect more than one function call in **lapply()**, namely **hist()** and **axis()**. Furthermore, the anonymous function allows us to use the first argument to **lapply()**, the gender levels *female* and *male*, in more than one position: to subset the SAT scores with sat[gender == X] and to set the main label of each figure with paste("SAT distribution", X).

5.4.3 UPS – Response profiles by subject   EXAMPLE!

The unpleasant sounds project collects many measurements per subject. *There is the chance that effects of position in the experiment affect these measurements, for example, that subjects are more vigilant at the beginning of the session and become more and more tired towards the end.* This is a well known problem in perception research. If we display all ratings of each subject against its point of measurement in the experiment, tendencies in the responses of each subject should stand out. For example, tendencies towards extreme judgements or average judgements should become apparent. Furthermore, measurements with narrow or wider ranges in rating variability should become visible. The following script creates the response profile of pleasantness rating for each subject. Because it is a big display, the script stores its output in a pdf, which you can inspect with a pdf viewer.

UPS_FIG_ResponseProfilesBySubject.R

```
## UPS - Response profiles by subject
source("Scripts/UPS_PRE_DataImport.R")

# Preparation
levels(UPS$sound) <- list("Fi" = "Fingernails",
                          "Vo" = "Vomiting",
                          "Ts" = "Tschiritsch",
                          "St" = "Styrofoam")
levels(UPS$variation) <- list("OR" = "Original",
                              "HP" = "High-pass",
                              "LP" = "Low-pass",
                              "BP" = "Band-pass",
                              "BS" = "Band-stop",
                              "OP" = "Only pitch",
                              "WP" = "Without pitch")
UPS$soundvariation <- interaction(UPS$sound, UPS$variation)
wUPS <- reshape(UPS,
                v.names = "rating",
                timevar = "soundvariation",
                idvar = "subject",
                drop = c("group", "age", "gender",
                         "sound", "variation"),
                direction = "wide")

# Create figure
pdf(file = "Figures/UPS_ResponseProfilesBySubject.pdf",
    width = 12, height = 16)
layout(matrix(1:96, nrow = 48))
par(mar = c(0,0,0,0), oma = c(6, 3, 3, 1))
lapply(as.list(wUPS$subject), function(X)
{
  plot(matrix(wUPS[wUPS$subject == X, 2:29]),
       xlim = c(-1, 28),
       ylim = c(.5, 6.5),
       type = "b",
       pch = 20,
       xaxt = "n",
       yaxt = "n")
  if(X %in% c(78, 130)) axis(1, at = 1:28,
                             labels = names(wUPS)[2:29],
                             las = 2)
  axis(2, at = 1:6, labels = 1:6, line = -4,
       tick = FALSE, cex.axis = .4, las = 1)
  text(-1, 3.5, labels = X, font = 2)
})
```

```
mtext("UPS - Response profiles by subject",
      side = 3, outer = TRUE, line = 1)
mtext("Rating of sound pleasantness (1: pleasant to 6: unpleasant)",
      side = 2, outer = TRUE, line = 1)
dev.off()
```

Output

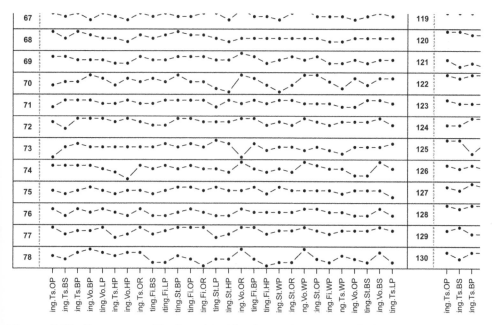

Figure 5.20 Response profiles by subject for the unpleasant sounds data.

Analysis

Figure 5.20 shows a small fraction of the figure. There are 96 profiles collected in this display – one for each subject. A single profile shows the subject ID, a rudimentary y-axis, and one point for each of the subject's ratings. A healthy profile should show variation between ratings, indicating that some stimuli are more unpleasant than others. Most subjects, even in this small section of the display, show such profiles. However, some are more restricted to smaller ranges of response categories than others, highlighting the importance in the instructions to use the complete range of options in the evaluation of the stimuli. Subject 68 is such an example: after half of the stimuli, his or her responses only slightly vary between categories 3 and 4. Considering the grand display in the pdf file, all data seems to be useful for further analysis. No subject produced a flat line or any other profile susceptible of ignorance regarding the experimental stimuli.

The R script consists of two sections to produce the figure: preparation of data and creation of the figure. Planning for a complex layout, we shorten the labels of the factors sound and variation using the **levels()** function. Because it is easier to work with only one factor variable than with two, **interaction()** then binds the two variables together to a new factor variable consisting of all the combinations of levels of sound and variation. Next, we reshape the data to wide format using **reshape()** to facilitate the subsequent call of **plot()**. Notice that the resulting data frame wUPS drops some variables because we do not need them for the display. After preparation we create the figure. For this, we initialise a pdf file and then divide the page into 96 figure regions using **layout()**. Afterwards, **lapply()** used with the subject numbers lets us define an anonymous function with **function()**. Inside the function definition demarcated with {}, the details of the single diagrams are defined. The combination of **lapply()** and **function()** allows us to let the coordinate systems and annotations for all diagrams look the same. Note, however, that we use **if()** to set an x-axis with **axis()** only for the two last diagrams in each column on the page, respectively. The two calls of **mtext()** lie outside the function definition because they add annotation to the whole page. Finally, **dev.off()** closes the pdf file, so that it can be opened in any pdf viewer.

R LANGUAGE! ## 5.4.4 Complex layouts in R

Complex layouts can even extend on multiple pages. For example, you may want to get a view of the empirical distributions of 30 items from a questionnaire. You could divide the graphical device into six plot regions and use five pages to present all distributions. Of course, you can store all resulting five pages in one pdf file. The code demonstration shows **layout()** in action together with some of its friends, namely, **pdf()**, **for()**, and **matrix()**.

Ch5_ComplexHistogramLayout.R

```
## A graphical layout of 30 histograms over 5 pages
# Create random data
set.seed(12345)
Data <- matrix(rnorm(3000, seq(10, 12, length.out = 30), sd = 2),
               nrow = 100,
               byrow = TRUE)

# Create graphical output
pdf(file = "Ch5_ComplexHistogramLayout.pdf", onefile = TRUE)
layout(matrix(1:6, nrow = 3, byrow = TRUE))
for(i in 1:30)
{
  hist(Data[, i],
       ylim = c(0, 40),
       breaks = -.5:20.5,
       xlab = paste("Var", i),
       main = NULL)
}
dev.off()
```

Output

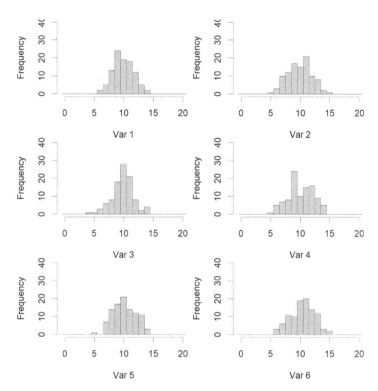

Figure 5.21 First page of a complex histogram layout comprising five pages.

Analysis

This demonstration uses random data from the normal distribution stored in the matrix Data. A pdf file receives all graphical output until **dev.off()** closes the file. The **layout()** function divides the pdf device region into six figure regions. Note that **layout()** goes after **pdf()**; so, we open the device first and divide it afterwards. Then, a **for()** loop generates histograms for all data columns.

The former demonstrations show **layout()** in action for routine data display. However, the function can do more and structure the graphical device with diagrams in different sizes and arbitrary arrangements. Suppose that a main diagram is to be created with some auxiliary diagrams that go to the bottom and to the right of the main diagram. For example, the scatter plot of the bivariate distributions of some scales X and Y should be displayed in the main plot region, and the frequency distributions of two sets of underlying item distributions of X1-X4 and Y1-Y4 should be the supplements. The following matrix can be used to divide the graphics device accordingly, and the following code shows how to create the display.

$$\begin{pmatrix} 1 & 1 & 1 & 1 & 2 \\ 1 & 1 & 1 & 1 & 3 \\ 1 & 1 & 1 & 1 & 4 \\ 1 & 1 & 1 & 1 & 5 \\ 6 & 7 & 8 & 9 & 10 \end{pmatrix}$$

Ch5_LayoutMainSupplement.R

```
## A layout of one main and several auxiliary diagrams
# Create correlated random data with two factors
set.seed(12345)
Lambda <- matrix(c(.3, .3, .3, .3, .2, .2, .2, .2,
                   .2, .2, .2, .2, .3, .3, .3, .3), nrow = 8)
Sigma <- Lambda %*% t(Lambda) + diag(rep(.3, nrow(Lambda)))
Data <- mvtnorm::rmvnorm(n = 100, sigma = Sigma)

# Set up layout
m <- matrix(rep(1,16), nrow = 4)
m <- cbind(m, 2:5)
m <- rbind(m, 6:10)
par(oma = c(1, 1, 1, 1))
layout(m)
layout.show(10)

# Create graphical output
plot(rowMeans(Data[,1:4]),
     rowMeans(Data[,5:8]),
     xlab = "Scale X",
     ylab = "Scale Y")
par(mar = c(3, 2, .5, .5))
apply(Data, MARGIN = 2, hist,
      main = NULL,
      ylab = NULL,
      ylim = c(0,40))
```

Output

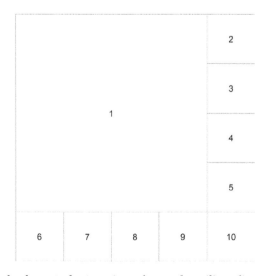

Figure 5.22 Complex layout of one main and several auxiliary diagrams.

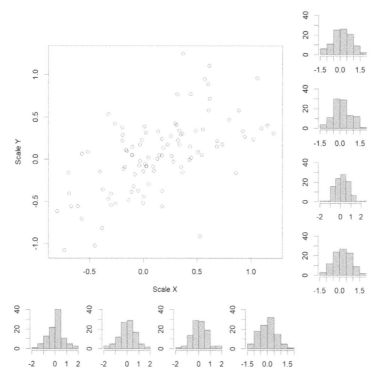

Figure 5.23 Use of **layout()** for complex graphical arrangements.

Analysis

As usual, in the first part of the script, random data is created from the multivariate normal distribution with **mvtnorm::rmvnorm()** for eight variables with expected means of 0 and a covariance matrix Sigma. Our approach to create the covariance matrix by matrix multiplication of factor loadings plus uniqueness might look a bit complicated; however, it saves us from typing in many numbers in a square matrix and it still captures the essence that the eight variables are based on two underlying factors. We simply reverse the operation of factor analysis to decompose the empirical covariance matrix.

After data creation we set up the graphical display with **layout()**. The matrix m is created in a few steps and contains the information where all diagrams should reside, respectively. We check the structure of the display with **layout.show()** (Figure 5.22). Finally, we use two calls to graphical functions, **plot()** and **hist()**, to create a total of nine figures (Figure 5.23). Note that we combined two different types of diagrams and also two different functions to create the display. First, one function call of **plot()** creates one figure; next, the combination of **apply()** and **hist()** creates eight figures in one go. Field number 10 of the graphics device is simply left empty; however, if we do not set up the graphic device anew, the next call to a graphical function will fill in the remaining field.

In sum, **layout()** and **matrix()** together with **par()** can create beautiful arrangements of diagrams. It requires some experimenting to find ones own preferred arrangement, especially regarding the proportions of margins and outer margins. Our examples here use one graphical device with multiple figures. There are also options to work with multiple devices at the same time, to switch between them back and forth, and to selectively open and close devices. The functions **windows()**, **dev.cur()**, **dev.list()**, and **graphics.off()** help with this.

What you have learned so far

1. To produce a complete diagram with only one function call
2. To manipulate the display by using graphical parameters
3. To add annotation and other elements to diagrams
4. To store diagrams in files
5. To check graphically whether the data meet the assumptions for inferential statistics
6. To create complex layout of several diagrams on one page

Exercises

Explore functions

1. There are several histogram functions available in different packages. Obtain an overview with ??histogram. Try such functions like **sfsmisc::histBxp()**, **lattice::histogram()**, **psych::multi.hist ()**, **plotrix::histStack()**, or **Hmisc::histbackback()** with the **MusicData.csv** data set. Compare the output with the output of **hist()**.
2. Create a dummy figure with plot(1:20, pch = 1:20). With this diagram, practise manipulating several parameters of **plot()**. Find all the parameters with ?plot. For example, try the parameters main, xlab, or ylab. Find more parameters for **plot()** with ?par; for example, las, cex, cex.axis, pch, bty, or col.
3. Use **layout()** with the music evaluations of the **MusicData.csv** data set. Create a layout of four rows and five columns of histograms for the 20 variables. Try to create the same page layout with two alternative solutions: use the parameters mfcol and mfrow of **par()** and use the **split.screen()** function.
4. Explore different functions to create error bars. Use the factor variables and dependent variables of **MusicPreferencesSchool.csv** for this. Try **gplots::plotCI()**, **Hmisc::errbar()**, **plotrix::plotCI()** or **psych::error.bars.by()**.

Solve tasks

1. Take the example graphics from sections 5.3 and 5.4 and store the graphics in files. Use functions like **png()** or **pdf()** and add them to the code at the appropriate place. Do not forget **dev.off()** in each example.
2. Install the **plotrix** package. Load the package with **library()**. Then run library(help = "plotrix") at the console and study the available functions. Open the help pages of some of them, for example, ?gantt.chart or ?dendroPlot. Copy-paste the example code at the bottom of the help pages to the console, run the code, and study the graphics display. Do you have use for some of the functions in **plotrix** for your own data?

Read code

1 The following script uses high-level and low-level functions to create an enhanced diagram. Read the code and describe in your own words how it works. Which high-level and which low-level functions come to work?

```
## High-level and low-level functions

# Preparation
d <- read.csv("MusicPreferencesSchool.csv")
levels(d$Musician) <- list("Inst." = "Plays instrument",
                           "n. Inst." = "Plays no instrument")

# Generate diagram
with(d, plot(Creativity, PreferenceSong1))
with(d[d$Creativity >= mean(d$Creativity),],
     {
             points(Creativity,
                    PreferenceSong1,
                    pch = 16)
             text(Creativity,
                  PreferenceSong1,
                  Musician,
                  adj = c(-.2, 1))
     })
abline(v = mean(d$Creativity))
```

2 The following script generates group means and error bars. Describe in your own words how descriptive statistics figure in the script.

```
## Display means and standard errors

# Preparation
d <- read.csv("MusicPreferencesSchool.csv")
library(gplots)

d <- within(d, SumPreference <- PreferenceSong1 + PreferenceSong2)
levels(d$School) <- list("GS" = "Grammar school",
                         "JHS" = "Junior high school",
                         "SMS" = "Secondary modern school")

descr <- with(d,
              {
                  list(means = tapply(SumPreference,
                                      School,
                                      mean,
                                      na.rm = TRUE),
                       se = tapply(SumPreference,
                                   School,
                                   function(X) sqrt(var(X) / length(X))))
              })
```

```
# Generate diagram
par(mar = c(5, 5, 2, 1))
plotCI(x = descr$means,
         uiw = descr$se,
         xaxt = "n",
         xlim = c(.5, 3.5),
         xlab = "School",
         ylab = "Music preference\n(Mean +- SE)")
axis(side = 1,
       at = 1:3,
       labels = names(descr$means),
       cex = 0.7)
```

3   The following script creates a neat layout of diagrams. Describe in your own words how
    we initialise the layout, how we assign a common x-axis to both diagrams, and how we
    manipulate the figure margins as needed.

```
## Layout of diagrams

# Preparation
d <- read.csv("MusicPreferencesSchool.csv")
XLIM <- c(49.5, 65.5)

# Generate diagram
layout(c(1, 2, 2))
with(d,
     {
             par(mar = c(0, 5, .5, .5))
             hist(Creativity,
                  breaks = seq(XLIM[1], XLIM[2], 2),
                  main = NULL, xlim = XLIM,
                  xaxt = "n")
             par(mar = c(5, 5, .5, .5))
             plot(Creativity, PreferenceSong1, xlim = XLIM)
     })
```

Analyse data

1   Import **MusicData.csv** and create box plots for the twenty music evaluations Ex1 to
    Ex20. Use the function **boxplot()**.

Apply in the real world

1   Explore your own data with graphics. Create histograms to assess univariate distributions,
    scatter plots to assess covariance and correlation, or error bars to assess group differences
    on your dependent variables.

Become a statistics programmer

1   What do the function calls plot(1:20, pch = 1:8, col = 1:3) and paste(c("a",
    "b", "c", "d"), 1:3) have in common? They use recycling. Explain the programming
    concept with the two function calls.

WBD

1   In the example about web-based cognitive bias modification, we created diagrams about the relation between independent and dependent variables. However, we also need to enquire whether covariates and demographics distribute evenly between experimental conditions. Explore graphically the distributions of years of education, employment status, and the number of past depression episodes between the two experimental conditions. Use bar diagrams, histograms, or mean diagrams.

2   In the main text, we programmed grouped frequency distributions and a mean diagram for the depression scores with `lattice::histogram()` and `interaction.plot()`. Extend this analysis for the variables anhedonia and prospective imagery test.

ITC

1   Go to the ITC example in Chapter 5 showing the distribution of SAT scores divided by gender. In this display replace gender by age. Use `cut()` to categorise age into four groups.

2   Create a scatter plot showing the SAT score on the x-axis and the binary variable `p1_pnotq_correct` on the y-axis. Add connected points to the diagram which show the means of the binary variable for the categorised SAT score. This display is the basis for the logistic regression in Chapter 7.

3   Create the same scatter plot as in the foregoing task with the binary responses for the selection tasks two and three. Then, use `layout()` to stack the displays for all three selection tasks on top of each other.

UPS

1   In the main text we created the interaction plot for all combinations of the factors sound and variation. Refine this graphical analysis to respect the two different conditions of the factor group: origin known and origin unknown. Create one interaction plot for each of the two conditions. Use data subsetting with `[]` and `%in%` to split the data set. If possible, use `layout()` to stack the two displays on top of each other.

2   Store the graphic of the foregoing task as pdf. Use the `pdf()` function for this.

SCB

1   To begin with the analysis, check the quality of the data by the calculation of descriptive statistics and the display of histograms for all of the nine groups. Unify the x-axes and the y-axes of the different histograms. Print all histograms on one page with the use of `layout()`.

2   Create three grouped bar diagrams for the dependent variables. The x-axis should show the different intervention programs, the bars should be grouped by type of school. Use `barplot()`.

3   Draw Q-Q-diagrams to check the normality assumption for the pretest-posttest differences for the three types of conflict. Use `qqnorm()`. Do you see gross deviations of the data from the normal distribution?

6

INFERENTIAL STATISTICS I: A COMPLETELY RANDOMISED FACTORIAL DESIGN

Chapter overview

This chapter describes statistical analysis with one of the most common designs used in psychological research: the analysis of variance (ANOVA). It shows how to calculate the ANOVA model and how to obtain the results. The chapter also includes factorial designs using more than one independent variable and the analysis of repeated measures. Programming features: formulas and statistical models, collection of information from complex R objects.

6.1 Review of analysis of variance (ANOVA)

This section introduces the basic one-way ANOVA: one dependent variable is related to one qualitative independent variable. The analysis shows whether the different conditions of the independent variable yield significant mean differences in the dependent variable.

6.1.1 One-way analysis of variance

Analysis of variance (ANOVA) is a versatile method to test statistical hypotheses in the social sciences. In its basic form of a completely randomised design, the influence of three or more conditions of a qualitative independent variable on a quantitative dependent variable is studied. The statistical null-hypothesis states that there are no mean differences in the dependent variable between conditions. The alternative hypothesis states that there is at least one pair of group means which has a difference greater than zero. The aim of ANOVA is to reject the null-hypothesis in favour of the alternative hypothesis and thus conclude a significant influence of the independent variable on the dependent variable. If the study yields no significant result, no conclusion is possible. A significant difference between conditions is empirical evidence in favour of an empirical hypothesis derived from theory.

For many researchers and students in the social sciences, analysis of variance lies at the heart of statistical analyses. In many journal publications, the significance test of an ANOVA is the crucial information for the evaluation of the data. We will calculate a one-way ANOVA, which means that one independent variable is related to one dependent variable. It requires only one or two lines of R code. However, we have to create data first.

Ch6_RandomData.R

```
## Random data for the analysis of variance
if(!exists('d'))
{
  set.seed(12345)
  d <- data.frame(A = gl(3, 50, labels = c("a1", "a2", "a3")),
                  B = gl(2, 25, 150, labels = c("b1", "b2")),
                  Y = round(rnorm(150,
                             rep(c(5, 5.5, 5, 4.8, 5.3, 6),
                                 each = 25)), 2))
}
```

Analysis

We save this script with the file name Ch6_RandomData.R so that the data can be used in the subsequent analyses. In the data frame, A and B are factors and the ANOVA will reveal their association with the dependent variable Y. The function **gl ()** – generate factor levels – generated the design formed by these factors. The function **rnorm ()** generates the dependent variable Y with random draws from normal distributions with six different means, corresponding with the combinations of factor levels formed by A and B. The data frame d consists of 150 lines.

At the console, we can now acquire a descriptive overview of the distribution of Y in the conditions formed by the levels of A. We then compare the group means for the three levels of A with an analysis of variance using **aov()**.

Ch6_OnewayANOVA.R

```
## One-way analysis of variance - Descriptives and analysis
source("./Ch6_RandomData.R")

table(d$A)
mean(d$Y)
with(d, tapply(Y, A, mean))
OnewayANOVA <- aov(Y ~ A, d)
OnewayANOVA
summary(OnewayANOVA)
```

Output

```
> table(d$A)

a1 a2 a3
50 50 50
> mean(d$Y)
[1] 5.426067
> with(d, tapply(Y, A, mean))
    a1     a2     a3
5.4302 5.2110 5.6370
> OnewayANOVA <- aov(Y ~ A, d)
> OnewayANOVA
Call:
   aov(formula = Y ~ A, data = d)

Terms:
                       A Residuals
Sum of Squares   4.53818 189.92780
Deg. of Freedom        2       147

Residual standard error: 1.136673
Estimated effects may be unbalanced
> summary(OnewayANOVA)
             Df Sum Sq Mean Sq F value Pr(>F)
A             2   4.54   2.269   1.756  0.176
Residuals   147 189.93   1.292
```

Analysis

The table shows us that we have a balanced design with equal numbers of subjects in each condition: **mean()** produces the grand mean of the design and **tapply()** produces the group means of Y for each level of A. Then **aov()** creates the object OnewayANOVA, which is more complex than the vectors and data frames that we discussed so far. The object has a complex structure with headings, text, numbers, and tables in the plain R style. The line starting with Call shows the function call. The expression Y ~ A in R is a formula, which links a dependent variable to an independent variable. To the left of the tilde operator ~ is the dependent variable, to the right is the independent variable. We applied such a formula already with graphics. Formulas are often used in R and can become very complex. In fact, any experimental design can be specified in a formula in R. An overview of the formula notation is provided by Crawley (2005). In factorial designs, we will later use formulas like Y ~ A * B, which contain A and B as independent variables. Under the heading Terms are sum of squares and degrees of freedom between the factor levels of A and for the residuals. The sum of squares for the levels of A comprises the variance in Y due to the categorisation of experimental subjects into different conditions. The sum of squares for the residuals comprises the error variance in Y, which is not due to any factor of influence included in the analysis. Finally, summary(OnewayANOVA) comprises the F-test and p-value, which we can determine on the basis of the sum of squares and degrees of freedom, which we will study now.

6.1.2 Sum of squares and F-test

The sum of squares associated with the levels of the independent variable are often called between-group variation or $SS_{between}$. The sum of squares associated with the residuals are often called within-group variation or SS_{within}. Let us inspect the formulas for both measures of variation.

$$SS_{between} = \sum_j n_j \left(\bar{Y}_j - \bar{Y} \right)^2$$

Factor A comprises three groups $j = 1$-3. The n_j are the three sample sizes, which are multiplied with the squared differences between the group means Y_j and the grand mean Y. These products are then summed up over all three groups.

$$SS_{within} = \sum_j \sum_i \left(Y_{ij} - \bar{Y}_j \right)^2$$

For SS_{within}, the squared differences between all measurements y_{ij} and their respective group mean y_j are summarised for the whole sample. We can calculate the two measures of variation with the information we already obtained with **table()**, **mean()**, and **tapply()**. That is, $y = 5.43$, $y_1 = 5.43$, $y_2 = 5.21$, and $y_3 = 5.64$. Sample sizes for all groups are $n_{1-3} = 50$. Using this information, we type in the formulas for $SS_{between}$ and SS_{within}: and further calculate all information of summary(OnewayANOVA).

Ch6_SumOfSquaresFtest.R

```
## Sum of squares and F-test
source("./Ch6_RandomData.R")

SSb <- sum(50 * (5.43 - 5.43)^2,
           50 * (5.21 - 5.43)^2,
           50 * (5.64 - 5.43)^2)
SSw <- sum(sum((d$Y[1:50] - 5.43)^2),
           sum((d$Y[51:100] - 5.21)^2),
           sum((d$Y[101:150] - 5.64)^2))

dfA <- 2
dfRes <- 147

MSb <- SSb / dfA
MSw <- SSw / dfRes

F <- MSb / MSw

Pr <- pf(F, dfA, dfRes, lower.tail = FALSE)
```

Output

```
> SSb
[1] 4.625
> SSw
[1] 189.9283
> MSb
[1] 2.3125
> MSw
[1] 1.292029
> F
[1] 1.78982
> Pr
[1] 0.1706099
```

Analysis

The results correspond well with `summary(OnewayANOVA)`, which we calculated earlier. The object `OnewayANOVA` also contains the degrees of freedom of factor A and for the residuals. Dividing the sum of squares by the respective degrees of freedom yields the mean squares. Division of the mean squares produces the F value, for which we can obtain a probability value, or p-value, from the F-distribution to test for significance. A significant result is often assumed if the probability falls below 0.05. Information from tables like `summary(OnewayANOVA)` are often reported in scientific journals.

6.2 Model specification for the ANOVA in R

After revisiting the statistics of ANOVA we now focus on the formula interface of R. We specify the relationship between independent and dependent variables with a formula.

R FUNCTION! ### 6.2.1 t.test()

The most straightforward association between a qualitative independent variable and a quantitative dependent variable is the mean difference in the dependent variable between two groups of the independent variable. In this situation, the *t*-test allows for the conclusion, whether the population means of the dependent variable differ between the two levels of the independent variable. The following script demonstrates the **t.test()** function.

Ch6_tTest.R

```
## t-test
source("Ch6_RandomData.R")

tt1 <- t.test(Y ~ B, d)
tt2 <- with(d, t.test(Y[A == "a1"], Y[A == "a2"]))
```

Output

```
> tt1

        Welch Two Sample t-test

data:  Y by B
t = -2.0597, df = 146.99, p-value = 0.04119
alternative hypothesis: true difference in means is not equal to 0
95 percent confidence interval:
 -0.74486796 -0.01539871
sample estimates:
mean in group b1 mean in group b2
        5.236000         5.616133

> tt2

        Welch Two Sample t-test

data:  Y[A == "a1"] and Y[A == "a2"]
t = 0.95283, df = 97.927, p-value = 0.343
alternative hypothesis: true difference in means is not equal to 0
95 percent confidence interval:
 -0.2373317  0.6757317
sample estimates:
mean of x mean of y
   5.4302    5.2110
```

Analysis

Two *t*-tests, `tt1` and `tt2`, show different ways to specify the *t*-test. Calls of `tt1` and `tt2` show the standard results display with *t*-value, degrees of freedom, and the *p*-value. Since factor B consists of only two levels, it suffices to specify the factor only. Since factor A consists of three levels, the levels of A to be compared need to be selected explicitly, in this case `a1` and `a2`. The *t*-test might look too simple to be useful; however, it is the one inferential statistical method which yields the most unambiguous conclusions.

If you want to calculate a *t*-test given only two empirical means, standard deviations, and sample sizes, R offers the **pt()** function to test an empirical *t*-value for significance. This is relevant if you use summary statistics from journal publications, where you do not have the data set. Simply use the formula for the *t*-value from a statistics textbook (e.g., Hays, 1994), then use **pt()** with this value and the appropriate degrees of freedom. The function returns the *p*-value for the test of significance.

6.2.2 aov()   R FUNCTION!

We use **aov()** to create an analysis of variance. The function uses the formula notation with the tilde operator `~` to separate dependent from independent variables: the dependent variable stands on the left-hand side, the independent variables on the right-hand side. The resulting `aov` object consists of sum of squares and the *F*-test to evaluate the significance of the mean differences between experimental conditions. However, the `aov` object consists of even more information, and we use specialised functions to access this information. Let us use the `aov` object we created above and uncover more of the information it comprises.

Ch6_aov.R

```
## Using aov() to generate an ANOVA object
source("./Ch6_RandomData.R")


# Create model and summary
OnewayANOVA <- aov(Y ~ A, d)
summary(OnewayANOVA)


# Enquire further information
is(OnewayANOVA)
methods(class = "aov")
methods(class = "lm")


# aov methods
model.tables(OnewayANOVA)
TukeyHSD(OnewayANOVA)
coef(OnewayANOVA)


# lm methods
model.matrix(OnewayANOVA)
confint(OnewayANOVA)
layout(matrix(1:4, 2))
plot(OnewayANOVA)
```

Output

```
> summary(OnewayANOVA)
          Df Sum Sq Mean Sq F value Pr(>F)
A           2   4.54   2.269   1.756  0.176
Residuals 147 189.93   1.292
>
> # Enquire further information
> is(OnewayANOVA)
[1] "aov"      "lm"       "oldClass"
> methods(class = "aov")
 [1] coef         coerce       extractAIC   initialize   model.tables print
 [7] proj         se.contrast  show         slotsFromS3  summary      TukeyHSD
[13] vcov
see '?methods' for accessing help and source code
> methods(class = "lm")
 [1] add1         alias         anova         case.names    coerce
 [6] confint      cooks.distance deviance      dfbeta        dfbetas
[11] drop1        dummy.coef    effects       extractAIC    family
[16] formula      hatvalues     influence     initialize    kappa
[21] labels       logLik        model.frame   model.matrix  nobs
[26] plot         predict       print         proj          qr
[31] residuals    rstandard     rstudent      show          simulate
[36] slotsFromS3  summary       variable.names vcov
see '?methods' for accessing help and source code
>
> # aov methods
> model.tables(OnewayANOVA)
Tables of effects

 A
A
      a1       a2       a3
 0.00413 -0.21507  0.21093
> TukeyHSD(OnewayANOVA)
  Tukey multiple comparisons of means
    95% family-wise confidence level

Fit: aov(formula = Y ~ A, data = d)

$A
        diff       lwr       upr     p adj
a2-a1 -0.2192 -0.7574587 0.3190587 0.6006127
a3-a1  0.2068 -0.3314587 0.7450587 0.6350661
a3-a2  0.4260 -0.1122587 0.9642587 0.1499128
```

```
> coef(OnewayANOVA)
(Intercept)            Aa2            Aa3
     5.4302        -0.2192        0.2068
>
> # lm methods
> model.matrix(OnewayANOVA)
     (Intercept) Aa2 Aa3
1              1   0   0
2              1   0   0
3              1   0   0
[...]
148            1   0   1
149            1   0   1
150            1   0   1
attr(,"assign")
[1] 0 1 1
attr(,"contrasts")
attr(,"contrasts")$A
[1] "contr.treatment"

> confint(OnewayANOVA)
                   2.5 %        97.5 %
(Intercept)    5.1125208 5.7478792
Aa2           -0.6684663 0.2300663
Aa3           -0.2424663 0.6560663
```

Analysis

We already inspected the model summary of this analysis consisting of the sum of squares and the F-test. To retrieve more information from the object, we should first check what kind of object OnewayANOVA is and what methods apply to it. We use **is ()** to find out that the object is of class aov and of class lm. We then use **methods ()** with these two class names to obtain all functions that apply to our analysis. Further functions to inspect an aov object are the following. A call of **model.tables ()** returns information regarding the estimated effects in the model. **TukeyHSD ()** returns all pairwise comparisons for all factor levels and interactions of factors in the design. Finally, **coef ()** returns all effect estimates in the plan. Moreover, the functions for objects of class lm – linear regression models – also apply. In fact, every ANOVA is a linear regression. It is only a matter of preference or convenience for the statistical display and for the format of the statistical report. Consequently, R allows for an easy switch between ANOVA and regression for a given model.

One-way ANOVA employs only one independent variable, whereas a so-called factorial design uses more than one independent variable, say, A and B. The operator that links these variables determines how they figure in the ANOVA model. The plus operator + uses the simple main effects of A and B. The star operator * includes the main effects of the two factors and their interaction. The colon operator : calculates the interaction without main effects. Factorial designs are standard in research publications and they often give interesting clues about relationships in the data. In fact, many research hypotheses include the idea of interaction between factors; for example, that a certain drug has a different effect on male and female participants. Table 6.1 contains different model formulations for often encountered ANOVA models. Regression analysis in Chapter 7 will use the same syntax. The table further provides the corresponding experimental design formulas, which statistics books often use (e.g., Kirk, 2012). We will use these formulas in the subsequent sections.

Table 6.1 Model specifications for the ANOVA.

Analysis	aov model in R[1]	Linear Model[2]
One-way analysis of variance. Effect model of factor A	`Y ~ A`	$Y_{ij} = \mu + \alpha_j + \epsilon_{ij}$
Factorial design, main effects only	`Y ~ A + B`	$Y_{ijk} = \mu + \alpha_j + \beta_k + \epsilon_{ijk}$
	`Y ~ A + B + C`	$Y_{ijkl} = \mu + \alpha_j + \beta_k + \gamma_l + \epsilon_{ijkl}$
Factorial design, interaction only	`Y ~ A : B`	$Y_{ijk} = \mu + (\alpha\beta)_{ij} + \epsilon_{ijk}$
Factorial design, main effects and interaction	`Y ~ A * B`	$Y_{ijk} = \mu + \alpha_j + \beta_k + (\alpha\beta)_{ij} + \epsilon_{ijk}$
	`Y ~ A * B * C`	$Y_{ijkl} = \mu + \alpha_j + \beta_k + \gamma_l + (\alpha\beta)_{ij} + (\alpha\gamma)_{jl} + (\beta\gamma)_{kl} + (\alpha\beta\gamma)_{jkl} + \epsilon_{ijkl}$
Repeated-measures analysis (randomised blocks), within-subjects factor A	`Y ~ A + Error(subj/A)`	$Y_{ij} = \mu + \alpha_j + \pi_i + \epsilon_{ij}$
Split-plot factorial design, within-subjects factor B	`Y ~ A * B + Error(subj/B)`	$Y_{ijk} = \mu + \alpha_j + \pi_{i(j)} + \beta_k + (\alpha\beta)_{jk} + (\beta\pi)_{ki(j)} + \epsilon_{ijk}$

[1] Y: dependent variable; A, B, C: independent variables; `subj`: subject ID coded as a factor.
[2] Y: Measurement of the dependent variable; i: subject number; j, k, l: conditions associated with factors A, B, and C; μ: grand mean of the design; α_j: treatment effect of condition j of Factor A; β_k: treatment effect of condition k of factor B; γ_l: treatment effect of condition l of factor C; ϵ: error component of the model with normal distribution.

EXAMPLE! ## 6.2.3 ITC – Calculation of the original statistical tests

Reproducible science not only means reproducibility of research methods but also of the statistics, which led to the conclusions. We now use the *t*-test to recalculate the results reported by Baranski (Open Science Collaboration, 2015b) in her replication of the original experiment by Stanovich and West (2008). Additionally, we propose an approach using an ANOVA to test the hypotheses.

ITC_MOD_tTest.R

```
## ITC - Recalculation of the t-test
if(!exists('ITC_tTest'))
{
  source("./Scripts/ITC_PRE_LoadData.R")

  ITC_tTest <- with(ITC$stvu,
              {
                    p1_pnotq_correct_relevelled <-
```

```
relevel(p1_pnotq_correct, "correct")
                    t.test(sat ~ p1_pnotq_correct_relevelled,
                            alternative = "two.sided",
                            var.equal = TRUE)
                })
}
```

Output

```
> ITC_tTest

        Two Sample t-test

data:  sat by p1_pnotq_correct_relevelled
t = 0.99484, df = 175, p-value = 0.3212
alternative hypothesis: true difference in means is not equal to 0
95 percent confidence interval:
 -31.50548  95.55087
sample estimates:
  mean in group correct mean in group incorrect
             1735.714                1703.692
```

Analysis

The object `ITC_tTest` stores the results of the *t*-test. To obtain the same sign of the *t*-value obtained by Baranski, we use **relevel()** to set "correct" as the reference category of the variable `p1_pnotq_correct`. Simply enter `ITC_tTest` at the console to see the results. The test yields almost but not exactly the same insignificant test result as obtained by Baranski in her replication report. Her replication and our analysis use the same degrees of freedom, indicating that both analyses used the same number of subjects. However, her average SAT scores of M = 1729 for subjects who answered correctly on the card selection task, and M = 1698 for those who answered incorrectly differ from the means returned by our *t*-test.

ITC_MOD_ANOVA.R

```
## ITC - ANOVA of the associations between SAT scores and
correct responses
if(!exists('ITC_ANOVA'))
{
  source("./Scripts/ITC_PRE_LoadData.R")
  ITC_ANOVA <- list(SAT_by_p1correct = aov(sat ~ p1_pnotq_correct,
                               data = ITC$stvu),
              SAT_by_p2correct = aov(sat ~ p2_pnotq_correct,
                               data = ITC$stvu),
              SAT_by_p3correct = aov(sat ~ p3_pnotq_correct,
                               data = ITC$stvu))
}
```

Output

```
> lapply(ITC_ANOVA, summary)
$SAT_by_p1correct
                 Df  Sum Sq Mean Sq F value Pr(>F)
p1_pnotq_correct   1   43393   43393    0.99  0.321
Residuals        175 7672831   43845

$SAT_by_p2correct
                 Df  Sum Sq Mean Sq F value Pr(>F)
p2_pnotq_correct   1   71987   71987   1.648  0.201
Residuals        175 7644238   43681

$SAT_by_p3correct
                 Df  Sum Sq Mean Sq F value Pr(>F)
p3_pnotq_correct   1   99805   99805   2.293  0.132
Residuals        175 7616419   43522
```

Analysis

The script returns the list ITC_ANOVA consisting of three analyses of variance of SAT scores by numbers of correct responses. To display the analyses at the console, **lapply()** uses the summary function for all three ANOVAS. None of the three ANOVAs returned a significant result. Because of the equivalence of the two group ANOVA with the two-sided *t*-test for the first card selection task, the *p*-values of 0.321 are the same.

To be honest, reading through the replication report and reprogramming the analysis, I was not really happy about the changed role of the variables between theory and analysis. In theory the cognitive ability as measured by the SAT score is treated as the potential factor of influence for doing the card selection task. In the analysis it is treated as the dependent variable. In fact, the correct or incorrect response to the card selection task is treated as the independent variable. Although it is clear that the *t*-test and the ANOVA do not prove causal relationships, I find it better if the role of the variables in theory and analysis correspond. For the given data, a logistic regression of a binary outcome on a continuous predictor variable could have been a better analysis (see Chapter 7).

R FUNCTION! ## 6.2.4 methods()

An analysis of variance is a complex R object with several components. There are functions which extract these components and present their information at the console. Complex objects in R usually have such functions and they can be enquired with **methods()**. For example, a call of methods(class = "lm") prints out a list of almost 40 functions that could all be applied on an lm object. The most prominent of these are **print()** and **plot()**. Try to look for functions that apply to other complex objects like aov objects in the analysis of variance. In turn, the functions themselves can go as arguments for **methods()**, which then returns all the objects to which a particular one applies. Let us try both options.

Ch6_methods.R

```
## Using methods() to retrieve available functions and classes
methods(class = "data.frame")
methods(plot)
```

Output

```
> methods(class = "data.frame")
 [1]  $<-               [            [[            [[<-           [<-
aggregate
 [7]  anyDuplicated as.data.frame as.list       as.matrix     by             cbind
[13]  coerce        dim                  dimnames       dimnames<-      droplevels
duplicated
[19]  edit          format       formula       head          initialize   is.na
[25]  Math          merge        na.exclude    na.omit       Ops          plot
[31]  print         prompt       rbind         row.names     row.names<-  rowsum
[37]  show          slotsFromS3  split         split<-       stack        str
[43]  subset        summary      Summary       t             tail         transform
[49]  type.convert  unique       unstack       within
see '?methods' for accessing help and source code
> methods(plot)
 [1]  plot.acf*          plot.data.frame*    plot.decomposed.ts* plot.default
 [5]  plot.dendrogram*   plot.density*       plot.ecdf           plot.factor*
 [9]  plot.formula*      plot.function       plot.hclust*        plot.histogram*
[13]  plot.HoltWinters*  plot.isoreg*        plot.lm*            plot.medpolish*
[17]  plot.mlm*          plot.ppr*           plot.prcomp*        plot.princomp*
[21]  plot.profile.nls*  plot.raster*        plot.spec*          plot.stepfun
[25]  plot.stl*          plot.table*         plot.ts             plot.tskernel*
[29]  plot.TukeyHSD*
see '?methods' for accessing help and source code
```

Analysis

The first call of **methods()** supplies data.frame as a class name, the second call supplies **plot()** as a generic method. For data.frame the console shows several functions that we already know and applied, like **merge()**, **by()**, or **cbind()**. For **plot()** the console shows a long list of classes to which the function applies. Of this list, we already know the classes table or data.frame. The **methods()** function is a strong aid to help you find the best functions to apply in the analysis. Moreover, some packages define their own classes and associated specialised functions. For example, the **tm** package for text mining defines the vCorpus class. The **methods()** function helps you to become familiar with such classes.

EXAMPLE! ## 6.2.5 SCB – ANOVA test for an intervention effect

The main analysis in the conflict reconciliator project is an analysis of variance to test whether conflict reconciliators entail less bullying and violence on schoolyards. For the three types of conflict, we first calculate the effect of factor A, the intervention program.

SCB_MOD_ANOVA_InterventionEffect.R

```
## SCB - ANOVA test for an intervention effect
source("./Script/SCB_PRE_DataImport.R")
source("./Script/SCB_PRE_VariableLabels.R")
source("./Script/SCB_PRE_DataAggregation.R")


M1 <- aov(PersonPerson_PREPOST ~ A, data = SCB)
M2 <- aov(PersonGroup_PREPOST ~ A, data = SCB)
M3 <- aov(GroupGroup_PREPOST ~ A, data = SCB)


summary(M1)
summary(M2)
summary(M3)
```

Output

```
> summary(M1)
             Df Sum Sq Mean Sq F value  Pr(>F)
A             2  282.2  141.09    10.6 4.5e-05 ***
Residuals   177 2356.6   13.31
---
Signif. codes:  0 '***' 0.001 '**' 0.01 '*' 0.05 '.' 0.1 ' ' 1
> summary(M2)
             Df Sum Sq Mean Sq F value Pr(>F)
A             2   84.1   42.05   3.609 0.0291 *
Residuals   177 2062.1   11.65
---
Signif. codes:  0 '***' 0.001 '**' 0.01 '*' 0.05 '.' 0.1 ' ' 1
> summary(M3)
             Df Sum Sq Mean Sq F value Pr(>F)
A             2  113.5   56.74    4.64 0.0109 *
Residuals   177 2164.6   12.23
---
Signif. codes:  0 '***' 0.001 '**' 0.01 '*' 0.05 '.' 0.1 ' ' 1
```

Analysis

We calculated the intervention effect for all three types of conflict using the differences between pre- and post-measurements of conflict counts. With the **aov()** function we define the analyses with a formula expression relating the dependent variable to the independent variable. The three model summaries then show the effectiveness of the intervention program. However, we cannot tell the direction of the effect, that is, whether conflict mediators lead to more or less conflicts on schoolyards. So, we need to gather more information from the model objects.

SCB_CON_ANOVA_InformationANOVA.R

```
## SCB - Gather information from aov object
source("./Script/SCB_MOD_ANOVA_InterventionEffect.R")


methods(class = "aov")


model.tables(M1)
model.tables(M2)
model.tables(M3)


TukeyHSD(M1)
TukeyHSD(M2)
TukeyHSD(M3)
```

Output

```
> methods(class = "aov")
  [1]  coef                    coerce            extractAIC      initialize
model.tables print        proj          qqnorm        se.contrast
[10] show            slotsFromS3  summary      TukeyHSD      vcov
see '?methods' for accessing help and source code
>
> model.tables(M1)
Tables of effects

 A
A
    Conflict reconciliators Enhanced teacher attendance
No intervention
                1.0778                   0.6778                    -1.7556
> model.tables(M2)
Tables of effects

 A
A
    Conflict reconciliators Enhanced teacher attendance
 No intervention
                0.4833                   0.4833                    -0.9667
> model.tables(M3)
Tables of effects
```

```
 A
 A
                 Conflict    reconciliators    Enhanced    teacher    attendance
No intervention
                      1.0278                    -0.1222                   -0.9056
>
> TukeyHSD(M1)
  Tukey multiple comparisons of means
    95% family-wise confidence level

Fit: aov(formula = PersonPerson_PREPOST ~ A, data = SCB)

$A
                                                            diff        lwr
upr      p adj
Enhanced    teacher    attendance-Conflict    reconciliators  -0.400000  -1.974580
1.1745798 0.8200464
No intervention-Conflict  reconciliators               -2.833333 -4.407913
-1.2587535 0.0001007
No intervention-Enhanced teacher attendance            -2.433333 -4.007913
-0.8587535 0.0009919


> TukeyHSD(M2)
  Tukey multiple comparisons of means
    95% family-wise confidence level

Fit: aov(formula = PersonGroup_PREPOST ~ A, data = SCB)

$A
                                                            diff        lwr
upr      p adj
Enhanced teacher attendance-Conflict reconciliators -2.553513e-15 -1.472927
1.47292698 1.0000000
No intervention-Conflict  reconciliators               -1.450000e+00 -2.922927
0.02292698 0.0547183
No intervention-Enhanced teacher attendance            -1.450000e+00 -2.922927
0.02292698 0.0547183


> TukeyHSD(M3)
  Tukey multiple comparisons of means
    95% family-wise confidence level

Fit: aov(formula = GroupGroup_PREPOST ~ A, data = SCB)

$A
                                                            diff        lwr
upr      p adj
```

Enhanced teacher attendance-Conflict reconciliators -1.1500000 -2.659084 0.3590843 0.1721493

No intervention-Conflict reconciliators -1.9333333 -3.442418 -0.4242490 0.0079266

No intervention-Enhanced teacher attendance -0.7833333 -2.292418 0.7257510 0.4389106

Analysis

The `model.tables()` function extracts the group means from the model objects. This shows that conflict reconciliators indeed lead to improved conflict counts for all three types of conflict. Additionally, the Tukey pairwise comparisons of group means show that the difference between conflict reconciliators and enhanced teacher attendance does not show any significant difference, suggesting a similar effectiveness of the two interventions. However, we find significant pairwise differences between the two interventions and the no intervention condition.

6.3 Contrasts for the ANOVA

This section introduces how to calculate planned comparisons or (a priori) contrasts. It is one of the most preferred ways to compare the means between experimental conditions. To use contrasts we need to handle factor variables.

6.3.1 Definition and application of contrasts

STATISTICAL ISSUE!

Because the ANOVA only tells us whether or not all group means are equal, and because this is usually less than we specify in our empirical hypothesis, we have to search for better methods. *We need a statistic that most closely fits with the hypotheses that we have in mind when we set up the experiment.* For example, we usually state that therapy A is better than therapy B and not that the therapies simply have different effects. Another example could be that social pressure leads to more conformity, and not only that social pressure leads to more or less conformity, compared with a control condition. Almost always we hypothesise about better or worse, about more or less. For the testing of such hypotheses, a priori contrasts are made. *A priori contrasts* also enhance the test power for hypothesised effects in the ANOVA.

A contrast is a linear combination of the expected means in a design. For example, $\psi_1 = 1\mu_1 - 1\mu_2 + 0\mu_3$ is a contrast for the comparison of the first and second group means in a design consisting of three groups. It expresses our hypothesis that $\mu_1 > \mu_2$ – that the expected mean of condition 1 is bigger than the expected mean of condition 2. Another contrast is $\psi_2 = .5\mu_1 + .5\mu_2 - 1\mu_3$. It compares the average of the first two group means with the third group mean. We collect the so-called contrast coefficients in a vector c with $c_{\psi_1} = (1, -1, 0)$ and $c_{\psi_2} = (.5, .5, -1)$. The numbers weights for the summation of all the means in an experimental design. R calculates a t-test for the significance of the contrast, which is, in fact, a test of the mean difference that we hypothesised.

6.3.2 Contrasts in R

In the analysis of variance, we employ the `glht()` function of the `multcomp` package (Hothorn, Bretz, & Westfall, 2008) to calculate contrasts. The function stands for general linear hypothesis testing. Let us try it in R with our small data simulation. We will first set up the contrast coefficients, then calculate the ANOVA and apply the contrast calculations with the resulting aov object.

Ch6_APrioriContrastsMultcomp.R

```
## A priori contrasts with multcomp::glht()
source("Ch6_RandomData.R")
library(multcomp)

# Preparation
ContrastCoefficients <- rbind("a1>a2" = c(1, -1, 0),
                              "a1a2>a3" = c(.5, .5, -1))

# Calculations
GrandMean <- mean(d$Y)
Means <- tapply(d$Y, d$A, mean)
m <- aov(Y ~ A, d)
mC <- glht(m,
           linfct = mcp(A = ContrastCoefficients),
           alternative = "greater")

# Results
GrandMean
Means
summary(m)
summary(mC)
```

Output

```
> # Results
> GrandMean
[1] 5.426067
> Means
    a1     a2     a3
5.4302 5.2110 5.6370
> summary(m)
             Df Sum Sq Mean Sq F value Pr(>F)
A             2   4.54   2.269   1.756  0.176
Residuals   147 189.93   1.292
> summary(mC)

        Simultaneous Tests for General Linear Hypotheses
```

Multiple Comparisons of Means: User-defined Contrasts

```
Fit: aov(formula = Y ~ A, data = d)

Linear Hypotheses:
              Estimate Std. Error t value Pr(>t)
a1>a2 <= 0      0.2192     0.2273   0.964  0.308
a1a2>a3 <= 0   -0.3164     0.1969  -1.607  0.997
(Adjusted p values reported -- single-step method)
```

Analysis

We attach **multcomp** at the beginning and then use two functions of it for the script: <u>glht()</u> and <u>mcp()</u>. We prepare the contrast coefficients as two named row vectors and bundle them in a matrix with <u>rbind()</u>. The analysis starts with the calculation of means and the analysis of variance object m with **aov()**. The function <u>glht()</u> then obtains m and the contrast coefficients as an argument for the `linfct` parameter. The parameter stands for linear function, in fact, for the linear combination of expected means. The results show that a summary of m produces the table that we already know. The summary of the object produced by <u>glht()</u> consists of estimates for the two contrasts that we specified. We can confirm that the contrasts are linear combinations of the empirical means:

$$\hat{\psi}_1 = 1 \cdot M_1 - 1 \cdot M_2 + 0 \cdot M_3 = 5.4302 - 5.2110 = 0.2192$$

$$\hat{\psi}_2 = 0.5 \cdot M_1 + 0.5 \cdot M_2 - 1 \cdot M_3 = 0.5 \cdot 5.4302 + 0.5 \cdot 5.2110 - 1 \cdot 5.6370 = -0.3164$$

The empirical contrasts are tested with a *t*-test against the test value 0. A significant contrast is equivalent to a significant mean difference as specified with the contrast coefficients. Contrasts most closely correspond with the ideas that we usually have about the order of means when we plan an experiment. We hypothesise that some means are either greater or smaller than other means, and not merely that the means differ. Contrasts directly turn such ideas in a statistical test. To calculate contrasts in R we can also use the contrasts attribute of factor variables. This is good if we want comparisons with reference categories in a linear regression. Another package that supplies more functions related to contrasts is **MBESS** (Kelley, 2007, 2021). Among other things, the package calculates effect sizes and confidence intervals for contrasts. So, use contrasts whenever you can. I really like them.

6.4 Two-factor ANOVA model

Experiments in the social or clinical sciences rarely apply only one independent variable. Usually, there are other independent variables, and the interaction between them is of theoretical interest. We will now practise how to incorporate more than one independent variable in the analysis.

6.4.1 The completely randomised factorial design

The obvious generalisation of the one-way ANOVA with one experimental factor is to include one or more other factors in the analysis. We can analyse their impact and their interaction with the factor of primary concern. For example, any therapy program would at least be evaluated regarding different effects for women and men. So, if we have factor A with levels a_1, a_2, and a_3, we add factor B, with levels b_1 and b_2, to the plan. In the simple case, all levels of A are crossed with all levels of B. The experimental conditions that emerge can be represented in a 3 × 2 table (Table 6.2).

Table 6.2 Completely randomised factorial design (CRF)

		B	
		b_1	b_2
A	a_1	a_1b_1	a_1b_2
	a_2	a_2b_1	a_2b_2
	a_3	a_3b_1	a_3b_2

All combinations of factor levels are realised in the experiment and yield data with several subjects. Summary statistics can be calculated and then an ANOVA can be calculated. In this case the summary table of the ANOVA not only contains two lines for between and within sources of variability but can contain several more lines, at least one for every experimental factor in the analysis. Moreover, interaction terms can be included and tested. A two-factor analysis of variance we calculate with the following script.

Ch6_TwoFactorANOVA.R

```
## Two-factor ANOVA
source("Ch6_RandomData.R")

# Setup model formula
AllEffects  <- formula(Y ~ A * B)
MainEffects <- formula(Y ~ A + B)
Interaction <- formula(Y ~ A : B)

# Generate analyses
ModelMainEffects <- aov(MainEffects, data = d)
ModelInteraction  <- aov(Interaction,  data = d)
ModelAllEffects  <- aov(AllEffects,  data = d)

# Display results
summary(ModelMainEffects)
summary(ModelInteraction)
summary(ModelAllEffects)
```

Output

```
> # Display results
> summary(ModelMainEffects)
            Df Sum Sq Mean Sq F value Pr(>F)
A            2   4.54   2.269   1.796 0.1697
B            1   5.42   5.419   4.288 0.0401 *
Residuals  146 184.51   1.264
---
Signif. codes:  0 '***' 0.001 '**' 0.01 '*' 0.05 '.' 0.1 ' ' 1
> summary(ModelInteraction)
            Df Sum Sq Mean Sq F value Pr(>F)
A:B          5   14.6   2.920   2.338 0.0448 *
Residuals  144  179.9   1.249
---
Signif. codes:  0 '***' 0.001 '**' 0.01 '*' 0.05 '.' 0.1 ' ' 1
> summary(ModelAllEffects)
            Df Sum Sq Mean Sq F value Pr(>F)
A            2   4.54   2.269   1.817  0.166
B            1   5.42   5.419   4.338  0.039 *
A:B          2   4.64   2.322   1.859  0.160
Residuals  144 179.87   1.249
---
Signif. codes:  0 '***' 0.001 '**' 0.01 '*' 0.05 '.' 0.1 ' ' 1
```

Analysis

The formula definitions with **formula()** specify three linear models. Here, Y is not only related to one factor but to two factors connected with one of three operators. A + B means that Y will be modelled with the main effects of the factors A and B. Furthermore, A:B specifies an interaction term, that is, the effect of all combinations of factor levels of A and B. Finally, A*B is an abbreviation of the commonly used combination of main effects and interaction, explicitly written as A + B + A : B.

The three formulas AllEffects, MainEffects, and Interaction, are then passed to the analysis of variance function **aov()**. Afterwards the results of the analyses are displayed with **summary()**. The first table consists of the results for the two main effects of A and B. We can observe significant effects for factor B because the empirical F-values fall in the rejection region with $p < .05$. In ModelInteraction we find the results of the interaction without consideration of the two main effects. Finally, ModelAllEffects shows all ANOVA tests which are possible with this design – main effects and interactions.

6.4.2 SCB – Two-factor ANOVA   EXAMPLE!

From the beginning, we included types of school in the experimental design, because it may modify the effectiveness of conflict reconciliators. So, we include factor B: type of school in the subsequent analysis and obtain a factorial ANOVA with two factors.

SCB_MOD_TwoFactorANOVA.R

```
## SCB - ANOVA test for an intervention effect
source("./Script/SCB_PRE_DataImport.R")
source("./Script/SCB_PRE_VariableLabels.R")
source("./Script/SCB_PRE_DataAggregation.R")


M1 <- aov(PersonPerson_PREPOST ~ A * B, SCB)
M2 <- aov(PersonGroup_PREPOST ~ A * B, SCB)
M3 <- aov(GroupGroup_PREPOST ~ A * B, SCB)


summary(M1)
summary(M2)
summary(M3)
```

Output

```
> summary(M1)
            Df Sum Sq Mean Sq F value  Pr(>F)
A            2  282.2  141.09  10.568 4.7e-05 ***
B            2   69.7   34.84   2.610  0.0765 .
A:B          4    4.0    1.01   0.075  0.9896
Residuals  171 2282.9   13.35
---
Signif. codes:  0 '***' 0.001 '**' 0.01 '*' 0.05 '.' 0.1 ' ' 1
> summary(M2)
            Df Sum Sq Mean Sq F value Pr(>F)
A            2   84.1   42.05   3.645 0.0282 *
B            2    1.9    0.95   0.082 0.9210
A:B          4   87.4   21.85   1.894 0.1137
Residuals  171 1972.8   11.54
---
Signif. codes:  0 '***' 0.001 '**' 0.01 '*' 0.05 '.' 0.1 ' ' 1
> summary(M3)
            Df Sum Sq Mean Sq F value Pr(>F)
A            2  113.5   56.74   4.601 0.0113 *
B            2   28.7   14.37   1.166 0.3142
A:B          4   27.3    6.82   0.553 0.6969
Residuals  171 2108.5   12.33
---
Signif. Codes:  0 '***' 0.001 '**' 0.01 '*' 0.05 '.' 0.1 ' ' 1
```

Analysis

The three analyses for the three different types of conflict, respectively, yield significant effects for the intervention programme. The suspected interaction between intervention programme and type of school, denoted as the term `A:B`, does not show significant effects. Use the function `model.tables()` with the three analysis objects to inspect the group means in the two-factor design.

6.5 More complex ANOVA models

We now turn to some classical ANOVA designs: randomised blocks, repeated measures, and split-plot factorial. The model formulas become more complex with these designs.

Thus far, we learned about the ANOVA with one or two experimental factors. And regarding experimental design, the inexperienced researcher is well advised to follow the command by Hays (1994, p. 585): "keep it simple!" However, experiments sometimes need more design options. For example, it may be more economic and experimentally feasible to apply all experimental conditions to each subject, if carry-over effects are not to be expected, or if each condition takes five minutes to conduct, as may be the case in perception studies with computer interfaces. These examples take the subject as a random factor of influence into account. For instance, in reaction time experiments, where differential effects of different conditions on reaction times are probed, one could expect that a person with quick reactions would be able to react quickly in all conditions, and a slower person would be expected to react slower in all conditions. These systematic influences of the person can be included in the statistical model to yield a more sensitive analysis, even for smaller effects. Moreover, many experiments combine this idea with the factorial design. That is, one factor may be a so-called within-subjects factor, meaning that each subject goes through all experimental conditions; another factor may be a between-subjects factor, meaning that each subject is allocated to only one of its conditions. For example, in a computer experiment each subject can go through two or three experimental tasks successively, making task a between-subjects factor. The experiment may further distinguish the gender of the participants, making this a between-subjects factor. Perhaps there is a hypothesis that males accomplish one particular task better than females and that females are better in another one of the tasks. Such interaction hypotheses can be analysed statistically, but require a more complex design.

We now turn to the randomised block design, which is sometimes called repeated-measures analysis. It employs a within-subjects factor. Then, turn to the split-plot factorial design, which combines between- and within-subjects factors. Kirk (2012) provides full mathematical descriptions of these designs. Here, we focus on how to write the model formulas in R.

6.5.1 The randomised block design

STATISTICAL ISSUE!

The randomised block design is appropriate if blocks of homogeneous experimental units are allocated to experimental conditions. Table 6.3 shows the data layout for this design. The blocks of homogeneous experimental units occupy the rows, and the experimental conditions occupy the columns. That is, each block of units is tested in all experimental conditions. In this case, we not only expect to find an effect of the experimental condition but also an effect of the blocks. In fact, the assumption is that units of the same block will yield more similar measurements than units of different blocks. For example, if all subjects of an experiment go through all experimental conditions, the subject yields a homogeneous block of measurements, and a systematic effect of the subject on the outcome of the experiment could be expected. If there is no interaction of personal characteristics with the experimental conditions, one would expect that persons scoring high in one condition should also be more likely to score high in the other conditions of the experimental design; persons scoring low in one condition are expected to score low in the other conditions. The experimental subject as a source of variation on the dependent variable is only one example of the more general proposition that experimental units stratified according to a certain variable are more similar with respect to the outcome of the experiment. This is the randomised block design. If one block of homogenous experimental units is in fact one subject, we speak of a repeated-measures design. We capture the idea of the randomised block design by adding another term to the linear model:

$$Y_{ij} = \mu + \alpha_j + \pi_i + \epsilon_{ij}$$

Each Y_{ij} is a measurement of person i in condition j. In this model the added term π_i stands for the effect of a block or person. Unlike α_j it is a random effect, because in replications of the experiment, we would expect to sample different blocks or subjects leading to different effects, which enter the model. In contrast, since α_j is bound to a defined experimental condition, its effect is considered to be fixed in the model. It is a fixed effect.

Table 6.3   Data model of the randomised block design.

	Treatment		
Group / Subject	a_1	a_2	a_3
1	Y_{11}	Y_{12}	Y_{13}
2	Y_{21}	Y_{22}	...
3	Y_{31}	...	...
4	Y_{41}	...	...
5	...	...	...
6	...	...	...
7	...	...	...
8	...	...	...

The analysis of variance function **aov()** also calculates the randomised block design. However, another term enters the model and uses the **Error()** function. Now, the model connecting the dependent variable (DV) with the independent variable (IV) and respecting blocks looks like this: Model <- aov(DV ~ IV + Error(Block)). The first part of the formula is the same as in the one-way ANOVA; however, the **Error()** function is added with ± as another term to the formula. The function includes the block effect (e.g., the effect of the experimental subjects) as a factor in the analysis and thus allows the testing of the effect of IV against a smaller residual. See Kirk (2012) for the statistical details of the randomised block design.

The randomised block or repeated measures design has *two main advantages*: (1) variation due to the factor block or experimental subject can be controlled and eliminated from the error variance; (2) the design is more efficient to conduct because one experimental unit goes through all conditions, thereby lowering the need for a large sample size. Kirk (2012) also defines a generalised randomised block design, which includes replications within each block × treatment combination. This design allows for the calculation of the block-treatment interaction effect.

6.5.2 UPS – Repeated-measures ANOVA   EXAMPLE!

Let us inspect the repeated-measures analysis (i. e., randomised block design) for the unpleasant sounds data set. We assume that each person brings his or her own tendency to the experiment to give either higher, medium, or lower responses and this tendency may be present in all experimental conditions.

UPS_MOD_RepeatedMeasuresANOVA.R

```
## UPS - Repeated measures analysis
if(!exists('RM_ANOVA'))
{
  source("Scripts/UPS_PRE_DataImport.R")
  UPS$subject_factor <- factor(UPS$subject, ordered = FALSE)
  RM_ANOVA <- with(UPS,
                  {
                            list(m_sound  =  aov(rating  ~  sound +
Error(subject_factor),
                              data = UPS),
                            m_soundXvar  =  aov(rating  ~  sound *
variation +
Error(subject_factor),
                                data = UPS),
                            Replications  =  replications(rating  ~
sound * variation +
Error(subject_factor),
                                  data = UPS))
                  })
}
```

Output

```
> summary(RM_ANOVA$m_sound)

Error: subject_factor
          Df Sum Sq Mean Sq F value Pr(>F)
Residuals 95   762.9    8.03


Error: Within
           Df Sum Sq Mean Sq F value    Pr(>F)
```

```
sound            3      50   16.659    15.71 4.03e-10 ***
Residuals 2589   2746   1.061
---
Signif. codes:   0 '***' 0.001 '**' 0.01 '*' 0.05 '.' 0.1 ' ' 1
> summary(RM_ANOVA$m_soundXvar)

Error: subject_factor
            Df Sum Sq Mean Sq F value Pr(>F)
Residuals 95   762.9    8.03

Error: Within
                Df Sum Sq Mean Sq F value   Pr(>F)
sound            3   50.0   16.66   20.39 4.66e-13 ***
variation        6  344.7   57.45   70.32  < 2e-16 ***
sound:variation 18  306.3   17.02   20.83  < 2e-16 ***
Residuals     2565 2095.4    0.82
---
Signif. Codes:   0 '***' 0.001 '**' 0.01 '*' 0.05 '.' 0.1 ' ' 1
> RM_ANOVA$Replications
          sound       variation sound:variation
            672             384              96
```

Analysis

The script creates two repeated-measures analyses and collects them in the object RM_ANOVA. The subject ID must enter as a factor in the analysis, so we convert it with **factor()** before the calculations. Then we create the two model objects m_sound and m_soundXvar. The analysis m_sound uses only sound as a within-subjects variable. By using Error(subject_factor) we include the subject effect in the model to remove its systematic effect from the residual variance and, in fact, lower the residual for the F-test for the factor sound. As can be seen in the output under Error: Within, sound shows a significant effect on the pleasantness ratings.

The analysis m_soundXvar further includes the sound variation as a further within-subject factor. It uses the same additional error term in the model. Here the results in the section Error: Within show three F-tests against the residual: sound, variation, and their interaction all show significant effects on the pleasantness ratings.

EXAMPLE! ## 6.5.3 UPS - Split-plot factorial analysis of variance

Let us further include between-subject factors in the analysis of the unpleasant sounds data. The study is, in fact, partly motivated by an interaction hypothesis; that is, the sound rating is assumed to vary depending on the interaction of the variables sound, whether the source of the sound is known to the subject, and variation of the sound. We use the so-called split-plot factorial design to analyse the effects of within-subject and between-subject factors in one analysis. Kirk (2012) provides a comprehensive statistical account of this design. The model specification is similar to the randomised block analysis.

UPS_MOD_SplitPlotFactorialANOVA.R

```
## UPS - Split-plot factorial ANOVA model
if(!exists('SPF_ANOVA'))
{
  source("Scripts/UPS_PRE_DataImport.R")
  UPS$subject_factor <- factor(UPS$subject)

  SPF_ANOVA <- with(UPS,
                 {
                             list(m_groupXsound = aov(rating ~ group *
sound + Error(subject_factor), data = UPS),
                                    m_groupXgenderXsoundXvariation =
aov(rating ~ group + gender +
sound + variation +
group:sound +
Error(subject_factor), data = UPS))
                 })
}
```

Output

```
> summary(SPF_ANOVA$m_groupXsound)

Error: subject_factor
          Df Sum Sq Mean Sq F value Pr(>F)
group      1    9.7   9.666   1.206  0.275
Residuals 94  753.2   8.013


Error: Within
              Df Sum Sq Mean Sq F value   Pr(>F)
sound          3   50.0  16.659  15.694 4.09e-10 ***
group:sound    3    1.4   0.458   0.431    0.731
Residuals   2586 2745.0   1.061
---
Signif. codes:  0 '***' 0.001 '**' 0.01 '*' 0.05 '.' 0.1 ' ' 1
> summary(SPF_ANOVA$m_groupXgenderXsoundXvariation)

Error: subject_factor
          Df Sum Sq Mean Sq F value   Pr(>F)
group      1    9.7    9.67   1.367 0.245339
gender     1   95.5   95.52  13.508 0.000397 ***
Residuals 93  657.7    7.07
---
```

```
Signif. codes:  0 '***' 0.001 '**' 0.01 '*' 0.05 '.' 0.1 ' ' 1


Error: Within
                Df Sum Sq Mean Sq F value   Pr(>F)
sound            3   50.0   16.66  17.906 1.68e-11 ***
variation        6  344.7   57.45  61.745  < 2e-16 ***
group:sound      3    1.4    0.46   0.492    0.688
Residuals     2580 2400.3    0.93
---
Signif. Codes:  0 '***' 0.001 '**' 0.01 '*' 0.05 '.' 0.1 ' ' 1
```

Analysis

The script creates a list object SPF_ANOVA containing two ANOVA models. The variable subject ID must again enter as a factor in the models and it is transformed with **factor()**. The first model m_groupXsound only uses the factors group and sound to account for the pleasantness ratings. The first factor distinguishes between whether or not the sound source is known to the subject. The results show that the F-test for sound uses the additional error term based on subject_factor, whereas the F-tests of the sound factor and the group-sound interaction use the residual. The analysis shows no significant effect of whether the sound source is known or unknown on the pleasantness ratings but an effect of the type of sound.

The analysis m_groupXgenderXsoundXvariation has a similar structure but further includes gender as a between-subject factor and variation as a within-subject factor. The variables group and gender use the additional error term based on the subject factor; sound and variation use the residual. We find a significant gender effect. At the console, with(UPS, tapply(rating, gender, mean)) shows that females on average gave higher ratings of unpleasantness than males.

Notice that summary(aov(rating ~ group, UPS)) yields a significant effect of group but the split-plot analysis did not. In fact, due to the mechanics of the split-plot factorial design, the between-subjects factor group has much less test power compared to the corresponding completely randomised design with this factor. Kirk (2012) described this in detail in his chapter about split-plot factorial design.

────────── **What you have learned so far** ──────────

1   To calculate a one-way analysis of variance and explore the results
2   To write a model formula in R for analysis of variance designs
3   To calculate and test statistical contrasts
4   To extend the model with further independent variables to create a factorial design
5   To calculate an analysis of variance in complex designs using repeated measures

===== **Exercises** =====

Explore functions

1 Use **methods()** to study all functions applicable to an `aov` object. Practise with an analysis of variance in the **MusicPreferencesSchool.csv** data set. First, calculate the `aov` object with `myANOVA <- aov(PreferenceSong1 ~ Musician, data = MusicPreferencesSchool)`; second, call `methods(class = "aov")`; third, apply the available methods with `myANOVA`.

Solve tasks

1 Calculate contrasts within the **MusicPreferencesSchool.csv** data set. Set up a contrast matrix to compare the preference ratings between any two types of school. Use **multcomp::glht()** for the calculation.

Read code

1 The following code fragment creates two simple ANOVA models with the **MusicPreferencesSchool.csv** data. Describe in your own words how we initialise and compare the models.

```
## Two-factor ANOVA

# Preparation
d <- read.csv("MusicPreferencesSchool.csv")
d$SumPreference <- d$PreferenceSong1 + d$PreferenceSong2

# ANOVA
m1 <- aov(SumPreference ~ School, d)
m2 <- update(m1, . ~ . * Musician)

summary(m1)
summary(m2)
anova(m1, m2)
```

2 The following script calculates a one-way ANOVA by hand using the **MusicPreferencesSchool.csv** data set. Check the calculations with an introductory statistics book and describe all steps of the calculations in your own words. Check the results against `summary(aov(Y ~ School, d))`.

```
## By-hand calculation of the ANOVA sum of squares

# Preparation
d <- read.csv("MusicPreferencesSchool.csv")
d$Y <- d$PreferenceSong1 + d$PreferenceSong2

# Calculate statistics
M            <- mean(d$Y)
```

(Continued)

```
M_Groups    <- tapply(d$Y, d$School, mean)
n_Groups    <- tapply(d$Y, d$School, length)
df_b <- nlevels(d$School) - 1
df_w <- length(d$Y) - nlevels(d$School)

# Calculate F- and p-value (apply formulas)
SS_b                  <- sum(n_Groups * (M_Groups - M)^2)
SS_w                  <- sum(unlist(lapply(levels(d$School),
                      function(X)
                      {
                          sum((d$Y[d$School %in% X] - M_Groups[X])^2)
                      })))
MS_b <- SS_b / df_b
MS_w <- SS_w / df_w
F <- MS_b / MS_w
p <- pf(q = F, df1 = df_b, df2 = df_w, lower.tail = FALSE)

# Return p-value
sprintf("%.10f", p)
```

Apply in the real world

1   Analyse your own data. If your data set consists of categorical independent variables and quantitative dependent variables, calculate an analysis of variance with **aov()** to test hypotheses regarding these variables.

EXAMPLE! **WBD**

1   Use the data set in wide format. Calculate difference scores between baseline measurements and all subsequent time points, respectively (i. e., post-treatment, 1-month, 3-month, and 6-month). Then, use a *t*-test to compare means of the difference scores between control condition and intervention. Use **t.test()** and supply a formula for this analysis. Is there an intervention effect in the hypothesised direction?
2   Recalculate the foregoing analysis with an analysis of variance. Use **aov()** for this analysis.
3   Use the data set in long format. Calculate a split-plot factorial analysis of variance for the depression outcome, respecting condition and measurement as independent variables.

EXAMPLE! **ITC**

1   Refine the ANOVAS from the main text for this example. Include gender and age categories as independent variables.
2   Use the count variable of correct responses we created in the exercises in Chapter 3. Calculate an ANOVA for this variable and use gender and age categories as independent variables. Is the task performance independent of gender and age?

UPS

1   Go back to the analyses for the UPS example in the main text in Chapter 6. Recreate the aov objects using the **update ()** function to enter the factors step-by-step to the model.
2   Calculate a priori contrasts to check whether the condition origin known led to higher ratings of unpleasantness. Use the function **glht ()** from the package **multcomp** for the calculations.
3   Invent and calculate further contrast analyses for the factors sound, variation, and gender.

SCB

1   Examine the influence of the three intervention programs on the reduction of conflict count at the schools. Use three one-way analyses of variance for all three types of conflict. Use the function **aov ()** for the calculations.
2   It is questionable whether the three intervention programs show different effects depending on the type of school. Formulate corresponding hypotheses and test the hypotheses with a two-factor analysis of variance. Calculate for all three types of conflict.
3   The analysis of variance tests whether there exists any difference between expected values in the design, but does not test for the differences between single expected values. Such specific hypotheses can be investigated with the use of contrasts and corresponding t-tests.
4   Use contrasts in the one-factor design to test whether conflict reconciliators are more effective than (a) the enhancement of teacher attendance and (b) the group with no intervention regarding the reduction in conflicts. Calculate two contrasts. What is your decision regarding the effectiveness of the conflict reconciliators?
5   Can you recalculate the contrast values with the descriptive statistics? Closely inspect the empirical means and the contrast values.
6   In the example SCB about conflict reconciliators at schools, prove the effect of conciliators against the other two conditions. Start with a one-way ANOVA.
7   In the SCB example, calculate two a priori contrasts to test the superiority of the conciliators program against the other two conditions regarding bullying.
8   Add the factor "school type" to the design and calculate the two-factor ANOVA. Can we conclude that there is an interaction present, that is, a different effect of the conciliators program depending on school?

7

INFERENTIAL STATISTICS II: A MULTIPLE REGRESSION ANALYSIS

━━━━━━━━━━━━━━━━━━━━━━━━━━━ **Chapter overview** ━━━━━━━━━━

This chapter broadens the scope of inferential statistics to linear regression. In regression analysis you will learn how to fit and test regression models. Furthermore, you will experience the ease of implementing interaction terms. The linear modelling approach is then extended to cover logistic regression. Programming features: simple and multiple linear regression, test of regression parameters, data imputation, logistic regression.

7.1 Linear regression analysis with one or two predictors

This section starts with a simple regression of one dependent variable on one predictor variable. We review the linear model equation from the statistics books and check where to find the regression coefficients in the R output. We will also familiarise ourselves with the idea of strength of association.

STATISTICAL
ISSUE!

7.1.1 The regression equation

Sometimes we believe in the linear relationship between two quantitative variables X and Y. For example, intelligence (X) could be associated with grades at school (Y), emotionality with empathy, or the time spend for relaxation with physical endurance. Linear regression is made for the study of such relationships. The method uses a mathematical equation for the relationship between the variables, the so-called linear model equation.

$$Y_i = a + bX_i + e_i$$

Y_i is the measured Y-value of person i; it is the sum of three terms: an intercept term a, the measurement X_i of variable X multiplied with regression coefficient b, and an error term e_i. The aim of regression analysis is to estimate the coefficients a and b from data. That is, we measure the X and Y variables in a sample of participants and then statistically estimate the two regression coefficients a and b.

To calculate our first regression analysis in R, we use a small example relating health-related quality of life (QOL) to monthly income (INC), security of the job (JOB), adverse life events (ADV), or quality of personal relationships (REL). We may hypothesise that QOL has an association with the other variables of the form that higher levels of INC, JOB, ADV, or REL correspond with higher levels of QOL, and lower levels of these variables correspond with lower levels of QOL. The linear model equation to calculate QOL from the values of INC looks like this:

$$QOL_i = a + bINC_i + e_i$$

We now calculate a and b from sample data. The first of the following scripts creates the data frame for this example. The second script consists of the regression analysis based on the function `lm()`.

Ch7_DataCreation.R

```
## Random data for linear regression
if(!exists('d'))
{
  Means <- c(QOL = 20,
             INC = 1500,
```

```
            JOB = 4,
            ADV = 3,
            REL = 5)
  Sigma <- matrix(c( 5,   30,   1, .2,   .3,
                    30,  500,  10,  0,  10,
                     1,   10, 1.5,  0,  .4,
                    .2,    0,   0,  1,   0,
                    .3,   10,  .4,  0, 1.5), nrow = 5)
  set.seed(12345)
  d <- data.frame(round(mvtnorm::rmvnorm(n = 100,
                                    mean = Means,
                                    sigma = Sigma)))
  rm(Means, Sigma)
}
```

Ch7_SimpleLinearRegression.R

```
## A simple linear regression
source("Ch7_DataCreation.R")

# Scatter plot
with(d, plot(INC, QOL, pch = 16))

# Descriptives
means <- apply(d, 2, mean)
means
r <- cor(d)
print(r, digits = 2)

# Regression equation
m <- lm(formula = QOL ~ INC, data = d)
summary(m)

# Regression line
abline(m)
abline(v = means["INC"], lty = "dashed")
abline(h = means["QOL"], lty = "dashed")

# Diagnostic diagrams
layout(matrix(1:4, 2))
plot(m)

# Regression with standardised variables
m_z <- lm(scale(QOL) ~ scale(INC), d)
summary(m_z)
```

Output

```
> # Descriptives
> means <- apply(d, 2, mean)
> means
    QOL     INC     JOB     ADV     REL
  20.30 1502.27    4.12    3.15    5.10
> r <- cor(d)
> print(r, digits = 2)
     QOL  INC   JOB   ADV  REL
QOL 1.00 0.73 0.357 0.267 0.29
INC 0.73 1.00 0.429 0.118 0.38
JOB 0.36 0.43 1.000 0.067 0.16
ADV 0.27 0.12 0.067 1.000 0.21
REL 0.29 0.38 0.163 0.210 1.00
>
> # Regression equation
> m <- lm(formula = QOL ~ INC, data = d)
> summary(m)

Call:
lm(formula = QOL ~ INC, data = d)

Residuals:
    Min      1Q  Median      3Q     Max
-3.8762 -1.0206 -0.0231  1.0280  4.1138

Coefficients:
              Estimate Std. Error t value Pr(>|t|)
(Intercept) -78.856744   9.504613  -8.297 5.88e-13 ***
INC           0.066005   0.006326  10.434  < 2e-16 ***
---
Signif. codes:  0 '***' 0.001 '**' 0.01 '*' 0.05 '.' 0.1 ' ' 1

Residual standard error: 1.534 on 98 degrees of freedom
Multiple R-squared:  0.5263,    Adjusted R-squared:  0.5214
F-statistic: 108.9 on 1 and 98 DF,  p-value: < 2.2e-16

> # Regression with standardised variables
> m_z <- lm(scale(QOL) ~ scale(INC), d)
> summary(m_z)

Call:
lm(formula = scale(QOL) ~ scale(INC), data = d)

Residuals:
```

```
      Min      1Q    Median      3Q       Max
-1.74768 -0.46018 -0.01043  0.46349  1.85482

Coefficients:
              Estimate Std. Error t value Pr(>|t|)
(Intercept) -8.283e-16  6.918e-02    0.00        1
scale(INC)   7.254e-01  6.953e-02   10.43   <2e-16 ***
---
Signif. codes:  0 '***' 0.001 '**' 0.01 '*' 0.05 '.' 0.1 ' ' 1

Residual standard error: 0.6918 on 98 degrees of freedom
Multiple R-squared:  0.5263,     Adjusted R-squared:  0.5214
F-statistic: 108.9 on 1 and 98 DF,  p-value: < 2.2e-16
```

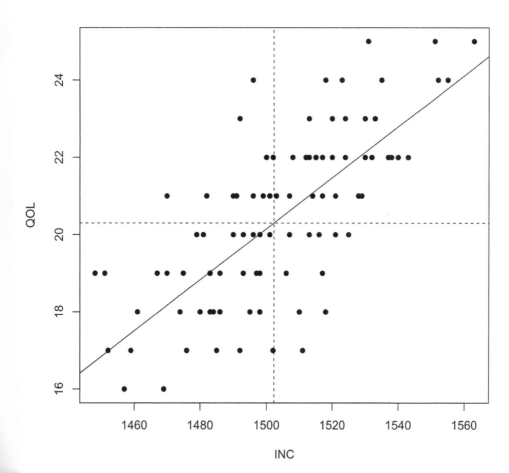

Figure 7.1 Scatter plot and solid regression line. Dashed lines mark the means of INC and QOL.

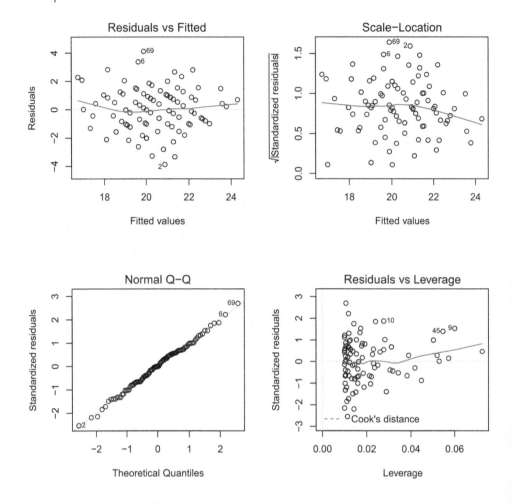

Figure 7.2 Diagnostic diagrams in linear regression.

Analysis

The data creation uses **rmvnorm()** from the **mvtnorm** package to generate correlating data (Genz et al., 2021). The matrix Sigma consists of the variances and covariances, which initialise the associations between all variables. A scatter plot then gives an idea about the relationship between INC and QOL (Figure 7.1). It shows a cloud of points, in which higher levels of INC correspond with higher levels in QOL, on average, and lower levels of INC correspond with lower levels in QOL, on average. The solid regression line corresponds with the yet unknown regression equation. As can be seen, prediction is never perfect and the data points do not lie on the regression line but are scattered around it. So, estimation always incorporates measurement error, which is the deviation of the data points from the regression line.

To arrive at the regression line, we calculate the coefficients a and b of the regression equation, made from the first two terms of the linear model equation.

$$\hat{Y}_i = a + bX_i$$

$\hat{Y}_i$ denotes the so-called predicted values for all participants. It means that all values of X translate in predicted values, which lie on the regression line. In our example the regression equation looks like this:

$$\widehat{QOL}_i = a + bINC_i$$

The regression analysis estimates the coefficients a and b following the principle of least squares. That is, the analysis selects a regression line to minimise the sum of squared differences between the data points and the regression line. The regression line then optimally represents the data points.

Although it requires differential calculus to actually do the analysis (Hays, 1994), we can fortunately find the coefficients a and b in the output of the **lm()** function in our R script. After some descriptives, the **lm()** function calculates a and b for the regression equation with the formula m <- lm(formula = QOL ~ INC, data = d). The syntax of this function closely corresponds with the syntax of the analysis of variance introduced in Chapter 6. That is, the tilde operator assigns the dependent variable on the left to the independent variable on the right. Comparable with ANOVA is also that **summary()** provides the summary table for the regression. The table contains estimates for the regression coefficient and the intercept term. So that we can now formulate the regression equation $\hat{Y} = a + bX$ as:

$$\widehat{QOL} = -78.86 + 0.07INC$$

The summary also provides a significance test of the hypothesis that the regression coefficient is zero. In the case of, INC a coefficient of zero would mean that there is no linear association between INC and QOL. A positive association like the one found would mean that higher values of INC correspond with higher values of QOL. The *p*-value *(Pr(>|t|))* of the according *t* value falls below the commonly used significance level of $\alpha = 0.05$. This result means that, given that the hypothesis of no linear association between INC and QOL is true, the obtained regression coefficients or higher coefficients occur in less than 5% of identical replications of the experiment. Following the logic of statistical inference, one concludes that one no longer assumes that there is no linear association between INC and QOL, but a positive linear association embodied by the regression equation.

In the script the function **abline()** then draws the regression line in the scatter plot that we created earlier in the script (Figure 7.1). The line follows the ascend of the cloud of data points. In fact, it must do so in order to minimise the sum of squared differences between regression line and all the data points. Now, take another look at the smart **abline()** function. Apparently it takes the complete model as an argument and selects the relevant information from the model itself. So, we must not index the two coefficients in the model ourselves to hand them to the function. However, **abline()** can also display many other kinds of straight lines depending on the arguments supplied, for example, auxiliary lines marking the means of QOL and INC.

The script then draws more diagrams to help us assess the prerequisites of the regression analysis. We simply create a layout for four diagrams and then call `plot(m)`. R returns four diagnostic diagrams (Figure 7.2). The top left diagram displays the residuals of the regression against the fitted values of the regression. In all sections of the fitted values, the residuals should show the same variance. The top right diagram shows another representation of the distribution of residuals displayed against the fitted values. The bottom left diagram shows the Q-Q diagram for an assessment of whether the data appear to be sampled from the normal distribution. Most data points should fall close to the straight line from the bottom left to the top right. For more information about diagnostic diagrams in regression analysis see Hays (1994).

The last part of the script does the regression with standardised variables. That is, INC and QOL are transformed with **scale()** to have a mean of 0 and a standard deviation of 1. In this case the regression coefficient equals the correlation coefficient between the variables.

STATISTICAL
ISSUE!

7.1.2 Explained variance and strength of association

The regression equation by itself only defines a straight line through the data cloud; it says nothing about the strength of association between X and Y. In fact, the data points may fall in the vicinity of the regression line or at a greater distance, indicating a higher or lower strength of association, respectively. In other words, our prediction is erroneous and the question is: by knowing the values of X, how sure can we be in knowing the values of Y? How much of the variance of Y is explained variance due to the regression equation? We use the so-called coefficient of determination R^2 (R-squared) as a measure of strength of association:

$$R^2 = r_{XY}^2$$

This is the squared correlation coefficient between X and Y. It is the proportion of the variance of Y accounted for by X. R^2 may vary between 0 and 1, with 0 indicating no association and 1 indicating a perfect linear association. An adjustment of the score to better estimate the population parameter of R^2 is the following (Harrell, 2015):

$$adjustedR^2 = 1 - \left(1 - R^2\right)\frac{N-1}{N-p-1}$$

In this formula, N is the sample size and p is the number of non-intercept parameters (i. e., regression coefficients) in the model. R^2 and the *adjusted* R^2 are part of the model summary in regression analysis. However, it can also be calculated by hand. Simply try the following at the console to verify the value in our regression summary.

Ch7_CoefficientOfDetermination.R

```
## R-squared coefficient of determination
source("Ch7_SimpleLinearRegression.R")

summary(m)
R2 <- (with(d, cor(INC, QOL)))^2
```

```
R2
var(predict(m)) / var(d$QOL)

# Adjusted R squared
N = length(d$QOL)
Radjusted <- 1 - (1-R2) * (N-1)/(N-1-1)
Radjusted
```

Output

```
> summary(m)

Call:
lm(formula = QOL ~ INC, data = d)

Residuals:
    Min      1Q  Median      3Q     Max
-3.8762 -1.0206 -0.0231  1.0280  4.1138

Coefficients:
             Estimate Std. Error t value Pr(>|t|)
(Intercept) -78.856744   9.504613  -8.297 5.88e-13 ***
INC           0.066005   0.006326  10.434  < 2e-16 ***
---
Signif. codes:  0 '***' 0.001 '**' 0.01 '*' 0.05 '.' 0.1 ' ' 1

Residual standard error: 1.534 on 98 degrees of freedom
Multiple R-squared:  0.5263,    Adjusted R-squared:  0.5214
F-statistic: 108.9 on 1 and 98 DF,  p-value: < 2.2e-16

> R2 <- (with(d, cor(INC, QOL)))^2
> R2
[1] 0.5262614
> var(predict(m)) / var(d$QOL)
[1] 0.5262614
>
> # Adjusted R squared
> N = length(d$QOL)
> Radjusted <- 1 - (1-R2) * (N-1)/(N-1-1)
> Radjusted
[1] 0.5214273
```

Analysis

The code shows different ways to arrive at R-squared. It is part of the regression summary together with the adjusted R-squared. We can calculate it as the squared correlation coefficient between QOL and INC. Furthermore, it is the quotient between the variance of the predicted data values and the variance of the dependent variable QOL. We use **predict()** to obtain the predicted values of the regression model. Finally, the formula given above produces the adjusted R-squared. Many statistical treatments explain the details of strength of association and the coefficient of determination (Hays, 1994; Kirk, 2012).

R FUNCTION! ## 7.1.3 lm()

At the heart of linear regression lies the **lm()** function. It works like **aov()** for ANOVA, but produces regression coefficients. It takes a formula relating a criterion or dependent variable to one or more predictor variables. Predictors can be quantitative or categorical; interaction terms are possible, too. The following script shows how to use **lm()** and demonstrates some technical aspects of its use. It uses the data set for this chapter relating health-related quality of life (QOL) to job security (JOB) and past experiences of adverse events (ADV).

Ch7_lm.R

```
## Linear regression with the lm() function
source("Ch7_DataCreation.R")

# Linear regressions
m1 <- lm(QOL ~ JOB, data = d)
m2 <- lm(QOL ~ JOB + ADV, data = d)
m3 <- lm(QOL ~ JOB * ADV, data = d)

# Model summaries
summary(m1)
summary(m2)
summary(m3)

# Predicted values
d$QOL_predict <- predict(m1)
head(d)

# Further information about lm objects
is(m3)
methods(class = "lm")
head(model.frame(m3))
head(model.matrix(m3))
```

Output

```
> # Model summaries
> summary(m1)

Call:
lm(formula = QOL ~ JOB, data = d)

Residuals:
     Min      1Q  Median      3Q     Max
 -4.8977 -1.5392  0.2815  1.4608  4.1023

Coefficients:
```

```
            Estimate Std. Error t value Pr(>|t|)
(Intercept)  17.5015      0.7683  22.779  < 2e-16 ***
JOB           0.6793      0.1795   3.784 0.000266 ***
---
Signif. codes:  0 '***' 0.001 '**' 0.01 '*' 0.05 '.' 0.1 ' ' 1

Residual standard error: 2.082 on 98 degrees of freedom
Multiple R-squared:  0.1275,    Adjusted R-squared:  0.1186
F-statistic: 14.32 on 1 and 98 DF,  p-value: 0.0002658

> summary(m2)

Call:
lm(formula = QOL ~ JOB + ADV, data = d)

Residuals:
    Min      1Q  Median      3Q     Max
-5.2594 -1.4052  0.1456  1.5045  3.7623

Coefficients:
            Estimate Std. Error t value Pr(>|t|)
(Intercept)  16.1880      0.8938  18.111  < 2e-16 ***
JOB           0.6479      0.1746   3.712 0.000344 ***
ADV           0.4579      0.1719   2.664 0.009048 **
---
Signif. codes:  0 '***' 0.001 '**' 0.01 '*' 0.05 '.' 0.1 ' ' 1

Residual standard error: 2.02 on 97 degrees of freedom
Multiple R-squared:  0.187,     Adjusted R-squared:  0.1702
F-statistic: 11.15 on 2 and 97 DF,  p-value: 4.37e-05

> summary(m3)

Call:
lm(formula = QOL ~ JOB * ADV, data = d)

Residuals:
   Min     1Q Median     3Q    Max
-5.272 -1.405  0.136  1.491  3.768

Coefficients:
            Estimate Std. Error t value Pr(>|t|)
(Intercept) 16.40757    2.05111   7.999 2.85e-12 ***
JOB          0.59122    0.50760   1.165    0.247
ADV          0.39050    0.59205   0.660    0.511
JOB:ADV      0.01733    0.14554   0.119    0.905
---
Signif. codes:  0 '***' 0.001 '**' 0.01 '*' 0.05 '.' 0.1 ' ' 1

Residual standard error: 2.031 on 96 degrees of freedom
Multiple R-squared:  0.1871,    Adjusted R-squared:  0.1617
F-statistic: 7.364 on 3 and 96 DF, p-value: 0.000171
```

```
>
> # Predicted values
> d$QOL_predict <- predict(m1)
> head(d)
  QOL  INC JOB ADV REL QOL_predict
1  22 1517   4   3   6    20.21849
2  17 1511   4   3   4    20.21849
3  22 1540   5   3   5    20.89774
4  20 1481   3   4   5    19.53924
5  23 1533   4   1   4    20.21849
6  23 1492   5   4   5    20.89774
>
> # Further information about lm objects
> is(m3)
[1] "lm"        "oldClass"
> methods(class = "lm")
 [1]  add1              alias           anova           case.names
coerce         confint
 [7]  cooks.distance  deviance         dfbeta          dfbetas
drop1          dummy.coef
[13]  effects                 extractAIC       family          formula
hatvalues      influence
[19]  initialize          kappa            labels          logLik
model.frame    model.matrix
[25] nobs               plot             predict         print           proj
qr
[31]  residuals              rstandard        rstudent        show
simulate       slotsFromS3
[37] summary          variable.names vcov
see '?methods' for accessing help and source code
> head(model.frame(m3))
  QOL JOB ADV
1  22   4   3
2  17   4   3
3  22   5   3
4  20   3   4
5  23   4   1
6  23   5   4
> head(model.matrix(m3))
  (Intercept) JOB ADV JOB:ADV
1           1   4   3      12
2           1   4   3      12
3           1   5   3      15
4           1   3   4      12
5           1   4   1       4
6           1   5   4      20
```

Analysis

After data creation, three calls of **lm()** show model formulas for a linear regression using one or two predictors. Model m1 is a simple linear regression with QOL as a dependent variable and JOB as a predictor. The tilde operator ~ connects both variables in the model. The data parameter tells R where to find the variables. The models m2 and m3 show multiple regressions with two predictors. The + operator tells R to use both predictors without interaction term; conversely, the * operator additionally includes the interaction term of both predictors. The model summaries then reveal the results of the three analyses. Model m1 includes one regression coefficient beside the intercept estimate, m2 includes two and m3 three coefficients. As can be seen, all models also include the coefficient of determination R-squared and adjusted R-squared. The regression models further include the predicted values, which **predict()** shows. That is, all data values of JOB enter the regression equation of m1 and so produce predicted values QOL_predict. These values lie on the regression line. We use **head()** to return only the top lines of the data frame.

Finally, we inspect some technical information about the regression models. The function **is()** shows us that m3 is an R object of class lm. Then, **methods()** shows all available functions for this class. Among these functions are **model.frame()** and **model.matrix()**. The first one shows the variables entering the regression model; the second one shows the design matrix including all model terms. The design matrix reveals that the interaction term of JOB and ADV comprises the products of the variables' data values.

7.1.4 SCB - Linear regression

EXAMPLE!

A secondary question in the student reconciliator project is whether the number of conflicts on schoolyards also depends on the amount of pocket money that students possess on average and on the condition of the school. Because these variables are quantitative predictors, a regression analysis may inform us about the associations.

SCB_MOD_LinearRegression.R

```
## SCB - Linear regression of pre-post-differences on covariables
source("./Script/SCB_PRE_DataImport.R")
source("./Script/SCB_PRE_VariableLabels.R")
source("./Script/SCB_PRE_DataAggregation.R")

M1 <- lm(PersonPerson_PREPOST ~ X1, SCB)
M2 <- lm(PersonGroup_PREPOST ~ X2, SCB)

summary(M1)
summary(M2)
```

Output

```
> summary(M1)

Call:
lm(formula = PersonPerson_PREPOST ~ X1, data = SCB)
```

```
Residuals:
    Min      1Q  Median      3Q     Max
-10.687  -2.734   0.194   2.313   9.409

Coefficients:
             Estimate Std. Error t value Pr(>|t|)
(Intercept) -0.02692    1.08864  -0.025    0.980
X1          -0.04774    0.21407  -0.223    0.824

Residual standard error: 3.85 on 178 degrees of freedom
Multiple R-squared:  0.0002793, Adjusted R-squared:  -0.005337
F-statistic: 0.04973 on 1 and 178 DF,  p-value: 0.8238

> summary(M2)

Call:
lm(formula = PersonGroup_PREPOST ~ X2, data = SCB)

Residuals:
    Min      1Q  Median      3Q     Max
-8.3867 -2.2879 -0.1892  2.7121  8.5146

Coefficients:
             Estimate Std. Error t value Pr(>|t|)
(Intercept)  0.97909    1.85015   0.529    0.597
X2          -0.04937    0.12128  -0.407    0.684

Residual standard error: 3.471 on 178 degrees of freedom
Multiple R-squared:  0.0009301, Adjusted R-squared:  -0.004683
F-statistic: 0.1657 on 1 and 178 DF,  p-value: 0.6844
```

Analysis

The analyses using the <u>lm()</u> function show no significant associations between the covariables and the pre-post-differences. Furthermore, R-squared shows very low proportions of explained variance.

7.2 Multiple linear regression

The idea of linear regression easily extends from one to multiple predictors. This enables comparisons between regression models using different sets of predictors.

7.2.1 Multiple regression

In multiple regression analysis we study the association between several predictors and one dependent variable. In our example, monthly income (INC), security of the job (JOB), adverse life events (ADV), or the quality of personal relationships (REL) may all impact on quality of life (QOL). These predictors may even interact; for example, the impact of adverse life events on quality of life may be different for persons who experience their personal relationships either as good or bad. In fact, regression analysis maintains the same concept of interaction as analysis of variance. We calculate a multiple regression by adding terms to the linear equation; usually, each predictor of interest obtains its own term in the model. So, the linear equation could look like this:

$$Y_i = b_0 + b_1 X_{i1} + b_2 X_{i2} + ... + b_j X_{ij} + e_i$$

In this equation the intercept term becomes b_0, and b_1 to b_j become the regression coefficients. The index i marks the subjects $i = 1$ to N. In our example the linear equation would be:

$$QOL_i = b_0 + b_1 INC_i + b_2 JOB_i + b_3 ADV_i + b_4 REL_i + e_i$$

Adding interaction terms as the product of predictors would be:

$$QOL_i = b_0 + b_1 INC_i + b_2 JOB_i + b_3 ADV_i + b_4 REL_i + b_5 ADV_i \times REL_i + e_i$$

This equation uses the interaction between ADV and REL and assigns the coefficient b_5 to it. In this way, we can extend the model in many different ways. The aim of multiple regression is still to estimate the b coefficients of the model, again, by minimising the sum of the squared differences between predicted and the actual values of Y. To calculate the coefficients in the multiple regression case you need differential calculus, but R gently takes care of this burden.

The linear regression equation is flexible enough to accept qualitative predictors, too. The conditions of an experimental intervention or other qualitative variable may be coded as so-called dummy predictors taking the value of 0 if a person does not and 1 if a person does belong to a certain condition. Regression analysis then estimates the regression coefficients of these predictors as the group effects, which we also study in analysis of variance.

7.2.2 Fitting a multiple regression in R   R LANGUAGE!

Let us calculate the multiple regression for health-related quality of life (QOL) and include all four predictors. In a second step we will add an interaction term as well.

Ch7_MultipleLinearRegression.R

```
## Multiple linear regression
source("Ch7_DataCreation.R")

# Descriptive statistics
apply(d, 2, mean)
apply(d, 2, sd)

print(cov(d), digits = 2)
print(cor(d), digits = 2)

# Multiple regression
```

```
m1 <- lm(QOL ~ INC + JOB + ADV + REL, data = d)
m2 <- update(m1, . ~ . + ADV:REL)

summary(m1)
summary(m2)
```

Output

```
> # Descriptive statistics
> apply(d, 2, mean)
    QOL     INC     JOB     ADV     REL
  20.30 1502.27    4.12    3.15    5.10
> apply(d, 2, sd)
       QOL        INC        JOB        ADV        REL
  2.217925 24.376614   1.165844   1.183856   1.159066
>
> print(cov(d), digits = 2)
       QOL    INC     JOB    ADV    REL
QOL   4.92   39.2   0.923  0.702   0.74
INC  39.22  594.2  12.200  3.404  10.81
JOB   0.92   12.2   1.359  0.093   0.22
ADV   0.70    3.4   0.093  1.402   0.29
REL   0.74   10.8   0.220  0.288   1.34
> print(cor(d), digits = 2)
      QOL  INC   JOB   ADV  REL
QOL  1.00 0.73 0.357 0.267 0.29
INC  0.73 1.00 0.429 0.118 0.38
JOB  0.36 0.43 1.000 0.067 0.16
ADV  0.27 0.12 0.067 1.000 0.21
REL  0.29 0.38 0.163 0.210 1.00
>
> # Multiple regression
> m1 <- lm(QOL ~ INC + JOB + ADV + REL, data = d)
> m2 <- update(m1, . ~ . + ADV:REL)
>
> summary(m1)

Call:
lm(formula = QOL ~ INC + JOB + ADV + REL, data = d)

Residuals:
    Min      1Q  Median      3Q     Max
-3.8372 -0.8691  0.0588  0.8844  3.7031

Coefficients:
              Estimate Std. Error t value Pr(>|t|)
(Intercept) -75.367822  10.509544  -7.171 1.61e-10 ***
INC           0.062837   0.007304   8.603 1.59e-13 ***
```

```
JOB              0.099042    0.142964    0.693   0.49014
ADV              0.351676    0.130164    2.702   0.00817 **
REL             -0.048397    0.142871   -0.339   0.73555
---
Signif. codes:  0 '***' 0.001 '**' 0.01 '*' 0.05 '.' 0.1 ' ' 1

Residual standard error: 1.498 on 95 degrees of freedom
Multiple R-squared:  0.5625,     Adjusted R-squared:  0.5441
F-statistic: 30.54 on 4 and 95 DF,   p-value: 2.442e-16

> summary(m2)

Call:
lm(formula = QOL ~ INC + JOB + ADV + REL + ADV:REL, data = d)

Residuals:
    Min      1Q  Median      3Q     Max
-3.8254 -0.9461  0.1722  0.8808  3.6512

Coefficients:
             Estimate Std. Error t value Pr(>|t|)
(Intercept) -79.195577  10.555120  -7.503 3.47e-11 ***
INC           0.062929   0.007203   8.736 8.92e-14 ***
JOB           0.095681   0.141003   0.679   0.4991
ADV           1.482641   0.603692   2.456   0.0159 *
REL           0.708692   0.419266   1.690   0.0943 .
ADV:REL      -0.227459   0.118637  -1.917   0.0582 .
---
Signif. codes:  0 '***' 0.001 '**' 0.01 '*' 0.05 '.' 0.1 ' ' 1

Residual standard error: 1.477 on 94 degrees of freedom
Multiple R-squared:  0.579,     Adjusted R-squared:  0.5566
F-statistic: 25.85 on 5 and 94 DF,   p-value: 2.494e-16
```

Analysis

The script first includes the data creation script from the beginning of this chapter. Afterwards it generates descriptive statistics – means, standard deviations, covariances, and correlations. Then $\underline{\text{lm}()}$ calculates the regression analysis. It relates the dependent variable QOL with the tilde operator $\sim$ to a linear combination of the predictors. In this case $\underline{+}$ operators go between the predictors. They indicate that only the effects of the predictors should be included without out interaction terms. Notice how closely the notation of the model follows the notation of the linear equation given above. Only the coefficients are missing in the R notation and they will be calculated by $\underline{\text{lm}()}$. The available operators in linear modelling are the same as those introduced in the analysis of variance in Chapter 6 and summarised in Table 6.1. They are $\underline{+}$ for the addition of model terms, $\underline{:}$ to specify interaction terms, and $\underline{*}$ to specify the simultaneous inclusion of main effects and interactions in the model. The function $\underline{\text{summary}()}$ applied with the linear model object m1 returns a table at the console which contains the coefficients of the model. The b coefficients of the model are summarised in the column `Estimate`, and the row `(Intercept)` contains the b_0 coefficient. With this information, we can write down the regression equation and replace the bs with the concrete numbers from the column `Estimate`. The table further shows the t-tests for all coefficients testing against the assumption of no association between predictor and dependent variable in the model. As can be seen, income (INC) and the experience of adverse life events (ADV) have a significant associations with health-related quality of life.

We use **`update()`** with a special formula notation to add the interaction between ADV and REL to the model. The first argument is the model to be updated, m1. The dot before ~ includes the original dependent variable in the update and the dot after ~ does the same for the predictors. Then we add the interaction term ADV:REL to the original model terms. The model m2, however, yields no significant association between the interaction term and QOL. Notice the changed coefficients and t-tests of ADV and REL if we include their interaction. We can see that inclusion or exclusion of a single predictor can lead to changes in all other coefficients and even cancel significant coefficients. The behaviour of the regression coefficients in different regression models with the same data depends on the correlation between the predictors. This problem is discussed in statistical textbooks about regression (Hays, 1994).

7.2.3 Testing regression parameters and the multiple R-squared

Each regression coefficient of a multiple regression may be tested for significance. We test the coefficients with the t-distribution and usually with a level of significance of $\alpha = 0.05$. In a regression equation including four predictors this yields four statistical tests. In the regression object m1 of the previous, example `summary(m1)` not only shows the regression coefficients but also the corresponding t-tests. As we have seen above, adding terms to the model, like the interaction term in m2, influences the coefficients in the model and the tests of significance. Besides interaction terms, polynomial terms may also be added to the model, like X^2 or X^3, so variable associations must not be linear. In fact, the number of possible model terms quickly grows with addition of predictor variables. Some algorithmic strategies for inclusion or exclusion of predictors have been proposed in the past (Draper & Smith, 1981). In contrast, researchers usually have clear ideas of which variables they want to include in a model. So, in regression analysis many authors recommend starting from theory and then deciding for a theoretically motivated set of predictors, interactions, and polynomials, and then calculating one regression equation including all terms of interest.

Another option in the situation with many predictors is to *shift focus from testing single coefficients in the regression equation to testing the model as a whole* and to statistically compare between different models. This means comparing R-squared between regressions. This is a good option if there are different sets of conceptually related predictors, for example, a group of socio-demographic variables and another group of subjective attitude scales. In a first step, one could include demographics in the design and see how much variance of the dependent variables this set of variables explains, that is, what the magnitude of R-squared is. In a second step, one adds subjective attitudes to the regression analysis and calculates R-squared again. A statistical comparison of the two R-squared values then shows whether inclusion of the additional variable yields a significant increase in explained variance. The multiple R-squared, and also differences between R-squared, values are tested in the F-distribution. R automatically supplies the R-squared statistic and the F-test with each model summary; however, the comparison between models needs another line of code.

Let us do this in our example to predict quality of life (QOL) with four predictors. Perhaps the two variables of monthly income (INC) and job security (JOB) may be conceived of as quasi-objective characteristics of a person, and the other two variables of adverse life events (ADV) and quality of personal relationships (REL) are more subjective. Our plan is to calculate the regression of QOL on the predictors INC and JOB first, and then to test whether the further inclusion of ADV and REL yield a significant increase in explained variance. We use the **anova()** function for this comparison.

Ch7_MultipleLinearRegression_ModelComparison.R

```
## Model comparisons in multiple linear regression
source("Ch7_MultipleLinearRegression.R")

m3 <- lm(QOL ~ INC + JOB, data = d)
summary(m3)
m4 <- update(m3, . ~ . + ADV + REL)
summary(m4)
anova(m3, m4)
```

Output

```
> m3 <- lm(QOL ~ INC + JOB, data = d)
> summary(m3)

Call:
lm(formula = QOL ~ INC + JOB, data = d)

Residuals:
    Min      1Q  Median      3Q     Max
-3.8444 -1.0465  0.1234  0.9321  4.0065

Coefficients:
             Estimate Std. Error t value Pr(>|t|)
(Intercept) -76.012671  10.303739  -7.377 5.51e-11 ***
INC           0.063820   0.007021   9.090 1.24e-14 ***
JOB           0.106431   0.146806   0.725     0.47
---
Signif. codes:  0 '***' 0.001 '**' 0.01 '*' 0.05 '.' 0.1 ' ' 1

Residual standard error: 1.538 on 97 degrees of freedom
Multiple R-squared:  0.5288,	Adjusted R-squared:  0.5191
F-statistic: 54.43 on 2 and 97 DF,  p-value: < 2.2e-16

> m4 <- update(m3, . ~ . + ADV + REL)
> summary(m4)

Call:
lm(formula = QOL ~ INC + JOB + ADV + REL, data = d)

Residuals:
    Min      1Q  Median      3Q     Max
```

```
-3.8372 -0.8691  0.0588  0.8844  3.7031

Coefficients:
              Estimate Std. Error t value Pr(>|t|)
(Intercept) -75.367822  10.509544  -7.171 1.61e-10 ***
INC           0.062837   0.007304   8.603 1.59e-13 ***
JOB           0.099042   0.142964   0.693  0.49014
ADV           0.351676   0.130164   2.702  0.00817 **
REL          -0.048397   0.142871  -0.339  0.73555
---
Signif. codes:  0 '***' 0.001 '**' 0.01 '*' 0.05 '.' 0.1 ' ' 1

Residual standard error: 1.498 on 95 degrees of freedom
Multiple R-squared:  0.5625,    Adjusted R-squared:  0.5441
F-statistic: 30.54 on 4 and 95 DF,  p-value: 2.442e-16

> anova(m3, m4)
Analysis of Variance Table

Model 1: QOL ~ INC + JOB
Model 2: QOL ~ INC + JOB + ADV + REL
  Res.Df    RSS Df Sum of Sq      F Pr(>F)
1     97 229.47
2     95 213.05  2     16.42 3.6609 0.0294 *
---
Signif. codes:  0 '***' 0.001 '**' 0.01 '*' 0.05 '.' 0.1 ' ' 1
```

Analysis

The two model summaries of m3 and m4 show the coefficients and tests for each model. The **anova()** function then compares between the models. The terms included are listed first, so we can check that the models differ only respecting the terms ADV and REL. The F-test for model comparison shows a p-value below $\alpha = 0.05$. So, we conclude that inclusion of the more subjective variables of ADV and REL in addition to INC and JOB in the model yields a significant increase in explained variance of QOL.

EXAMPLE! ## 7.2.4 SCB - Multiple linear regression

Earlier, we used the two covariables of renovation status of the school and average pocket money of the school students as covariables in simple linear regressions with only one predictor. Now, we enter both covariables in one regression model and obtain a multiple regression analysis with two predictors.

SCB_MOD_MultipleLinearRegression.R

```
## SCB - Multiple linear regression
if(!exists('MultipleLinearRegression'))
{
  source("./Script/SCB_PRE_DataImport.R")
```

```
source("./Script/SCB_PRE_VariableLabels.R")
source("./Script/SCB_PRE_DataAggregation.R")

MultipleLinearRegression <- list(M1 = lm(PersonPerson_PREPOST ~ X1 + X2, SCB),
                                 M2 = lm(PersonGroup_PREPOST ~ X1 * X2, SCB))
}
```

SCB_OUT_MultipleLinearRegression_ConsoleOutput.R

SCB – Multiple linear regression - Console output
```
source("./Script/SCB_MOD_MultipleLinearRegression.R")

summary(MultipleLinearRegression$M1)
summary(MultipleLinearRegression$M2)

head(model.frame(MultipleLinearRegression$M2))
head(model.matrix(MultipleLinearRegression$M2))
```

Output

```
> summary(MultipleLinearRegression$M1)

Call:
lm(formula = PersonPerson_PREPOST ~ X1 + X2, data = SCB)

Residuals:
    Min      1Q   Median      3Q      Max
-10.5847  -2.5685  -0.1326   2.5668   9.8998

Coefficients:
            Estimate Std. Error t value Pr(>|t|)
(Intercept)   4.2488     2.0578   2.065   0.0404 *
X1            0.1404     0.2248   0.624   0.5331
X2           -0.3442     0.1413  -2.436   0.0159 *
---
Signif. codes:  0 '***' 0.001 '**' 0.01 '*' 0.05 '.' 0.1 ' ' 1

Residual standard error: 3.797 on 177 degrees of freedom
Multiple R-squared:  0.0327,    Adjusted R-squared:  0.02177
F-statistic: 2.992 on 2 and 177 DF,  p-value: 0.05273

> summary(MultipleLinearRegression$M2)

Call:
lm(formula = PersonGroup_PREPOST ~ X1 * X2, data = SCB)

Residuals:
   Min     1Q Median     3Q    Max
-8.400 -2.564 -0.135  2.614  8.243

Coefficients:
            Estimate Std. Error t value Pr(>|t|)
(Intercept)  2.76450    7.19154   0.384    0.701
```

```
X1              -0.19414    1.40649  -0.138     0.890
X2              -0.24723    0.47829  -0.517     0.606
X1:X2            0.02871    0.09111   0.315     0.753

Residual standard error: 3.476 on 176 degrees of freedom
Multiple R-squared:  0.009424,  Adjusted R-squared:  -0.007461
F-statistic: 0.5581 on 3 and 176 DF,  p-value: 0.6434

>
> head(model.frame(MultipleLinearRegression$M2))
  PersonGroup_PREPOST X1 X2
1                   0  5 16
2                   4  5 17
3                   4  8 16
4                   0  5 18
5                  -1  5 17
6                   3  8 16
> head(model.matrix(MultipleLinearRegression$M2))
  (Intercept) X1 X2 X1:X2
1           1  5 16    80
2           1  5 17    85
3           1  8 16   128
4           1  5 18    90
5           1  5 17    85
6           1  8 16   128
```

Analysis

The regression model m1 shows a significant effect of X2, the amount of pocket money, on the pre-post-differences of number of person-against-person conflicts. So, perhaps pocket money could serve as a covariate in an analysis of covariance to give more statistical test power to the analysis of the intervention effect. Furthermore, m2 shows the regression of both predictors and their interaction regarding person-against-group conflicts. This regression shows no significant effect. The model matrix for m2 shows that the interaction term in this model consists of the products of the X1 and X2 value pairs. You may add the regression for the group-against-group conflicts to the script.

R LANGUAGE! ## 7.2.5 Packages specialised on regression analysis

Some packages help with regression analysis. Here are some of the popular ones: **nlme** (Pinheiro, Bates, DebRoy, Sarkar, & Team, 2021), **car** (Fox & Weisberg, 2019), **lme4** (Bates, Mächler, Bolker, & Walker, 2015), and **rms** (Harrell Jr, 2021b). Use library(help="PackageName") to obtain all the functions each one comprises. The package **nlme** provides nonlinear mixed effects modelling with many functions. Its main interface is the modelling functions **gls()**, **gnls()**, **lme()**, and **nlme()**. Knowing how to write formulas for regression analysis, you can easily adapt to the functions of this package. The functions follow the convention of generic functions to extract data from the model objects; that is, functions like **coef()**, `summary()`, `print()`, `residuals()`, and some more can be applied to the model objects. The package further contains graphic functions for the comprised models and examples of split-plot experiments. The `car` package enhances your work with regression models. Try out the functions `Anova()` or `vif()` with regression model objects to calculate the ANOVA and variance inflation factors, respectively. The package also consists of practical graphics functions for regression analysis. The package `lme4` specialises in mixed effects modelling. It consists of an elaborate vignette showing how to initialise formula for the analyses. Use `vignette("lmer")` at the console to open the vignette. The package `rms` further enhances regression analysis, especially regarding survival analysis. A code demonstration collects all its capabilities and can be retrieved with `demo(all, package="rms")`.

7.3 Further topics in regression

In this section, we study regression with data imputation in the face of missing values. We further show heuristics for how to proceed with the results for publication.

7.3.1 Regression analysis with imputed data   R LANGUAGE!

Real-life data often contains missing data. Often it is not problematic to simply accept this and analyse the available data. Other situations, however, require accounting for missing data and imputing data. The R package `mice` contains functions for data imputation and analysis (Van Buuren & Groothuis-Oudshoorn, 2011). You simply need to give `mice` a data frame with missing data and it returns a complete data set. That's it. The function `mice()` makes many smart guesses about the nature of the data and tries to employ as much information from the data frame as possible for the imputation. You can find the theoretical foundations and a step-by-step example of data imputation in the highly recommended paper by Van Buuren and Groothuis-Oudshoorn (2011). In the following, we run a minimal example of data imputation to get started. We use the data from the beginning of this chapter about health-related quality of life (QOL) and its covariates INC, JOB, ADV, and REL. The script consists of three parts: data deletion, data imputation, and analysis. The most important functions will be `mice()`, `with()`, and `pool()`.

Ch7_MultipleRegressionWithDataImputation.R

```
## Multiple regression with data imputation
library(mice)
source("Ch7_DataCreation.R")

# Delete data
set.seed(12345)
DeletionIndices <- sample(400, 50)
d <- as.matrix(d)
d[100 + DeletionIndices] <- NA
d <- as.data.frame(d)
view(d)
```

```
# Generate imputations
start_time <- Sys.time()
d_imp <- mice(d,
                seed = 12345,
                m = 20,
                maxit = 5)
end_time <- Sys.time()
print(end_time - start_time)

# Analysis
m_imp1 <- with(d_imp, lm(QOL ~ INC + JOB + ADV + REL))
m_imp2 <- with(d_imp, lm(QOL ~ INC + JOB + ADV * REL))
pool(m_imp1)
pool(m_imp2)
# If package texreg is available
texreg::screenreg(lapply(list(m_imp1, m_imp2), pool),
                  single.row = TRUE)
```

Output

```
> # Generate imputations
> start_time <- Sys.time()
> d_imp <- mice(d,
+                 seed = 12345,
+                 m = 20,
+                 maxit = 5)

 iter imp variable
  1   1  INC  JOB  ADV  REL
  1   2  INC  JOB  ADV  REL
  1   3  INC  JOB  ADV  REL
  [...]
  5  18  INC  JOB  ADV  REL
  5  19  INC  JOB  ADV  REL
  5  20  INC  JOB  ADV  REL
> end_time <- Sys.time()
> print(end_time - start_time)
Time difference of 3.95 secs
>
> # Analysis
> m_imp1 <- with(d_imp, lm(QOL ~ INC + JOB + ADV + REL))
> m_imp2 <- with(d_imp, lm(QOL ~ INC + JOB + ADV * REL))
> pool(m_imp1)
Class: mipo     m = 20
        term m estimate     ubar        b        t dfcom   df   riv lambda    fmi
1 (Intercept) 20 -77.7635 1.14e+02 1.10e+01 1.26e+02    95 81.5 0.1011 0.0919
0.1134
```

```
2              INC 20    0.0643 5.45e-05 5.40e-06 6.02e-05     95 81.1 0.1039 0.0941
0.1157
3              JOB 20    0.2086 1.84e-02 2.72e-03 2.12e-02     95 74.8 0.1555 0.1346
0.1568
4              ADV 20    0.3881 1.69e-02 1.27e-03 1.83e-02     95 84.2 0.0786 0.0729
0.0941
5              REL 20   -0.1055 2.05e-02 1.43e-03 2.20e-02     95 84.9 0.0735 0.0684
0.0896
> pool(m_imp2)
Class: mipo    m = 20
          term m estimate      ubar        b        t dfcom   df    riv lambda    fmi
1 (Intercept) 20 -81.0433 1.15e+02 1.06e+01 1.26e+02     94 81.2 0.0967 0.0882
0.110
2          INC 20   0.0643 5.33e-05 4.98e-06 5.85e-05     94 81.0 0.0980 0.0892 0.111
3          JOB 20   0.2202 1.80e-02 2.67e-03 2.08e-02     94 74.0 0.1563 0.1351 0.158
4          ADV 20   1.4441 3.57e-01 5.44e-02 4.14e-01     94 73.5 0.1601 0.1380 0.161
5          REL 20   0.5670 1.58e-01 3.18e-02 1.91e-01     94 67.7 0.2122 0.1750 0.198
6      ADV:REL 20  -0.2095 1.34e-02 2.24e-03 1.58e-02     94 71.7 0.1753 0.1492 0.172
> # If package texreg is available
> texreg::screenreg(lapply(list(m_imp1, m_imp2), pool),
+                   single.row = TRUE)

=======================================================
                   Model 1               Model 2
-------------------------------------------------------
(Intercept)   -77.76 (11.22) ***   -81.04 (11.23) ***
INC             0.06  (0.01) ***     0.06  (0.01) ***
JOB             0.21  (0.15)         0.22  (0.14)
ADV             0.39  (0.14) **      1.44  (0.64) *
REL            -0.11  (0.15)         0.57  (0.44)
ADV:REL                             -0.21  (0.13)
-------------------------------------------------------
nimp            20                    20
nobs           100                   100
R^2             0.57                  0.58
Adj. R^2        0.55                  0.56
=======================================================
*** p < 0.001; ** p < 0.01; * p < 0.05
```

Analysis

The script starts with data creation and with the deletion of random data points. We delete no data from the dependent variable QOL but from the four covariates. To replace the data with missing values NA, I found it easier to convert the data frame to a matrix, and then use matrix indices for the replacement. Afterwards, I converted it back into a data frame.

The data imputation with `mice()` comes next. It obtains our data frame with missing values, a seed to make the imputation replicable, a parameter m holding the number of complete data frames, and the number of iterations `maxit` for the estimation of each missing data value. The idea to impute multiple data frames and the idea to impute each data value several times are explained in detail in Van Buuren and Groothuis-Oudshoorn (2011). Important for us now is that `mice()` creates the object d_imp holding twenty data frames; it is an object of class `mids` standing for multiply imputed data set. The function `complete()` provides access to any one of the twenty data frames. Try `complete(d_imp, 1)`, for example. Before going to the analysis, notice that imputation with `mice()` usually takes some time. I enclosed the imputation with two calls of **`Sys.time()`** and then printed the duration of the imputation. Changing the parameters of **`mice()`** can lead to imputations taking seconds or several hours.

To analyse all twenty data frames simultaneously, we use **`with()`** with the `mids` object and then write the modelling function, in this case **`lm()`**. This creates m_imp1, which is a `mira` object, standing for multiply imputed repeated analysis. It consists of the twenty analyses. We also write a second model m_imp2 using one interaction term. Finally, **`pool()`** bundles and prints the results as a `mipo` object, standing for multiple imputed pooled outcomes. The pooling of analyses follows the rules by Rubin (1987). The pooling aggregates the twenty coefficients for each predictor to a single number, making it easy to report the analysis. Furthermore, **`screenreg()`** from the **`texreg`** package is able to combine the results of the two models in a nice table.

We practised data imputation with **`mice`**, but other packages can help with missing data, too. Try **`impute()`** from the **`Hmisc`** package or study the vignette of the **`mi`** package with `vignette("mi_vignette")`.

TEAMWORK! ### 7.3.2 Strategies to select the appropriate regression to publish

Regression analysis provides many options for data analysis. It requires heuristics or strategies to yield the results for your publication. In the following, we do not focus on R functions or the statistical foundations of regression, but rather on strategies you and your team could follow to finish the results section of your publication.

- Identify examples of regression analyses and results tables in your target journal and discuss them with your team. Annotate the printouts for interesting features you want to include yourself. For example, check whether the journal prefers statistical tests or confidence intervals for the regression coefficients? How elaborate should you describe issues relating to the correlations between predictors? Also refer to the statistical guidelines many journals offer for required information in the results.
- Identify statistical textbooks about regression analysis, which offer an account of the statistical formulas that are used in the procedure. My favourites are Draper and Smith (1981), Hays (1994), and Harrell (2015). Sometimes, it helps to cite sources like this to justify a particular manoeuvre that you apply to the data.
- Locate functions in R packages which could help you in the analysis. As shown above, several packages enhance regression analysis. Study their main functions and vignettes; then, ask your team whether someone has experience with the packages. Find out whether the results output fits with functions for beautiful tabular output. For example, `lm` objects are easily

turned into publication-ready tables with **stargazer()** from the stargazer package. This is much easier than copy-pasting the results from the R console to the word processor. We will focus on this issue in Chapter 8.

- Produce models in several of the functions you discovered and compare the numerical results. If you applied the correct functions and correctly specified the parameters, the results should be almost identical. This step greatly enhances your confidence with the calculations, especially if a statistical method is new to you. Furthermore, you can spot undesired parameter settings in R functions and correct them.
- Finalise a first version of the results tables and add prose for the results section in short time. Then ask your team or supervisor for feedback. Revise the analysis and prepare the next version of the results and tables. If necessary, ask for another feedback round.

The foregoing steps help you to finish your manuscript. However, regression can still produce nonsense if it does not relate to the research question or if the data is deficient. Regression analysis is simply a statistical method that almost always leads to a regression equation on the basis of data, regardless of the presence of meaning or causality. The many options of multiple regression seduce us to consider more and more complex models. Therefore, perhaps an admonition from Jacob Cohen (1990) is appropriate: "Less is more."

7.4 Logistic regression

In logistic regression the dependent variable has only two values, usually 0 and 1. This may be the presence or absence of a certain characteristic or diagnosis. To analyse such binary response variables we use logistic regression. This section introduces the model equation of logistic regression and its calculation in R.

7.4.1 The regression equation in logistic regression

STATISTICAL ISSUE!

A linear regression of a binary response variable (i. e., a variable with only two possible values) could easily be calculated using the standard regression equation. The regression calculations would simply yield the regression coefficients. However, it is difficult to interpret the regression line, because the regression would not predict the two possible data values. Much better is to find an appropriate transformation of the dependent variable to make it continuous and then apply linear regression. Agresti (2002, 2007) comprehensively introduces the mathematical derivation of logistic regression and we will go through the basic assumptions quickly. Hosmer and Lemeshow (2000) provide another comprehensive and readable account of logistic regression.

In logistic regression, our aim is to model the probability of a binary response $Y = 1$ given a certain value x of the predictor variable X:$\pi(x) = P\,(Y = 1|X = x)$. We relate this probability to the predictor variable with the so-called logit transformation of $\pi(x)$.

$$\text{logit}\left[\pi(x)\right] = \log\left(\frac{\pi(x)}{1-\pi(x)}\right) = \alpha + \beta x$$

The formula shows that logit $[\pi(x)]$ is a linear function of X. Solving the equation for $\pi(x)$ shows its relation to x.

$$\pi(x) = \frac{\exp(\alpha + \beta x)}{1 + \exp(\alpha + \beta x)}$$

The aim in logistic regression is to estimate the coefficients α and β. In logistic regression, we use the maximum likelihood (ML) method for estimation of the coefficients. The method has the following aim: it finds the regression coefficients that maximise the probability of the obtained data. This is unlike the least-squares principle, which strives to find parameter estimates that minimise error between observed data and predicted values. We will not study the details of maximum likelihood here; the method is described in detail in Timm (1975) and Casella and Berger (2002). Let us calculate an example of a logistic regression.

R LANGUAGE! ## 7.4.2 Fitting a logistic regression in R

The example for this chapter, testing the association between health-related quality of life (QOL) and four covariates, may serve to illustrate logistic regression. Consider a cut-off value of 21 in QOL would usually be associated with happiness in life. We can recode the variable in the values 0 and 1, whereby 0 means at most a value of 20 in QOL, and 1 means 21 or higher in QOL. In this situation, logistic regression may reveal predictors for this binary response variable. We will use the function **stats::glm()** to generate the regression model.

Ch7_LogisticRegression.R

```
## Logistic regression
library(memisc)
source("Ch7_DataCreation.R")

# Data preparation
table(d$QOL)
d$QOL_cat <- recode(d$QOL,
                    0 <- range(min, 20),
                    1 <- range(21, max))

# Descriptive statistics
with(d, table(QOL_cat, JOB))
with(d, tapply(QOL_cat, JOB, mean))
with(d,
     {
         plot(jitter(JOB), QOL_cat, pch = 16)
         points(tapply(QOL_cat, JOB, mean), pch = 18, cex = 2)
     })

# Logistic regression model
m <- glm(QOL_cat ~ JOB,
         data = d,
         family = binomial)
```

```
summary(m)

# Visualise predictions
curve(exp(coef(m)[1] + coef(m)[2] * x) / (1 + exp(coef(m)[1] + coef(m)[2] * x)),
      from = 1, to = 7, add = TRUE, lwd = 2)
points(d$JOB, predict(m, type = "response"),
       lwd = 3, lty = "dotted")
```

Output

```
> # Data preparation
> table(d$QOL)

16 17 18 19 20 21 22 23 24 25
 3  8 13 15 13 16 17  6  6  3
> d$QOL_cat <- recode(d$QOL,
+                     0 <- range(min, 20),
+                     1 <- range(21, max))
>
> # Descriptive statistics
> with(d, table(QOL_cat, JOB))
       JOB
QOL_cat  1  2  3  4  5  6  7
      0  1  7 12 22  6  4  0
      1  0  2  3 17 20  4  2
> with(d, tapply(QOL_cat, JOB, mean))
    1     2     3     4     5     6     7
0.000 0.222 0.200 0.436 0.769 0.500 1.000
> with(d,
+     {
+        plot(jitter(JOB), QOL_cat, pch = 16)
+        points(tapply(QOL_cat, JOB, mean), pch = 18, cex = 2)
+     })
>
> # Logistic regression model
> m <- glm(QOL_cat ~ JOB,
+          data = d,
+          family = binomial)
> summary(m)

Call:
glm(formula = QOL_cat ~ JOB, family = binomial, data = d)

Deviance Residuals:
   Min      1Q  Median      3Q     Max
-1.752  -1.099  -0.588   0.953   1.918

Coefficients:
            Estimate Std. Error z value Pr(>|z|)
```

```
(Intercept)    -3.146      0.931    -3.38  0.00073 ***
JOB             0.740      0.217     3.41  0.00066 ***
---
Signif. codes:  0 '***' 0.001 '**' 0.01 '*' 0.05 '.' 0.1 ' ' 1

(Dispersion parameter for binomial family taken to be 1)

    Null deviance: 138.47  on 99  degrees of freedom
Residual deviance: 123.97  on 98  degrees of freedom
AIC: 128

Number of Fisher Scoring iterations: 4
```

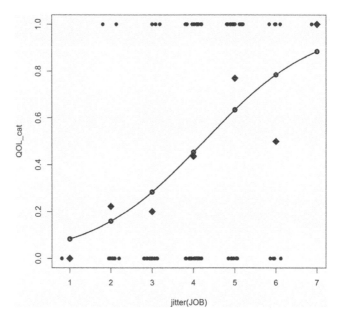

Figure 7.3 Logistic regression example of the categorised quality of life variable as a function of job security.

Analysis

The script calls the script for data generation first and loads the **memisc** package containing the **recode()** function. Then, **table()** provides the frequencies of the QOL categories, which shows that a cut off at value 21 would split the sample almost in half. Then, **recode()** does the recoding of QOL into the binary variable QOL_cat (i. e., categorised). The frequency table suggests that values of 1 in QOL_cat are associated with higher values of JOB than values of 0. Furthermore, the means of QOL_cat for the levels of JOB show an almost ascending trend. Figure 7.3 illustrates the distribution of QOL_cat as jittered points at 0 and 1 on the y-axis. The figure also shows the means of QOL_cat for the levels of JOB as diamonds. Our aim is to retrace the trend of the means with the S-shaped regression line characteristic for logistic regression. Therefore, **glm()** produces the logistic regression model and stores the result in the object m. Then, `summary(m)` yields the coefficients of the model, necessary to calculate expected frequencies and to generate the S-shaped curve. Finally, **`curve()`** adds the S-shaped regression line to Figure 7.3 by using the model coefficients with the model equation to calculate the expected mean frequencies. With **`points()`**, we mark the predicted frequencies with thick circles on the regression line.

The idea of logistic regression extends to ordinal regression for response variables with more than two ordered levels or categories. Ordinal regression is, in fact, a collection of repeated logistic regressions of adjacent levels of the response variable. The function **`clm()`** of the package **ordinal** (Christensen, 2019) calculate such regressions.

7.4.3 ITC – Reanalysis with a logistic regression   EXAMPLE!

In the Wason card selection task example, relating to the testing of a rule by turning over some cards, the proposed association is that between cognitive ability and correct responses. The analysis calculated by the authors was a *t*-test with correct or non-correct as the independent variable and cognitive ability as the dependent variable. There was no experimental manipulation and the two variables were measured; so, a causal influence of one variable on the other cannot be proven in the design, and we should at most talk about an association between the variables. However, in theory we guess that the authors actually mean that cognitive ability influences how a cognitive task is solved. This makes cognitive ability the designated independent variable and the statistical analysis should reflect this. So, the conducted *t*-test with task performance as the independent variable may prove the association between variables, but it reverses the thinking about the experiment. We suggest an alternative that treats SAT as a quantitative independent variable and task performance as a dichotomous dependent variable: it is the logistic regression. Logistic regression is appropriate for the analysis of dichotomous outcomes, and yields a function for prediction of probabilities of the two outcome values. The following shows two regression analyses, one for the solutions to the card selection problem 1, the other for the dichotomous variable of correct responses in all three card selection tasks. The script employs the **`glm()`** function using the logit link for the association between these dependent variables and independent variable SAT score.

ITC_MOD_LogisticRegression.R

```
## ITC - Logistic regression of correct responses
if(!exists('LogisticRegressionModels'))
{
  source("./Scripts/ITC_PRE_LoadData.R")

  LogisticRegressionModels <- list(m1 = glm(p1_pnotq_correct ~ sat,
                                 data = ITC$stvu,
                                 family = binomial(link="logit")),
                            m2 = glm(all_problems_correct ~ sat,
                                 data = ITC$stvu,
                                 family = binomial(link="logit")))
```

Output

```
> lapply(LogisticRegressionModels, summary)
$m1

Call:
glm(formula = p1_pnotq_correct ~ sat, family = binomial(link = "logit"),
    data = ITC$stvu)

Deviance Residuals:
    Min      1Q   Median      3Q      Max
-1.2426  -1.0151  -0.9405   1.3293   1.5341

Coefficients:
             Estimate Std. Error z value Pr(>|z|)
(Intercept) -1.6946213  1.2891444  -1.315    0.189
sat          0.0007387  0.0007429   0.994    0.320

(Dispersion parameter for binomial family taken to be 1)

    Null deviance: 237.58  on 176   degrees of freedom
Residual deviance: 236.58  on 175   degrees of freedom
AIC: 240.58

Number of Fisher Scoring iterations: 4


$m2

Call:
glm(formula = all_problems_correct ~ sat, family = binomial(link = "logit"),
    data = ITC$stvu)

Deviance Residuals:
    Min      1Q   Median      3Q      Max
-1.4003  -1.0879  -0.9694   1.2534   1.4423

Coefficients:
             Estimate Std. Error z value Pr(>|z|)
(Intercept) -1.8112134  1.2743377  -1.421    0.155
sat          0.0009286  0.0007354   1.263    0.207

(Dispersion parameter for binomial family taken to be 1)

    Null deviance: 243.33  on 176   degrees of freedom
Residual deviance: 241.71  on 175   degrees of freedom
AIC: 245.71

Number of Fisher Scoring iterations: 4
```

Analysis

The logistic regression model m1 relates to the association between solutions to the first card selection problem; the model m2 relates to whether or not all three problems were solved correctly. The list LogisticRegressionModels collects the two models and is the output of the script. At the console, then, lapply(LogisticRegressionModels, summary) yields the regression summaries for the two analyses. The coefficient estimate for the SAT scores does not reach a *p*-value (Pr in this case) lower than 0.05, leading us to the conclusion that we cannot assume cognitive ability as measured with the SAT score to be a predictor of performance in the card selection task.

━━━━━━━━━━━━ **What you have learned so far** ━━━━━━━━━━━━

- To calculate simple and multiple linear regressions
- To assess explained variance and strength of association
- To write the model formula in R associated with linear regression
- To calculate a regression analysis with imputed data
- To calculate a logistic regression

━━━━━━━━━━━━ **Exercises** ━━━━━━━━━━━━

Explore functions

1   Practise the functions associated with linear regression models with the data set **MusicData.csv**. Create a linear regression to predict the scores of one musical example from the scores of other musical examples. Then use methods(class = "lm") to obtain all functions related to lm objects. Use these functions with the lm object and figure out their meaning.

2   Use the linear regression model of the foregoing task. Call **plot()** with the model to obtain some diagnostic diagrams for the regression. Figure out how these diagrams may help you assess the quality of the regression analysis.

Solve tasks

1   Read the data set **MusicData.csv** into R. The data set contains preference ratings for 20 musical examples in a group of adolescents. Calculate covariance and correlation matrices for the 20 preference ratings. Select the appropriate variables with **grep()** or **paste()**. Limit the display to two decimal places. Calculate a display for the correlations between Ex01, Ex02, Ex03, and Ex20 with **psych::pairs.panels()**.

2   With **MusicData.csv**, calculate three linear regressions with the function **lm()**. Stepwise include Ex01, Ex02, and Ex03 as predictors for the criterion Ex20. What can you observe regarding the explained variance of Ex20?

3   Calculate the multiple R-squared of the first regression by division of two variances. Use the variance of the fitted values included in the regression model and the criterion variance.

(Continued)

4   The regression results are disappointing; we can only explain R-squared = .09 of the variance of Ex20. Include two more predictors to improve the regression, perhaps `Ex06` and `Ex09` after a close inspection of the correlation matrix.

Read code

1   With the `MusicPreferencesSchool.csv` data, the following code uses contrasts between treatments in a linear regression analysis to predict song preferences. Note how we manipulate the treatment contrasts and how we reset them to their default after the calculations. Describe all steps of the analysis in your own words and check the contrasts with `contrasts(d$School)`.

```
## Regression with user-defined contrasts

# Preparation
d <- read.csv("MusicPreferencesSchool.csv")

# Regression model
f <- formula("PreferenceSong1 ~ Creativity + School")
m1 <- lm(formula = f, data = d)
contrasts(d$School,
          how.many = 2) <- cbind(JHSvsSMS = c(-1, 1, 0),
                                 GSvsSMS = c(-1, 0, 1))
m2 <- lm(formula = f, data = d)
contrasts(d$School) <- NULL

# Model summaries
summary(m1)
summary(m2)
```

2   With the `MusicData.csv` data set, we practise how to efficiently aggregate a predictor set with several predictors in a linear regression. The functions **paste()** and **formula()** help us here. Describe in your own words for what particular purpose we need each call of **paste()**.

```
## Regression with multiple predictors

# Preparation
d <- read.csv("MusicData.csv")

# Calculate statistics
Predictors <- paste0("Ex", 1:5)

# Regression analysis
f <- formula(paste("Ex20 ~", paste(Predictors, collapse = " + ")))
m1 <- lm(formula = f, data = d)
m2 <- update(m1, . ~ . + Ex1:Gender)

# Results
summary(m1)
summary(m2)
```

Analyse data

1   Use the `MusiciansComplaints.csv` data set. Calculate a logistic regression to test whether back pain can be predicted from type of instrument. Notice that the data needs rearrangement after import and that the complaint variable needs to be dichotomised with a function related to factors.

Apply in the real world

1   Analyse your own data with a regression analysis. If your data set consists of quantitative independent and dependent variables, use `lm()` to calculate simple or multiple linear regressions. Try different predictor sets and use `update()` to efficiently structure the analyses.

WBD   EXAMPLE!

1   Replicate the primary analysis of the study with a multiple regression for the three outcomes depression, anhedonia, and prospective imagery test. Use condition and measurement as predictors.
2   Update the foregoing multiple regression and add other person characteristics like gender, employment status, or years of education to the model.

ITC   EXAMPLE!

1   Use the count variable of correct responses we created in the exercises in Chapter 3. Calculate a linear regression for this variable and use the SAT score as an independent variable. Is the task performance affected by the SAT score? Use `lm()` to calculate the regression.
2   Refine the linear regression of the foregoing task by adding gender as another independent variable. Also add the interaction between SAT and gender to the model. Inspect the results and conclude whether SAT affects task performance differently for females and males. Use `update()` to refine the model.

UPS   EXAMPLE!

1   Recreate the analyses of variance from Chapter 6 as linear regressions using dummy predictors. Furthermore, modify the contrast attributes of the factors in the design to test trend hypotheses.

SCB   EXAMPLE!

1   Perhaps the number of conflicts in a school simply depends on how well school students feel in their school and how well they can live with their allocated pocket money. Scrutinise this hunch with multiple regression analyses. Create a model which predicts the number of conflicts before the interventions by renovation status of the school (X1) and the average allocation of pocket money (X2). Use the function `lm()` and calculate the three regression equations for the three types of conflict.

8

PREPARING TABLES FOR PUBLICATION

━━━━━━━━━━━━━━━━━━━━━━━━ Chapter overview ━━━━━━━

This chapter shows how to produce tables for use in research publications. You will learn about R's facilities to export tables to be used in a word processor. So, you can finalise the display of your research results with R, which further enhances productivity. Programming features: file output of results, export of formatted tables, report generation.

8.1 Results tables for publication

The R console only shows ugly numbers as output. But what you publish are good-looking and well formatted tables. This section shows how to use R packages to create publication-ready tables.

PLANNING ISSUE!

8.1.1 Common situations which require a statistical results table

Here are some common situations which require a well formatted results table.

Table in a journal article. In this case you want to program the fine-tuned layout, which follows the journal guidelines. You should be able to copy the final display directly into your manuscript without any further changes; or even better, the results table should somehow automatically appear in your manuscript, and should also be automatically updated if the data change.

Table in a conference presentation. In this case, you also want to program a fine-tuned end product. However, you usually do not need to follow journal guidelines, but follow your own preferences for layout and formatting. It would, nevertheless, be nice for a research team to have a common style across all presentations of the team.

Quick and dirty table for your team meeting or for a colloquium. In this case, speed is of foremost importance. Sometimes your professor asks you to present results the next day. In this case, most researchers fall back to the copy-paste strategy just to fulfil the scheduled presentation. How nice it would be to let the results display still look clean and consistent.

To copy-paste the results from the console is truly quick and dirty. Most complex objects of R results already have a certain formatting for the display at the console. For example, simply call `summary()` with an aov object and obtain a table which looks almost like the display of ANOVA results you know from scientific journals. You could simply select the output at the console, copy it in the word processor, and do all the formatting there. However, the results are not copied as a table, but as mere text. Sure, it is easy to format the table in the word processor, but you have to repeat the steps for each table you create. In the following, we will look for better options to produce well formatted tables.

R FUNCTION!

8.1.2 capture.output()

R contains some built-in functions to export results to text files, and you have some control about what is exported. We use the functions `capture.output()`, `print()`, and `cat()` here. The first script in this chapter creates simulation data for all examples that follow and is stored in the file Ch8_DataCreation.R. The next script then shows how to simply write results in a plain text file.

Ch8_DataCreation.R

```
## Data creation
if(!exists("d"))
{
  set.seed(12345)
  d <- data.frame(id = 1:120)
```

```
with(d,
    {
        # Define true effects
        N <- 120
        Mu <- rep(5, N)
        Alpha <- rep(c(-1,1), each = N/2)
        Beta <- rep(c(-.5, .3, .2), times = 2, each = N/6)
        AlphaBeta <- rep(c(0, 0, -.1, .6, -.5, 0), each = N/6)
        Epsilon <- rnorm(N, 0, 5)

        # Linear equation
        Y <- Mu + Alpha + Beta + AlphaBeta + Epsilon
        A <- factor(rep(c("a1", "a2"), each = N/2))
        B <- factor(rep(c("b1", "b2", "b3"), times = 2, each = N/6))

        d$A <<- A
        d$B <<- B
        d$Y <<- Y
        d$C1 <<- sample(c("C1_1","C1_2"), size = 120, replace = TRUE)
        d$C2 <<- sample(c("C2_1",
                          "C2_2",
                          "C2_3",
                          "C2_4"), size = 120, replace = TRUE)
        d$X1 <<- rnorm(N, rep(c(5, 5, 4.5, 5, 4.5, 5.5), each = N/6), 2)
        d$X2 <<- rnorm(N, rep(c(11, 12, 13, 12, 12, 12), each = N/6), 3)
    })
}
```

Analysis

The script supplies data frame d. It consists of factor variables A, B, C1, and C2, of a dependent variable Y, and of two covariates X1 and X2. Variable Y is linked to A and B via a linear function, which also consists of a random component Epsilon sampled with **rnorm()**. The other variables are just added for demonstration purposes of this chapter and are all mutually independent.

Ch8_BasicOutputRObjectsTextFile.R

```
## Basic output of R objects in a text file.
source("./Ch8_DataCreation.R")
library(psych)

# Objects containing results
Descr <- describeBy(d$Y, list(d$A, d$B))
TableAB <- addmargins(table(d$A, d$B))
a <- aov(Y ~ A * B, data = d)
```

```
# Results output
capture.output(
  {
    cat("Descriptive statistics and ANOVA results\n",
        "=======================================\n\n")
    cat("\n\nStructure of the data frame\n\n")
    str(d)
    cat("\n\nDescriptive statistics\n\n")
    print(Descr)
    cat("\n\nConfirm balanced design\n\n")
    print(TableAB)
    cat("\n\nANOVA summary\n\n")
    summary(m)
  }, file = "Descriptives.txt"
)
```

Output

```
Descriptive statistics and ANOVA results
========================================


Structure of the data frame

'data.frame':   120 obs. of  8 variables:
 $ id: int  1 2 3 4 5 6 7 8 9 10 ...
 $ A : Factor w/ 2 levels "a1","a2": 1 1 1 1 1 1 1 1 1 1 ...
 $ B : Factor w/ 3 levels "b1","b2","b3": 1 1 1 1 1 1 1 1 1 1 ...
 $ Y : num  6.43 7.05 2.95 1.23 6.53 ...
 $ C1: chr  "C1_2" "C1_2" "C1_2" "C1_1" ...
 $ C2: chr  "C2_3" "C2_4" "C2_3" "C2_4" ...
 $ X1: num  3.6 3.87 4.48 2.87 4.79 ...
 $ X2: num  9.87 5.56 11.87 10.43 11.05 ...


Descriptive statistics

 Descriptive statistics by group
 : a1
 : b1
    vars  n mean   sd median trimmed  mad   min   max range  skew kurtosis   se
X1     1 20 3.88 4.17   3.97    3.92 3.88 -5.59 12.59 18.18 -0.17    -0.27 0.93
----------------------------------------------------------------------------
 : a2
 : b1
```

Figure 8.1 Part of a results file written with `capture.output()`.

Analysis

The script sources the data creation script and thus uses data frame d. It then loads package `psych` to have `describeBy()` available. Then, the script creates three objects containing results: descriptive statistics by group, a cross tab to check whether the experimental design is balanced, and an ANOVA object for testing the associations between A, B, and Y. The script finally calls `capture.output()` to write the results in a txt-file. The first argument of this function is a code block surrounded by braces { and }. Inside the braces, results annotation and results objects are placed one after the other. Annotation is practical with **cat()**, which simply prints strings. Notice the comma after the closing brace of **capture. output()**; it means that the function call is not complete. In fact, we have to supply a file name for the results. Opening the results file in a text editor looks like Figure 8.1. It is simply the console output with some added headings. I use **capture.output()** very often when I only need the numbers, for example, for the results section of a conference abstract that I have to submit the next day.

8.1.3 Formatted tables using the xtable package

R LANGUAGE!

If you want to present results to your colleagues, you would rather not give them a plain text printout and let them search for the important bits. Even for a team meeting you should prefer to distribute adequate results tables. The tables need not be beautiful, but they must be readable. The **xtable** package (Dahl, Scott, Roosen, Magnusson, & Swinton, 2019) with its main function **xtable()** exports tables to LaTeX or HTML (http://cran.r-project.org/web/packages/xtable/index.html). It produces HTML tables that you can copy-paste to your results presentation. The default output format, however, is LaTeX, but the type parameter can be changed to HTML. A table output requires two function calls in sequence: (1) the object has to be passed to **xtable()** and become an xtable object; then (2) this object has to be passed to **xtable**'s **print()** method. With the second function call there are many parameters available to change the appearance of the output. Before using the functions, we inspect with **methods()** for which types of objects **xtable()** applies.

Ch8_xtableObjects.R

```
## Objects for xtable()
library(xtable)
methods(xtable)
 [1] xtable.anova*           xtable.aov*
 [3] xtable.aovlist*         xtable.coxph*
 [5] xtable.data.frame*      xtable.glm*
 [7] xtable.lm*              xtable.matrix*
 [9] xtable.prcomp*          xtable.summary.aov*
[11] xtable.summary.aovlist* xtable.summary.glm*
[13] xtable.summary.lm*      xtable.summary.prcomp*
[15] xtable.table*           xtable.ts*
[17] xtable.zoo*
see '?methods' for accessing help and source code
```

Not all types of objects are listed here among the methods of **xtable()**. Such types need conversion into one of the listed types. For example, a numeric vector could be converted with **as.matrix()** and then be translated as an xtable. Let us test this at the console and convert a simple numeric vector into an xtable and return it in LaTeX and HTML code.

Ch8_xtableNumericVector.R

```
## Object transformation for xtable()
x <- rnorm(10)
x_m <- as.matrix(x)
x_m_xt <- xtable(x_m)
print(x_m_xt)
print(x_m_xt, type = "html")
```

Output

```
% latex table generated in R 4.1.1 by xtable 1.8-4 package
% Sat Nov 06 17:47:19 2021
\begin{table}[ht]
\centering
\begin{tabular}{rr}
  \hline
 & x \\
  \hline
1 & -0.45 \\
  2 & -1.42 \\
  3 & 0.39 \\
  4 & -1.18 \\
  5 & -0.49 \\
  6 & -0.86 \\
  7 & 1.45 \\
  8 & -0.61 \\
  9 & -1.70 \\
  10 & -0.59 \\
   \hline
\end{tabular}
\end{table}
> print(x_m_xt, type = "html")
<!-- html table generated in R 4.1.1 by xtable 1.8-4 package -->
<!-- Sat Nov 06 17:47:19 2021 -->
<table border=1>
<tr> <th>  </th> <th> x </th>  </tr>
  <tr> <td align="right"> 1 </td> <td align="right"> -0.45 </td> </tr>
  <tr> <td align="right"> 2 </td> <td align="right"> -1.42 </td> </tr>
  <tr> <td align="right"> 3 </td> <td align="right"> 0.39 </td> </tr>
  <tr> <td align="right"> 4 </td> <td align="right"> -1.18 </td> </tr>
  <tr> <td align="right"> 5 </td> <td align="right"> -0.49 </td> </tr>
  <tr> <td align="right"> 6 </td> <td align="right"> -0.86 </td> </tr>
  <tr> <td align="right"> 7 </td> <td align="right"> 1.45 </td> </tr>
  <tr> <td align="right"> 8 </td> <td align="right"> -0.61 </td> </tr>
  <tr> <td align="right"> 9 </td> <td align="right"> -1.70 </td> </tr>
  <tr> <td align="right"> 10 </td> <td align="right"> -0.59 </td> </tr>
   </table>
```

Analysis

As can be seen in the code, the ten random numbers are converted into a matrix first and are then converted into an xtable. We then use **print()** to return the table at the console, first as LaTeX and second as HTML. The function **xtable()** returns LaTeX code by default (or HTML if desired) which you can paste into your LaTeX or HTML document. In the compiled LaTeX document or in the web browser (in the case of HTML) the result would look approximately as shown in Table 8.1. To compare your result, copy the HTML output from the console into a plain text file and save it with the file extension .html. Then open this file in a web browser. Alternatively, run the line print(x_m_xt, type = "html") again and add the parameter file together with a file name to the function call to print the HTML code directly into an HTML file.

Table 8.1 Numeric vector shown as an xtable.

	x
1	-1.04
2	-0.21
3	1.78
4	1.46
5	-0.36
6	-0.61
7	0.50
8	0.25
9	-0.13
10	-1.25

The type argument of xtables **print()** (actually **print.xtable()**) determines what kind of table to produce, either "latex", or "html". The function has many more parameters to fine-tune the table's layout, for example, whether to place the caption above or below the table. Let us now use the data frame d, which we created above, and produce some results tables with the variables of d. We write all tables in separate HTML files and play with the function parameters a little bit. A detailed demonstration of **xtable**'s capabilities is given by Jonathan Swinton (2019) in his xtable Gallery.

Ch8_xTables.R

```
## Some xtables
source("Ch8_DataCreation.R")

descr <- with(d,
              {
                  n <- table(B)
                  m <- tapply(Y, B, mean)
                  print(m)
                  sd <- tapply(Y, B, sd)
                  descr <- cbind(n, m, sd)
```

```
                    dimnames(descr) <- list(c("b1", "b2", "b3"),
                                            c("N", "Mean", "SD"))
                    return(descr)
                  })

descr_xt <- xtable(descr,
                   caption = "Descriptive statistics")
print(descr_xt,
      type = "html",
      file = "DescriptiveStatistics.html",
      caption.placement = "top",
      html.table.attributes = "width=50%,
      style='color:red; background-color:lightyellow;
      border: 5px solid; padding:10px;
      font-family:calibri'")



ANOVA <- aov(Y ~ A * B, data = d)
ANOVA_xt <- xtable(summary(ANOVA),
                   caption = "ANOVA summary")
print(ANOVA_xt,
      type = "html",
      file = "ANOVA_Summary.html",
      caption.placement = "top",
      html.table.attributes = "style='border-top:2px solid;
      border-bottom:2px solid; border-spacing:15px'")
```

Output

Table 8.2 An xtable showing descriptive statistics. CSS was used for visual appearance.

Descriptive statistics

	N	Mean	SD
b1	40.00	5.57	5.25
b2	40.00	7.08	5.75
b3	40.00	5.54	5.76

Table 8.3 An xtable showing an ANOVA summary. Border formatting uses CSS.

ANOVA summary

	Df	Sum Sq	Mean Sq	F value	Pr(>F)
A	1	150.91	150.91	4.94	0.0283
B	2	62.01	31.00	1.01	0.3660
A:B	2	19.80	9.90	0.32	0.7241
Residuals	114	3485.82	30.58		

Analysis

The script creates two HTML files: DescriptiveStatistics.html, which consists of some descriptive statistics by group, and ANOVA_Summary.html, which consists of the results of an ANOVA. Screenshots of the tables are shown in Table 8.2 and Table 8.3. For those interested in HTML and CSS, I want to highlight one aspect in the HTML code of Table 8.3, which is given below.

Output

```
<!-- html table generated in R 3.3.3 by xtable 1.8-2 package -->
<!-- Thu Jul 26 21:33:26 2018 -->
<table style='border-top:2px solid;
      border-bottom:2px solid; border-spacing:15px'>
<caption align="top"> ANOVA summary </caption>
<tr> <th>      </th> <th> Df </th> <th> Sum Sq </th> <th> Mean Sq
</th> <th> F value </th> <th> Pr(&gt;F) </th> </tr>
  <tr> <td> A             </td> <td align="right"> 1 </td> <td align="right">
150.91 </td> <td align="right"> 150.91 </td> <td align="right"> 4.94 </td> <td
align="right"> 0.0283 </td> </tr>
  <tr> <td> B             </td> <td align="right"> 2 </td> <td align="right">
62.01 </td> <td align="right"> 31.00 </td> <td align="right"> 1.01 </td> <td
align="right"> 0.3660 </td> </tr>
  <tr> <td> A:B           </td> <td align="right"> 2 </td> <td align="right">
19.80 </td> <td align="right"> 9.90 </td> <td align="right"> 0.32 </td> <td
align="right"> 0.7241 </td> </tr>
  <tr> <td> Residuals   </td> <td align="right"> 114 </td> <td align="right">
3485.82 </td> <td align="right"> 30.58 </td> <td align="right">  </td> <td
align="right">  </td> </tr>
   </table>
```

Analysis

Inside the `<table>` element of HTML we find the parameter `style`, which holds specifications like border-top, border-bottom, and border-spacing. This setting of the style parameter we exactly provide as an argument to the parameter `html.table.attributes` of **print()**. So, this parameter offers us an interface to use HTML and CSS to modify the appearance of our xTable. However, although it provides many options, it is restricted to the `<table>` element of the HTML code and cannot modify other elements within the table, for example, table row elements `<tr>`. The description of how to use HTML and CSS together with xTables is not covered in the xTable gallery. The gallery only shows options for LaTeX output. The HTML tables created with xTable can be opened in a web browser and then be copy-pasted to Word or OpenOffice for use in a manuscript.

8.1.4 More helpful R packages to create and export tables   R LANGUAGE!

Besides **xtable**, other packages produce and export nice tables. We will visit some of them in the following and then have a minimal example for each.

- **R2HTML**: HTML export for R objects

The **R2HTML** package (Lecoutre, 2003) creates full reports in HTML using text, tables, and figures. It basically employs the cat function to write text into an HTML file. It can be mainly used in two modes. First, you can create an HTML file in one function call and pass the object to be printed to the function call. Second, you can initialise an HTML file at the beginning of your R session and then send all output of calculations to the file, which would otherwise go to the console. This is a helpful option to create reports. Moreover, CSS can be used with this package to improve the display. The package is helpful and comes in a mature version. However, its last update was a few years ago.

- **htmlTable**: advanced tables for markdown/HTML

The package **htmlTable** (Gordon, Gragg & Konings, 2021) is a powerful tool to create tables in rectangular format (e.g., for data frames or correlation matrices). It allows fine-tuning like grouping of rows and columns, colouring parts of the table, and use of CSS to further improve the appearance. It is not so good for model summaries, which often are stored as lists. Because **htmlTable** produces raw HTML markup without saving it to a file, it works best in a report generated with **knitr**.

- **tableone**: tabular comparison of the characteristics of intervention and control group in experiments

The results section of many clinical journal reports starts with a table to compare personal and other characteristics of intervention and control groups. This is mainly used to demonstrate baseline comparability between groups. Besides descriptive statistics, **tableone** (Yoshida & Bartel, 2021) also provides statistical tests to compare between groups regarding quantitative and qualitative variables.

- **tables**: formula-driven table generation

The **tables** package (Murdoch, 2020) uses formula notation to define tables. Its main function is **tabular()**. One of its strengths is the integration of CSS.

- **sjPlot**: data visualisation for statistics in social science

Although **sjPlot** (Lüdecke, 2020) is mainly for graphical display, it also produces HTML tables for common statistical results. The **tab model()** function creates complete model summaries. This function is comparable with **stargazer** (Hlavac, 2018) and **texreg** (Leifeld, 2013), because it also consists of many parameters to modify the appearance of the results table.

Let us practise some minimal examples using these packages.

Ch8_R2HTML_MinimalExample.R

```
## R2HTML minimal example
library(R2HTML)
source("Ch8_DataCreation.R")
```

```
# Statistics
Descriptives <- tapply(d$Y, d$A, mean)
Correlations <- cor(d[c("X1", "X2", "Y")])
M1 <- aov(Y ~ A, d)
M2 <- update(M1, . ~ . + B)
M3 <- update(M2, . ~ . + X1)


# HTML Output R2HTML
HTMLSetFile(file = "Ch8_R2HTML_MinimalExample")
HTMLInitFile(outdir = getwd(),
              filename = HTMLGetFile(),
              Title = "R2HTML minimal example")
HTML(Descriptives)
HTML(Correlations)
HTML(summary(M1))
HTML(summary(M2))
HTML(summary(M3))
HTML(anova(M1, M2))
HTMLEndFile(file = HTMLGetFile())
```

Output

a1	a2
4.9	7.2

	X1	X2	Y
X1	1.0000	-0.0095	0.0411
X2	-0.0095	1.0000	-0.0359
Y	0.0411	-0.0359	1.0000

	Df	Sum Sq	Mean Sq	F value	Pr(>F)	
A	1	150.9	150.9	4.9913	0.02736	*
Residuals	118	3567.6	30.2			

--- Signif. codes: 0 `***' 0.001 `**' 0.01 `*' 0.05 `.' 0.1 ` ' 1

	Df	Sum Sq	Mean Sq	F value	Pr(>F)	
A	1	150.9	150.9	4.9935	0.02736	*
B	2	62.0	31.0	1.0259	0.36170	
Residuals	116	3505.6	30.2			

--- Signif. codes: 0 `***' 0.001 `**' 0.01 `*' 0.05 `.' 0.1 ` ' 1

	Df	Sum Sq	Mean Sq	F value	Pr(>F)	
A	1	150.9	150.9	4.9505	0.02803	*
B	2	62.0	31.0	1.0171	0.36489	
X1	1	0.005957	0.005957	0.0002	0.98887	
Residuals	115	3505.6	30.5			

--- Signif. codes: 0 `***' 0.001 `**' 0.01 `*' 0.05 `.' 0.1 ` ' 1

Analysis of Variance Table

Model 1: Y ~ A Model 2: Y ~ A + B

	Res.Df	RSS	Df	Sum of Sq	F	Pr(>F)
1	118	3567.6				
2	116	3505.6	2	62.0	1.0	0.3617

Ch8_HtmlTable_MinimalExample.Rhtml

Save this script with the Rhtml extension and compile it with **knitr** (see Chapter 4).

```
<!--## htmlTable minimal example-->
<html>
<head>
<title>htmlTable minimal example</title>
</head>
<body>
<h1>htmlTable minimal example</h1>
<!--begin.rcode, echo=FALSE, warning=FALSE
rm(list=ls())
library(htmlTable)
library(psych)
source("Ch8_DataCreation.R")

Descriptives <- with(d, describeBy(Y,
                            group = list(A, B),
                            mat = TRUE,
                            digits = 2)[c("group1",
                                          "group2",
                                          "n",
                                          "mean",
                                          "sd")])
names(Descriptives)[1:2] <- c("A", "B")
```

```
htmlTable(Descriptives,
          caption = "Descriptive statistics",
          rnames = FALSE,
          cgroup = c("Factors",
                     "Statistics of Y"),
          n.cgroup = c(2,3))
end.rcode-->
</body>
</html>
```

Output

htmlTable minimal example

| Descriptive statistics | | | | |
| Factors | | Statistics of Y | | |
A	B	n	mean	sd
a1	b1	20	3.88	4.17
a2	b1	20	7.26	5.76
a1	b2	20	6.32	6.04
a2	b2	20	7.84	5.49
a1	b3	20	4.63	6.48
a2	b3	20	6.45	4.92

Ch8_tableone_MinimalExample.R

```
## tableone minimal example
source("Ch8_DataCreation.R")
library(tableone)

to <- CreateTableOne(vars = c("A", "Y", "C1", "C2", "X1", "X2"),
                strata = "B",
                data = d)
toExport <- print(to,
                  printToggle = FALSE,
                  noSpaces = TRUE,
                  exact = "C1",
                  nonnormal = "X2")
write.csv(toExport,

          file = "TableOneMinimalExample.csv")
```

Output

Table 8.4 Minimal example for *tableone* package.

	b1	b2	b3	p	test
n	40	40	40		
A = a2 (%)	20 (50.0)	20 (50.0)	20 (50.0)	1.000	
Y (mean (SD))	5.57 (5.25)	7.08 (5.75)	5.54 (5.76)	0.374	
C1 = C1_2 (%)	24 (60.0)	14 (35.0)	20 (50.0)	0.085	exact
C2 (%)				0.126	
C2_1	11 (27.5)	8 (20.0)	13 (32.5)		
C2_2	7 (17.5)	11 (27.5)	12 (30.0)		
C2_3	15 (37.5)	7 (17.5)	7 (17.5)		
C2_4	7 (17.5)	14 (35.0)	8 (20.0)		
X1 (mean (SD))	5.31 (2.07)	5.07 (1.77)	4.78 (2.01)	0.473	
X2 (median [IQR])	12.81 [10.53, 14.42]	12.85 [10.73, 13.98]	11.61 [9.98, 13.05]	0.138	nonnorm

Ch8_tables_MinimalExample.R

```
## tables minimal example
library(tables)
source("Ch8_DataCreation.R")

tabular(A*B+1 ~ (n = 1) + (Format(digits = 2) *
                           (Y + X1 + X2) *
                           (mean + sd)), data = d)
```

Output

```
        Y         X1        X2
A  B   n   mean sd  mean sd  mean sd
a1 b1  20  3.9  4.2 4.9  1.9 11.9 3.2
   b2  20  6.3  6.0 4.9  1.8 12.3 2.4
   b3  20  4.6  6.5 4.2  1.8 12.4 2.3
a2 b1  20  7.3  5.8 5.7  2.2 13.3 2.7
   b2  20  7.8  5.5 5.2  1.8 12.6 2.9
   b3  20  6.5  4.9 5.4  2.1 10.7 2.6
   All 120 6.1  5.6 5.1  1.9 12.2 2.7
```

Ch8_sjPlot_MinimalExample.R

```
## sjPlot minimal example
library(sjPlot)
```

```
source("Ch8_DataCreation.R")
m1 <- lm(Y ~ A + B, d)
m2 <- lm(Y ~ A * B, d)
m3 <- lm(Y ~ A * B + X1 + X2, d)
tab_model(m1, m2, m3)
```

Output

Table 8.5 Minimal example `sjPlot` package.

Predictors	Y Estimates	CI	p	Y Estimates	CI	p	Y Estimates	CI	p
(Intercept)	4.45	2.46 – 6.44	<0.001	3.88	1.43 – 6.33	**0.002**	5.44	−0.41 – 11.28	0.068
A [a2]	2.24	0.25 – 4.23	**0.027**	3.38	−0.09 – 6.84	0.056	3.57	0.00 – 7.13	**0.050**
B [b2]	1.51	−0.93 – 3.94	0.222	2.44	−1.03 – 5.90	0.166	2.50	−1.00 – 5.99	0.159
B [b3]	−0.03	−2.47 – 2.40	0.979	0.74	−2.72 – 4.21	0.672	0.80	−2.71 – 4.32	0.652
A [a2] * B [b2]				−1.86	−6.75 – 3.04	0.455	−2.01	−6.98 – 2.95	0.424
A [a2] * B [b3]				−1.55	−6.45 – 3.35	0.532	−1.94	−7.02 – 3.14	0.451
X1							−0.01	−0.55 – 0.52	0.963
X2							−0.13	−0.51 – 0.26	0.519
Observations	120			120			120		
R^2 / R^2 adjusted	0.057 / 0.033			0.063 / 0.021			0.066 / 0.008		

Table 8.6 Useful packages to generate tables for publication.

Package	Main functions to generate tables	Strengths	Exports
R2HTML	HTMLSetFile() HTMLInitFile() HTML() HTMLEndFile()	Useful to aggregate reports with several results	html
htmlTable	htmlTable()	Matrix like tables	html
tableone	CreateTableOne() print.TableOne()	Multivariable descriptives by Groups	csv
tables	tabular()	Multivariable descriptives by Groups	console
sjPlot	tab_model()	Model summaries	html

(Continued)

Table 8.6 (Continued)

Package	Main functions to generate tables	Strengths	Exports
xtable	xtable() print.xtable()	ANOVA and linear model summaries, flat tables	html, latex
stargazer	stargazer()	Linear model summaries	html, latex
texreg	Texreg(), htmlreg(), screenreg()	Linear model summaries	html, latex, console

EXAMPLE!
8.1.5 WBD - Patient characteristics by study arm

For our clinical example, the primary publication requires a table showing patient characteristics for all study arms. This demonstrates that groups are comparable and it suggests that allocation to study arm was not associated with any of these characteristics. The table usually is the first one in the publication manuscript, so that is why we use the package named **tableone** to create it.

Let us learn about the participants who took part in the study. Usually, clinical studies begin the results section with a table that summarises demographical information and important clinical variables between conditions. The package **tableone** creates this table with a few function calls.

WBD_DES_TableOnePatientCharacteristics.R

```
## WBD - Patient characteristics - table one
if(!exists("TableOne"))
{

  source("Scripts/WBD_PRE_DataImport.R")
  library(tableone)
  library(labelled)

  TableOne <- with(WBD,
                   {
                     CPS_TO <- set_variable_labels(CPS,
                                                   .labels =
attr(CPS, "variable.labels"))
                     var_label(CPS_TO$Gender) <- "Gender"

                                        # Distributions of demographic variables
between conditions: Table 1
                                        CreateTableOne(vars = c("Gender",
"BDIcategory",
                                            "Employmentstatus",
                                               "YrsEducation",
"NPastEpisodes",
"Comorbidcurrentanxiety",
                                                  "OnantiD",
"EverantiD",
```

```
                                                      "Past_psych_tx",
"Everpastcontact",
                                        "Current_contact",
                                        "Howfoundout",
                                            "BDI_1", "STAIT_1",
"EQ5DVAS_1", "SUIS_1",
                                                        "PITPV_1",
"PITPL_1", "PITPE_1",
                                                        "PITNV_1",
"PITNL_1", "PITNE_1",
                                            "SST_1_negativity",
                                            "ANH_1",
"EQ_Total"),
                                        strata = "Condition",
                                            data = CPS_TO)
                })
}
```

WBD_CON_PrintTableOnePatientCharacteristics.R

```
## WBD - Patient characteristics - print table one
source("./Scripts/WBD_DES_TableOnePatientCharacteristics.R")
library(tableone)
library(htmlTable)

htmlTable(print(TableOne,
    showAllLevels = TRUE,
    test = FALSE,
    varLabels = TRUE))
```

Output

Table 8.7 Table one of web-based cognitive bias modification.

		Stratified by Condition	
	Level	Imagery	Control
n		76	74
Gender (%)	Male	24 (31.6)	23 (31.1)
	Female	52 (68.4)	51 (68.9)
BDI-II category at baseline (%)	Mild to moderate	35 (46.1)	32 (43.2)
	Severe	41 (53.9)	42 (56.8)
Employment status (%)	In paid employment (full- or part-time)	46 (60.5)	39 (52.7)
	Student	18 (23.7)	24 (32.4)
	Not in employment	12 (15.8)	11 (14.9)

(Continued)

Table 8.7 (Continued)

		Stratified by Condition	
	Level	Imagery	Control
Years of education (%)	11 years or less	5 (6.6)	6 (8.1)
	12–15 years	24 (31.6)	19 (25.7)
	>= 16 years	47 (61.8)	49 (66.2)
Number of past MDEs (not including current) (%)	0–1	19 (25.0)	18 (24.3)
	2–3	20 (26.3)	13 (17.6)
	4+	37 (48.7)	43 (58.1)
Current comorbid anxiety disorders? (%)	No	34 (44.7)	34 (45.9)
	Yes	42 (55.3)	40 (54.1)
Currently taking licensed anti-depressant (%)	No	43 (56.6)	43 (58.1)
	Yes	33 (43.4)	31 (41.9)
Ever taken an anti-depressant (%)	No	23 (30.3)	22 (29.7)
	Yes	53 (69.7)	52 (70.3)
Psychological therapy/counselling in the past (%)	No	25 (32.9)	33 (44.6)
	Yes	51 (67.1)	41 (55.4)
Ever spoken to health professional about mood (%)	No	5 (6.6)	12 (16.2)
	Yes	71 (93.4)	62 (83.8)
Currently have contact with a health professional about mood (%)	No	53 (69.7)	49 (66.2)
	Yes	23 (30.3)	25 (33.8)
How did they find out about the study? (%)	Poster advert	5 (6.7)	11 (14.9)
	Radio/Newspaper advert	22 (29.3)	17 (23.0)
	Internet	32 (42.7)	28 (37.8)
	Other	16 (21.3)	18 (24.3)
BDI-II Baseline (mean (SD))		29.96 (8.63)	31.14 (10.17)
STAIT Baseline (mean (SD))		61.00 (6.33)	61.59 (6.87)
Eurqol VAS Baseline (mean (SD))		61.38 (20.17)	58.88 (18.91)
SUIS Baseline (mean (SD))		38.50 (9.90)	40.34 (7.91)
PIT Positive Vividness - Baseline (mean (SD))		2.82 (0.88)	2.87 (0.81)
PIT Positive Likelihood - Baseline (mean (SD))		2.54 (0.64)	2.52 (0.66)
PIT Positive Experiencing - Baseline (mean (SD))		2.62 (0.81)	2.46 (0.79)
PIT Negative Vividness - Baseline (mean (SD))		3.16 (0.89)	3.52 (0.80)

	Level	Stratified by Condition	
		Imagery	Control
PIT Negative Likelihood - Baseline (mean (SD))		3.14 (0.64)	3.30 (0.62)
PIT Negative Experiencing - Baseline (mean (SD))		3.09 (0.88)	3.13 (0.74)
SST negativity baseline (mean (SD))		0.57 (0.23)	0.60 (0.24)
Anhedonia Baseline (mean (SD))		3.18 (1.36)	3.47 (1.35)
Expectancy Questinonaire (mean (SD))		0.01 (2.59)	-0.01 (2.91)

Analysis

The table shows characteristics and clinical variables of the intervention group and the control group. The table is useful to check whether both groups are similar enough to warrant valid conclusions regarding the effectiveness of the intervention program.

8.1.6 Finalise ANOVA and regression results tables with stargazer and texreg

R LANGUAGE!

For many results displays, the packages **stargazer** (Hlavac, 2018) (http://cran.r-project.org/web/packages/stargazer/index.html) and **texreg** (Leifeld, 2013) (http://cran.r-project.org/web/packages/texreg/index.html) aggregate ready-to-publish tables with a single function call; for example, for descriptive statistics or results of a regression analysis. Besides tex markup, both packages produce HTML, which can be read by Microsoft Word and Open Office. The two packages consist of few functions only, but these functions consist of many parameters to modify the appearance of the table. However, most parameters have default values, so usually simply passing an object consisting of the results to one of the functions already produces a beautiful table. In many instances, nevertheless, you would first produce an initial version of the table, then decide on the required amendments, then find the correct parameters of the table functions to achieve the changes, and finally run a complex function call with results objects and specifications for layout parameters as arguments.

In the following, we will view a few small examples of the type of tables **stargazer** and **texreg** can handle. Next we go through a table of a comparison of some selected parameters of the two packages. We can see the differences between the packages and get a feeling about which function is more suitable in which situation. Finally, we practise the more complex task of creating a table for pooled results of a multiple imputed data set. Both packages contain vignettes to make table creation as easy as possible. Our aim here is to complement the documentation already given with some practical additions. Try the following results tables.

Ch8_StargazerTexregTables.R

```
# stargazer and texreg tables
source("./Ch8_DataCreation.R")
library(stargazer)
library(texreg)
```

```
# Statistical and graphical analysis
psych::describeBy(d$Y, group = list(d$A, d$B))
m1 <- aov(Y ~ A, data = d)
m2 <- aov(Y ~ A + B, data = d)
m3 <- aov(Y ~ A * B, data = d)

# Data frame and summary statistics with stargazer
stargazer(d,
          type = "html",
          out = "DataFrame_SG.html",
          summary = FALSE)

stargazer(d,
          type = "html",
          out = "Descriptives_SG.html",
          summary.stat = c("mean", "n", "sd", "min", "max"))

# Model summaries with stargazer
stargazer(lm(m1),
          type = "html",
          out = "LinearModelTable_m1_SG.html")

stargazer(lm(m1), lm(m2), lm(m3),
          type = "html",
          out = "LinearModelTable_m1m2m3_SG.html")

# Customise stargazer table
stargazer(lm(m1), lm(m2), lm(m3),
          type = "html",
          out = "LinearModelTable_m1m2m3_SG_customised.html",
          column.labels = c("Simple effect",
                            "Main effects",
                            "Main effects + interactions"),
          covariate.labels = c("A", "B 1-2", "B 1-3",
                                "A2:B2", "A2:B3",
                                "Intercept"),
          dep.var.caption = "Response variable",
          dep.var.labels = "Variable Y",
          model.numbers = FALSE,
          single.row = TRUE,
          star.cutoffs = c(.05, .01, .001),
          ci = TRUE,
          digits = 2)

# Model summaries with htmlreg
htmlreg(lm(m1),
        file = "LinearModelTable_m1_HR.html")
```

```
htmlreg(list(lm(m1), lm(m2), lm(m3)),
        file = "LinearModelTable_m1m2m3_HR.html")

# Customise htmlreg table
htmlreg(list(lm(m1), lm(m2), lm(m3)),
        file = "LinearModelTable_m1m2m3_HR_customised.html",
        single.row = TRUE,
        stars = c(.01, .05, .1),
        custom.model.names = c("Simple effect",
                               "Main effects",
                               "Main effects + interactions"),
        ci.force = TRUE,
        groups = list("Main factors" = 2:4,
                      "Interactions" = 5:6),
        bold = .05,
        caption = "Linear model summaries",
        caption.above = TRUE)
```

Analysis

The script first loads our simulation data for this chapter and the packages **stargazer** and **texreg**. Next, it creates three analyses of variance of the dependent variable Y against the independent variables A and B. The models are hierarchically arranged according to the inclusion of terms: m1 has only one term relating to the independent variables, m3 has three terms relating to main effects and an interaction. Then, the script directly proceeds to the output tables. The function **stargazer()** creates an HTML file containing the data frame (Table 8.8) and another file containing descriptive statistics (Table 8.9). Although its primary focus lies on the tables for statistical models, **stargazer()** is also capable of creating data frames and tables with summary statistics. The function guesses that if supplied with a data frame, the user might want a table with summary statistics; so, in the first call the parameter summary is set to FALSE in order to create the data frame in HTML. In the second call we again supply **stargazer()** with the data frame, but now we additionally specify some statistics for the parameter summary.stat. Consequently, the function returns a table containing summary statistics.

Table 8.8 First 20 data rows after export with **stargazer()**.

	id	A	B	Y	C1	C2	X1	X2
1	1	a1	b1	6.428	C1_2	C2_3	3.600	9.873
2	2	a1	b1	7.047	C1_2	C2_4	3.865	5.561
3	3	a1	b1	2.953	C1_2	C2_3	4.477	11.866
4	4	a1	b1	1.233	C1_1	C2_4	2.872	10.431
5	5	a1	b1	6.529	C1_2	C2_1	4.787	11.054

(Continued)

Table 8.8 (Continued)

6	6	a1	b1	-5.590	C1_1	C2_2	6.542	12.951
7	7	a1	b1	6.650	C1_1	C2_4	10.495	11.931
8	8	a1	b1	2.119	C1_1	C2_3	4.832	16.005
9	9	a1	b1	2.079	C1_2	C2_4	6.087	13.018
10	10	a1	b1	-1.097	C1_2	C2_1	6.506	10.167
11	11	a1	b1	2.919	C1_1	C2_4	3.383	10.562
12	12	a1	b1	12.587	C1_2	C2_1	7.002	16.104
13	13	a1	b1	5.353	C1_2	C2_4	5.912	12.414
14	14	a1	b1	6.101	C1_1	C2_3	2.131	12.745
15	15	a1	b1	-0.253	C1_2	C2_2	4.469	12.998
16	16	a1	b1	7.584	C1_1	C2_1	6.284	8.663
17	17	a1	b1	-0.932	C1_1	C2_2	4.170	14.490
18	18	a1	b1	1.842	C1_2	C2_1	4.081	5.106
19	19	a1	b1	9.104	C1_2	C2_2	3.415	13.308
20	20	a1	b1	4.994	C1_2	C2_1	2.683	17.779

Table 8.9 Descriptive statistics export with `stargazer()`.

Statistic	Mean	N	St. Dev.	Min	Max
id	60.500	120	34.785	1	120
Y	6.064	120	5.590	-7.802	18.186
X1	5.053	120	1.948	0.128	10.495
X2	12.188	120	2.745	5.106	18.726

The next three calls of **stargazer()** supply our three statistical models as arguments. In the first case we supply the linear model version of m1, that is, `lm(m1)`. Unfortunately, to my knowledge, **stargazer()** cannot do the sum of squares results table of an ANOVA, which psychologists commonly use. So, the output (Table 8.10) shows one column of regression coefficients and the respective standard errors. The regression constant in this case is the mean of Y in factor level a1; the coefficient Aa2 is the mean difference of Y between the levels a1 and a2. So, the Y mean of factor level a2 is 4.943 + 2.243 = 7.186, which we can check with a quick call of `tapply(d$Y, d$A, mean)` at the console. The table further consists of the sample size, R squared as a measure of strength of association, and the F-statistic from the analysis of variance. Now, think about what we did again: we wrote one line of code to define the statistical model and then used a few more lines for the call of **stargazer()**. And our result is a readily formatted and printable results table that we could directly copy-paste into a results section of a research paper. There is no need to tinker with table layout in the text editor any more or to move single numbers around in a results table, which is very error-prone. Moreover, assume that another 100 subjects are added to this data frame. It takes only a few seconds to run the analysis again and to produce the final results table. Let us now proceed to the next table.

Table 8.10 Linear model summary of an ANOVA, created with **stargazer()**.

	Dependent variable
	Y
Aa2	2.243**
	(1.004)
Constant	4.943***
	(0.710)
Observations	120
R^2	0.041
Adjusted R^2	0.032
Residual Std. Error	5.499 (df = 118)
F Statistic	4.991** (df = 1;118)
Note:	*$p<0.1$; **$p<0.05$; ***$p<0.01$

The next **stargazer** table (Table 8.11) simultaneously presents all three models that we created in a multi-column layout that we find very often in the scientific journals for the report of statistical models. This kind of presentation allows for the easy comparison of coefficients, strength of association, and F-statistics between models. As can be seen, with only three lines of code an almost publication-ready table can be created. If we add some parameter specifications, as we do in the next table, we can fine-tune the same results table even more to our needs. Table 8.12 shows the output of the next **stargazer()** call. In this function call, we used several parameters to change the row and column annotation. We further switched from the standard errors to confidence intervals using the parameter ci. Finally, we also changed the meanings of the * indicators of statistical significance. The package **stargazer** applies to many statistical models as you can find out with ?"stargazer models". Let us now compare with the equally powerful package **texreg** (Leifeld, 2013).

Table 8.11 Linear model summaries of three ANOVAS created with **stargazer()**.

	Dependent variable		
	Y		
	(1)	(2)	(3)
Aa2	2.243**	2.243**	3.378*
	(1.004)	(1.004)	(1.749)

(Continued)

Table 8.11 (Continued)

Bb2		1.509	2.437
		(1.229)	(1.749)
Bb3		-0.032	0.743
		(1.229)	(1.749)
Aa2:Bb2			-1.856
			(2.473)
Aa2:Bb3			-1.550
			(2.473)
Constant	4.943***	4.450***	3.883***
	(0.710)	(1.004)	(1.236)
Observations	120	120	120
R^2	0.041	0.057	0.063
Adjusted R^2	0.032	0.033	0.021
Residual Std. Error	5.499 (df = 118)	5.497 (df = 116)	5.530 (df = 114)
F Statistic	4.991** (df = 1;118)	2.348* (df=3;116)	1.522 (df = 5;114)

Note: $^*p<0.1; ^{**}p<0.05; ^{***}p<0.01$

Table 8.12 Linear model results table customised with `stargazer()`.

	Response variable		
	Variable Y		
	Simple effect	Main effects	Main effects + interactions
A	2.24* (0.28, 4.21)	2.24* (0.28, 4.21)	3.38 (-0.05, 6.81)
B 1-2		1.51 (-0.90, 3.92)	2.44 (-0.99, 5.86)
B 1-3		-0.03(-2.44, 2.38)	0.74 (-2.68, 4.17)
A2:B2			-1.86 (-6.70, 2.99)
A2:B3			-1.55 (-6.40, 3.30)
Intercept	4.94*** (3.55, 6.33)	4.45*** (2.48, 6.42)	3.88** (1.46, 6.31)
Observations	120	120	120
R^2	0.04	0.06	0.06
Adjusted R^2	0.03	0.03	0.02
Residual Std. Error	5.50 (df = 118)	5.50 (df = 116)	5.53 (df = 114)
F Statistic	4.99* (df = 1; 118)	2.35 (df = 3; 116)	1.52 (df = 5; 114)

Note: $^*p<0.05; ^{**}p<0.01; ^{***}p<0.001$

The package `texreg` distinguishes between LaTeX and HTML output by using different function names: `texreg()` and `htmlreg()`. If you are interested in a nice results table at the console, the package further offers `screenreg()` for this. However, in our examples, we use `htmlreg()`.

To use `htmlreg()` is comparable to the use of `stargazer()`: both functions return publication-ready results tables and both functions have many parameters to change the appearance of the tables. The first call of `htmlreg()` only displays the results for model m1 (Table 8.13), the next call displays the model comparison, similar to the output of `stargazer()` (Table 8.14). Note that models must be passed to the function as a list. Finally, we use several parameters of `htmlreg()` to change the appearance of the table (Table 8.15). Both packages, `stargazer` and `texreg`, are powerful aids for results display. To work with them most efficiently, one often needs their help pages to learn to employ all the available parameters.

Table 8.13 ANOVA summary created by `htmlreg()`.

	Model 1
(Intercept)	4.94***
	(0.71)
Aa2	2.24*
	(1.00)
R_2	0.04
Adj. R_2	0.03
Num. obs.	120
RMSE	5.50

***p<0.001, **p<0.01, *p<0.05

Statistical models

Table 8.14 Model comparison between three linear models created by `htmlreg()`.

	Model 1	Model 2	Model 3
(Intercept)	4.94***	4.45***	3.88**
	(0.71)	(1.00)	(1.24)
Aa2	2.24*	2.24*	3.38
	(1.00)	(1.00)	(1.75)
Bb2		1.51	2.44
		(1.23)	(1.75)
Bb3		-0.03	0.74
		(1.23)	(1.75)
Aa2: Bb2			-1.86
			(2.47)
			-1.55
Aa2: Bb3			(2.47)
R_2	0.04	0.06	0.06
Adj. R_2	0.03	0.03	0.02
Num. obs.	120	120	120
RMSE	5.50	5.50	5.53

***p<0.001, **p<0.01, *p<0.05

Statistical models

Table 8.15 Customised `htmlreg()` table.

	Simple effect	Main effects	Main effects + interactions
(Intercept)	**4.94** [3.55; 6.33]	**4.45** [2.48; 6.42]	**3.88** [1.46; 6.31]
Main factors			
Aa2	**2.24** [0.28; 4.21]	**2.24** [0.28; 4.21]	3.38 [−0.05; 6.81]
Bb2		1.51 [−0.90; 3.92]	2.44 [−0.99; 5.86]
Bb3		−0.03 [−2.44; 2.38]	0.74 [−2.68; 4.17]
Interactions			
Aa2:Bb2			−1.86 [−6.70; 2.99]
Aa2:Bb3			−1.55 [−6.40; 3.30]
R^2	0.04	0.06	0.06
Adj. R^2	0.03	0.03	0.02
Num. obs.	120	120	120

*Null-hypothesis value outside the confidence interval.

EXAMPLE! ### 8.1.7 SCB - Model summaries

The many results we calculated for the conflict reconciliator data were all mere console output and not adequate to be included in a publication. Therefore, we bring some beauty and a nice format to the results with **stargazer**.

SCB_OUT_ModelSummaries.R

```
## SCB - Model summaries
source("./Script/SCB_PRE_DataImport.R")
source("./Script/SCB_PRE_VariableLabels.R")
source("./Script/SCB_PRE_DataAggregation.R")
source("./Script/SCB_MOD_MultipleLinearRegression.R")
library(stargazer)

stargazer(MultipleLinearRegression,
          type = "html",
          out = "./Results/SCB_LinearModelSummaries.html",
          single.row = TRUE)
```

Output

Table 8.16 Results of a multiple linear regression of numbers of conflict on covariates renovation status of the school (X1) and average pocket money of school students (X2).

	Dependent variable:	
	PersonPerson_PREPOST	PersonGroup_PREPOST
	(1)	(2)
X1	0.140 (0.225)	−0.194 (1.406)
X2	−0.344** (0.141)	−0.247 (0.478)

X1:X2		0.029 (0.091)
Constant	4.249** (2.058)	2.764 (7.192)
Observations	180	180
R^2	0.033	0.009
Adjusted R^2	0.022	-0.007
Residual Std. Error	3.797 (df = 177)	3.476 (df = 176)
F Statistic	2.992* (df = 2; 177)	0.558 (df = 3; 176)
Note:		*p<0.1; **p<0.05; ***p<0.01

Analysis

The results (Table 8.16) show a significant influence of average pocket money on the counts of person-against-person conflicts. This kind of tabular display of regression coefficients and other model statistics for different models is often part of research papers.

8.2 **Report generation**

The results section of a research paper usually consists of several tables and figures; for example, a table to describe the sample, another one for results of statistical models, and figures to highlight key results. We now practise how to place all these elements of a results section in an HTML report.

8.2.1 Report generation with the knitr package   R LANGUAGE!

The **knitr** package creates full reports for us including text, tables, and figures. It was introduced by Yihui Xie (Xie, 2015; Xie, 2021a) and is titled "A General-Purpose Package for Dynamic Report Generation in R" (http://cran.r-project.org/web/packages/knitr/index.html). This is how it works: First, we write our report in an Rhtml file, which is a text file consisting of hypertext markup language (HTML) and R code. The file is structured so that the R code basically goes inside a modified HTML comment, which starts with `<!--begin.rcode` and ends with `end.rcode-->`. After we have written the file and saved it with the file extension Rhtml, we run **knit()** with the file, which produces an HTML file. This file consists of the HTML parts of the Rhtml file and the HTML representation of our calculations in R. A web browser can now open the HTML file and display the report.

We now practise two things with **knitr**: (1) to simply print console output of R calculations in an HTML file and add some annotation, and (2) to include beautiful table output that is itself rendered as HTML in the report and combine it with annotation. The first is quick and dirty, but it helps you to structure and comment your results. The second brings you closer to a report that you can publish. So, let us begin with the minimal example of printing console output in an HTML file. The example loads data from our simulation for this chapter and displays some statistics and a figure. Notice the ending Rhtml in the file name.

Ch8_knitrMinimalExample.Rhtml

```
<!--## Print console output in an HTML file-->
<html>
<head>
<title>Print console output in an HTML file</title>
</head>

<body>

<h1>Minimal knitr example</h1>

<p>Some lines of R functions with console output follow.</p>

<!--begin.rcode
rm(list=ls())
source("Ch8_DataCreation.R")
summary(d)
tapply(d$Y, d$B, mean)
m <- aov(Y ~ A * B, data = d)
summary(m)
end.rcode-->

<p>Diagnostic diagrams for the ANOVA</p>

<!--begin.rcode fig.width=7, fig.height=6
plot(m)
end.rcode-->

</body>
</html>
```

This script can be written in any text editor, but it cannot be executed by R directly. It must be sent to the **knit()** function of the knitr package. After you have stored the script in an Rhtml file, you can pass it as an argument to **knit()** and it will generate the HTML output file to be opened in a web browser. Use this code at the console:

```
library(knitr)
knit(input = "Ch8_knitrMinimalExample.Rhtml")
```

The resulting HTML file looks like Figure 8.2. As can be seen, the output starts with the heading and a small introductory text. Then follows a copy of the executed R code together with the console output. The bottom of the HTML file contains the four diagnostic diagrams for the ANOVA object (not shown here). We now inspect the Rhtml file to learn how knitr creates the output.

Output

Minimal knitr example

Some lines of R functions with console output follow.

```
rm(list=ls())
source("Ch8_DataCreation.R")
summary(d)
```

```
##          id            A        B           Y                C1
## Min.    :  1.00    a1:60    b1:40    Min.    :-7.802    Length:120
## 1st Qu.: 30.75    a2:60    b2:40    1st Qu.: 2.070    Class :character
## Median : 60.50             b3:40    Median : 6.400    Mode  :character
## Mean    : 60.50                     Mean    : 6.064
## 3rd Qu.: 90.25                      3rd Qu.: 9.444
## Max.    :120.00                     Max.    :18.186
##          C2                 X1                  X2
## Length:120         Min.    : 0.1281    Min.    : 5.106
## Class :character   1st Qu.: 3.6701    1st Qu.:10.342
## Mode  :character   Median : 5.2066    Median :12.442
##                    Mean    : 5.0533    Mean    :12.188
```

Figure 8.2 Part of the output of a minimal example created with `knitr`.

Analysis

The example starts with HTML tags, which structure the file. These tags are <html>, <head>, <title>, <body>, <h1>, and <p> in this example. Each tag consists of an opening and a closing part and holds the respective HTML element's content in between. For example, <html> begins the document and declares that everything that follows will be HTML until the closing tag </html>. So, the whole document is embraced by <html> and </html> as must be the case in every valid HTML file. The HTML element consists of two elements, a head and a body. These are marked by the tags <head> </head> and <body> </body>, respectively. As can be seen already, the document consists of a nested structure, and the relationships between elements are often called parent for containing elements, child for contained elements, or sibling for adjacent elements. This structure must be maintained: the opening and closing tag of a child element must both reside within the boundaries of the parent element. For example, incorrect would be <body><p>some content</body></p>; correct would be <body><p>some content</p></body>. In fact, this rule of enclosing elements within the boundaries of other elements corresponds with the nested function calls in R, for example round(mean(1:10)). Everything that belongs to **mean()** lies inside the parentheses of **round()**. The further element tags that we have in the example are <title>, <h1>, and <p>. The title element declares the title of the document, <h1> marks a heading of level 1, and the <p> tag marks a paragraph of text. As you may expect, there is much more to learn about HTML and there are more types of elements, but a small set of elements suffices to create nice reports in Rhtml. Let us now proceed to the embedded R code.

The R code resides inside an element of the HTML body. The element is tagged with <!--begin.rcode and end.rcode-->. In **knitr** this element is called a code chunk. The tags <!-- and --> actually mark comments in HTML, which are not rendered by the web browser; it is basically the same as # in R. Inside these tags, we can write the R code of our analysis. In the example, the first element containing R code contains a few lines: the first one secures that we start with an empty environment, the second one sources the script for data creation, and the subsequent ones calculate statistics. Then the code chunk ends and some more HTML paragraph follows. The second code chunk takes the ANOVA object and creates four diagrams associated with the model. Note that the R objects exist across the boundaries of a code chunk; so objects created in an earlier chunk can be used in a subsequent one. In the opening tag of code chunks, parameters about how **knitr** should handle the chunk can be placed. In the current chunk, which creates the figure, the parameters fig.width and fig.height are specified to set the size of the figure in inches. There are more parameters to be practised below. In sum, the minimal example shows how to basically combine HTML and R code in an Rhtml document, how to create the HTML report, and how to inspect it in a web browser.

Our next example is not minimal, but a full report of our simulation data consisting of several different output tables and figures. We use several functions for table generation and see how they fit in with the whole report. The following is the Rhtml file to be compiled at the console with **knit()**.

Ch8_HtmlReport.Rhtml

```
<!--## An HTML report combining text and results output-->
<html>
<head>
<title>Results of a simulation study</title>
</head>

<body>
<!--begin.rcode Read data, echo=FALSE, include=FALSE
rm(list=ls())
source("Ch8_DataCreation.R")
library(tableone)
library(psych)
library(xtable)
library(texreg)
library(stargazer)
library(htmlTable)
set.seed(12345)
ls()
end.rcode-->

<h1>A comprehensive HTML report of statistical results</h1>
<p>This report consists of a data overview, descriptive statistics, a linear
model, and a figure.</p>
```

```
<h2>Data inspection</h2>
<p>A small sample of lines from the data frame (N = <!--rinline nrow(d)-->)
gives a rough idea about the levels of factor variables and about the value
ranges of quantitative variables.</p>

<!--begin.rcode Inspect data, echo=FALSE, results="asis"
xt_data <- xtable(d[sort(sample(1:120, 10)),],
                  caption = "Random lines taken from the data frame")
print(xt_data,
      type = "html",
      caption.placement = "top",
      html.table.attributes = "border=0")
end.rcode-->

<h2>Descriptive statistics</h2>
<p>Variable distributions for the three levels of factor B is shown in table
one. The next table shows more detailed descriptives of quantitative variables
for the three levels of B. Finally, a cross table shows the frequencies of
combinations of the levels of factors A and C2.</p>

<!--begin.rcode Descriptive statistics, echo=FALSE, results="asis"
to <- CreateTableOne(vars = c("A", "Y", "C1", "C2", "X1", "X2"),
                     strata = "B",
                     data = d)
pr_to <- print(to,
               printToggle = FALSE)

stargazer(pr_to,
          type = "html",
          title = "Table one: Descriptives by groups")
end.rcode-->
<BR>
<!--begin.rcode Detailed descriptives, echo=FALSE, results="asis"
descr<- describeBy(d[c("Y", "X1", "X2")],
                   list(d$A, d$B),
                   mat = TRUE)

xt_descr <- xtable(descr,
                   caption = "Detailed descriptives by groups")
print(xt_descr,
      type = "html",
      caption.placement = "top",
      html.table.attributes = "border=0")
end.rcode-->
<BR>
<!--begin.rcode Cross tabs of factors, echo=FALSE
t_AC2 <- with(d, addmargins(table(A, C2)))
```

```
htmlTable(ct_AC2,
          caption = "Counts of A-C2 combinations",
          ctable = c("solid", "double"))
end.rcode-->
```

<h2>Statistical models</h2>
<p>The analysis starts with an ANOVA of the main effects and interaction of A
and B on Y. Next, the quantitative covariates X1 and X2 are included. Model
comparisons between the influences of factor variables only and the additional
influences of the covariates are possible.</p>

```
<!--begin.rcode ANOVA results, echo=FALSE, results="asis"
m1 <- lm(Y ~ A * B, data = d)
m2 <- lm(Y ~ A * B + X1 + X2, data = d)
m3 <- lm(Y ~ A * B + X1 * X2, data = d)

xt_ANOVA <- xtable(anova(m1),
                   caption = "Anova of main effects and interactions")
print(xt_ANOVA,
      type = "html",
      caption.placement = "top",
      html.table.attributes = "border=0")
end.rcode-->
<BR>
<!--begin.rcode Linear model summaries, echo=FALSE, results="asis"
htmlreg(list(m1, m2, m3),
        single.row = TRUE,
        center = FALSE,
        caption = "Summaries of linear models",
        caption.above = TRUE,
        custom.model.names = c("Factor influences",
                               "+ Covariates",
                               "+ Covariate interaction"))
end.rcode-->
```

<h2>Diagnostic diagrams for linear models</h2>
<p>Four diagnostic diagrams of model 1 testing influences of factors A and B
on Y.</p>

```
<!--begin.rcode, echo=FALSE
layout(matrix(1:4, 2))
plot(m1)
end.rcode-->
```

```
</body>
</html>
```

Use the function **knit()** again to turn this Rhtml file into an HTML file.

Output

A comprehensive HTML report of statistical results

This report consists of a data overview, descriptive statistics, a linear model, and a figure.

Data inspection

A small sample of lines from the data frame (N = 220) gives a rough idea about the levels of factor variables and about the value ranges of quantitative variables.

Random lines taken from the data frame

id	A	B	Y	C1	C2	X1	X2
14	14	a1	b1	6.10	C1_1 C2_3	2.13	12.75
24	24	a1	b2	-3.47	C1_1 C2_2	3.97	10.39
51	51	a1	b3	1.40	C1_2 C2_2	3.64	10.09
58	58	a1	b3	8.22	C1_2 C2_1	3.25	10.21
75	75	a2	b1	13.11	C1_2 C2_3	9.33	13.16
80	80	a2	b1	11.21	C1_1 C2_2	3.46	17.81
90	90	a2	b2	9.92	C1_1 C2_2	6.47	8.91
92	92	a2	b2	1.52	C1_1 C2_2	5.47	15.08
93	93	a2	b3	15.23	C1_1 C2_4	5.20	10.29
96	96	a2	b2	9.24	C1_2 C2_2	5.23	15.61

Descriptive statistics

Variable distributions for the three levels of factor B is shown in table one. The next table shows more detailed descriptives of quantitative variables for the three levels of B. Finally, a cross table shows the frequencies of combinations of the levels of factors A and C2.

Table one: Descriptives by groups

	b1	b2	b3	p	test
n	40	40	40		
A = a2 (%)	20 (50.0)	20 (50.0)	20 (50.0)	1.000	
Y (mean (SD))	5.57 (5.25)	7.08 (5.75)	5.54 (5.76)	0.374	
C1 = C1_2 (%)	24 (60.0)	14 (35.0)	20 (50.0)	0.079	
C2 (%)				0.126	
C2_1	11 (27.5)	8 (20.0)	13 (32.5)		
C2_2	7 (17.5)	11 (27.5)	12 (30.0)		
C2_3	15 (37.5)	7 (17.5)	7 (17.5)		
C2_4	7 (17.5)	14 (35.0)	8 (20.0)		
X1 (mean (SD))	5.31 (2.07)	5.07 (1.77)	4.78 (2.01)	0.473	
X2 (mean (SD))	12.56 (3.01)	12.46 (2.59)	11.54 (2.56)	0.191	

Detailed descriptives by groups

	item	group1	group2	vars	n	mean	sd	median	trimmed	mad	min	max	range	skew	kurtosis	se
Y1	1	a1	b1	1.00	20.00	3.88	4.17	3.97	3.92	3.88	-5.59	12.59	18.18	-0.17	-0.27	0.93
Y2	2	a2	b1	1.00	20.00	7.26	5.76	8.58	7.32	7.37	-3.06	17.75	20.81	-0.13	-1.17	1.29
Y3	3	a1	b2	1.00	20.00	6.32	6.04	7.06	6.52	7.18	-4.01	15.28	19.29	-0.23	-1.13	1.35
Y4	4	a2	b2	1.00	20.00	7.84	5.49	8.75	7.51	7.70	0.90	18.19	17.29	0.36	-1.24	1.23
Y5	5	a1	b3	1.00	20.00	4.63	6.48	6.40	4.96	6.19	-7.80	14.83	22.63	-0.43	-0.87	1.45
Y6	6	a2	b3	1.00	20.00	6.45	4.92	5.90	6.18	3.55	-0.42	17.10	17.52	0.39	-0.72	1.10

X11	7	a1	b1	2.00	20.00	4.88	1.93	4.47	4.71	1.88	2.13	10.49	8.36	1.05	1.12	0.43
X12	8	a2	b1	2.00	20.00	5.75	2.15	5.05	5.67	2.07	2.19	9.54	7.35	0.30	-1.23	0.48
X13	9	a1	b2	2.00	20.00	4.91	1.77	4.87	4.95	1.73	0.77	7.97	7.19	-0.24	-0.40	0.40
X14	10	a2	b2	2.00	20.00	5.22	1.79	5.45	5.50	0.38	0.13	7.60	7.47	-1.44	1.62	0.40
X15	11	a1	b3	2.00	20.00	4.17	1.79	4.41	4.23	1.93	0.20	7.69	7.49	-0.21	-0.48	0.40
X16	12	a2	b3	2.00	20.00	5.39	2.07	5.44	5.47	1.69	0.34	9.65	9.32	-0.39	0.13	0.46
X21	13	a1	b1	3.00	20.00	11.85	3.17	12.17	12.03	2.48	5.11	17.78	12.67	-0.35	-0.18	0.71
X22	14	a2	b1	3.00	20.00	13.27	2.74	13.32	13.19	2.84	8.32	18.72	10.39	0.10	-0.77	0.61
X23	15	a1	b2	3.00	20.00	12.34	2.35	12.57	12.31	1.76	7.88	18.73	10.85	0.46	0.90	0.53
X24	16	a2	b2	3.00	20.00	12.58	2.86	13.25	12.82	2.68	5.41	16.49	11.08	-0.73	-0.28	0.64
X25	17	a1	b3	3.00	20.00	12.40	2.29	12.63	12.25	3.36	9.39	16.61	7.23	0.31	-1.24	0.51
X26	18	a2	b3	3.00	20.00	10.69	2.60	10.84	10.78	2.15	5.29	15.90	10.61	-0.21	-0.21	0.58

Counts of A-C2 combinations

	C2				
	C2_1	C2_2	C2_3	C2_4	Sum
A					
a1	16	16	9	19	60
a2	16	14	20	10	60
Sum	32	30	29	29	120

Statistical models

The analysis starts with an ANOVA of the main effects and interaction of A and B on Y. Next, the quantitative covariates X1 and X2 are included. Model comparisons between the influences of factor variables only and the additional influences of the covariates are possible.

Anova of main effects and interactions

	Df	Sum Sq	Mean Sq	F value	Pr(>F)
A	1	150.91	150.91	4.94	0.0283
B	2	62.01	31.00	1.01	0.3660
A:B	2	19.80	9.90	0.32	0.7241
Residuals	114	3485.82	30.58		

Summaries of linear models

	Factor influences	+ Covariates	+ Covariate interaction
(Intercept)	3.88 (1.24)**	5.44 (2.95)	0.88 (5.92)
Aa2	3.38 (1.75)	3.57 (1.80)*	3.59 (1.80)*
Bb2	2.44 (1.75)	2.50 (1.76)	2.57 (1.77)
Bb3	0.74 (1.75)	0.80 (1.77)	0.71 (1.78)
Aa2:Bb2	-1.86 (2.47)	-2.01 (2.51)	-2.25 (2.52)
Aa2:Bb3	-1.55 (2.47)	-1.94 (2.56)	-1.82 (2.57)
X1		-0.01 (0.27)	1.00 (1.17)
X2		-0.13 (0.19)	0.25 (0.47)
X1:X2			-0.08 (0.09)
R²	0.06	0.07	0.07
Adj. R²	0.02	0.01	0.01
Num. obs.	120	120	120

***p < 0.001, **p < 0.01, *p < 0.05

Diagnostic plots for linear models

Four diagnostic plots of model 1 testing influences of factors A and B on Y.

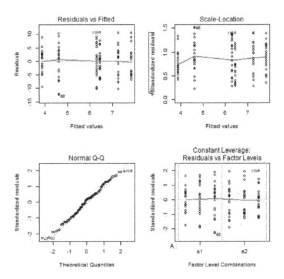

Figure 8.3 HTML report showing a full statistical analysis.

Analysis

This HTML file looks like Figure 8.3. It is a structured document containing title, headings, prose, statistics, and figures. The contents of the tables are described within the file and we do not need to repeat this here. Let us instead focus on how we produced the results with R and **knitr**. To get nice tables with **knitr**, the table-generating function inside the R code chunk must itself produce HTML. If we only use functions which print out something at the console, the HTML output would merely look as ugly as console output. So, to generate beautiful tables for our HTML report, we use functions like **xtable()**, **stargazer()**, **htmlreg()**, and **htmlTable()**. In fact, I only used so many diverse functions to demonstrate R's capabilities. In my own scripts I usually stick to one function for my tables throughout an analysis, for example, **xtable()**.

The Rhtml file starts with the HTML tags <html>, <head>, <title>, and <body>. This is standard in HTML files. The first chunk of R code is the first element of the body element. Its task is to source the data creation script, to include libraries into the search path, and to initialise the random number generator of R. The code chunk is named Read data as the second element of the <!--begin.rcode tag. As can be seen, code chunks offer parameters – the chunk options. Here, echo=FALSE sets that the R code will not be seen in the report and include=FALSE sets that the output of the chunk will not be included in the report either.

The first code chunk is followed by the HTML element tags <h1>, <p>, <h2>, and another <p>. The second code chunk then prints some random lines from the data frame using the **xtable()** and its associated print function. In **print()** we see, again, how the html.table. attributes parameter is used to turn off the border of the table in the report. Note that we used the parameter results="asis" in the opening tag of the chunk. This is necessary to prevent **knitr** from printing R output as R comments with ## at the beginning of each line (go back to our minimal example to check this). Because **print()** in this case delivers proper HTML, it should be placed "asis" in the HTML report.

The following code chunks all work in a similar fashion: they calculate descriptive statistics or results of an ANOVA or other linear models and present the results as HTML tables. Think about the script as a template and play with the parameters of the table-generating functions to find a layout that you like most.

To conclude, let us take another look at the structure of the Rhtml script. Although all elements are tagged appropriately, there is some prose here and some code there, some calculations and some output presentation. If the report were to grow, its appearance as code would look even more fragmented. So, wouldn't it be better to separate results calculation from results presentation? Perhaps it is better to put the calculations in one R script and then to source this script just before it is needed in the Rhtml report. This would remove all calculations from the report and leave only the presentation of the results. However, it requires the organisation of the statistical analysis into different scripts. We will practise this in the following example.

EXAMPLE! ## 8.2.2 ITC - Full report of results

In the reanalysis of the study testing for an influence of cognitive ability on logical reasoning, we calculated results and created figures in the foregoing chapters. Let us collect them all in a comprehensive report, which we could distribute in our team or research community. We use an Rhtml document to structure the report and include chunks of R code. We then compile the source document with **knitr**. Furthermore, we add a scatterplot showing the SAT score distributions for the different numbers of correct card selections.

ITC_FIG_ScatterplotSATByNumberCorrect.R

```
## ITC - Scatterplot SAT score by number of correct responses
source("./Scripts/ITC_PRE_LoadData.R")

with(ITC$stvc,
     {
         NumberCorrect <- rowSums(ITC$stvc[c("p1pnotq(correct)",
         "p2pnotq(correct)",
         "p3pnotq(correct)")])
         Means <- tapply(sat, NumberCorrect, mean)
         plot(NumberCorrect, sat,
              xaxt = "n",
              xlab = "Number of correct responses",
              las = 1)
         axis(1, at = 0:3)
         points(0:3, Means,
                cex = 2, pch = 18)
         text((0:3), Means,
              labels = round(Means, digits = 2),
              pos = c(4, 4, 4, 2))

     })
```

ITC_OUT_RecalculationReport.Rhtml

```
<!--## ITC results report-->
<html>

<head>
<title>ITC results report</title>
</head>

<body>
<!--begin.rcode Preparation, echo=FALSE, message=FALSE, warning=FALSE
rm(list=ls())
library(htmlTable)
library(xtable)
library(stargazer)
source("Scripts/ITC_PRE_LoadData.R")
source("Scripts/ITC_DES_TablesCorrectIncorrect.R")
source("Scripts/ITC_MOD_tTest.R")
```

```
source("Scripts/ITC_MOD_ANOVA.R")
source("Scripts/ITC_MOD_LogisticRegression.R")
end.rcode-->
```

<h1>Independence of thinking bias and cognitive ability - recalculation of results</h1>

<p>The report shows the recalculation of results from the replication study about the independence of thinking bias and cognitive ability, conducted by Baranski (2015).</p>

<h2>Frequency tables and figures</h2>

<p>The table shows the frequency of correct and incorrect solutions for the three card selection problems.</p>

```
<!--begin.rcode Frequency table, results="asis", echo=FALSE
m <- as.matrix(TablesCorrectIncorrect)
dimnames(m) <- list(`P1, P2` = c("inc, inc", "inc, co", "co, inc", "co, co"),
                P3 = c("inc", "co"))
htmlTable(m,
        caption = "Distribution of responses")
end.rcode-->
```

<p>As can be seen in the following histogram, the average SAT scores seem to differ between genders. The scatter plot shows a slight increase in average SAT scores with more correct card selections. However, different variances between groups and one outlier seem to be evident from this view of the data.</p>

```
<!--begin.rcode Histograms, echo=FALSE
source("Scripts/ITC_FIG_HistogramSATByGender.R")
end.rcode-->
<!--begin.rcode Scatter plot, echo=FALSE
source("Scripts/ITC_FIG_ScatterplotSATByNumberCorrect.R")
end.rcode-->
```

<h2>Model summaries</h2>

<p>The following t-test summary is the recalculation of the test by Baranski (2015). The subsequent analyses of variance show an alternative analyses for the hypotheses. Finally, the logistic regression summaries respect the experimental meaning of the variables, setting the proportion of correct responses as the dependent and SAT scores as the independent variable.</p>

```
<h3>Summary of the t-test.</h3>
<!--begin.rcode t-test, results="asis", echo=FALSE
ITC_tTest
end.rcode-->


<h3>Summaries of ANOVA models for the three card selection tasks.</h3>
<!--begin.rcode ANOVA, results="asis", echo=FALSE
print(xtable(summary(ITC_ANOVA[[1]])), type = "html")
print(xtable(summary(ITC_ANOVA[[2]])), type = "html")
print(xtable(summary(ITC_ANOVA[[3]])), type = "html")
end.rcode-->


<h3>Summaries of logistic regression models (estimate and standard error).
</h3>
<!--begin.rcode Logistic regression, results="asis", echo=FALSE
stargazer(LogisticRegressionModels,
          type = "html",
          single.row = TRUE,
          dep.var.labels = c("Problem 1 correct",
                             "All problems correct"))
end.rcode-->
</body>
</html>
```

I placed this script file one folder level above all other script files for the project, that is, in the main project folder. The script then finds all required R scripts in the scripts directory.

- IndependenceThinkingBiasCognitiveAbility
- ITC_OUT_RecalculationReport.Rhtml

- 📁 Data
- 📁 Scripts
    - 📄 ITC_PRE_LoadData.R
    - 📄 ITC_DES_TablesCorrectIncorrect.R
    - 📄 ITC_DES_CorrelationLogicIndexSAT.R
    - 📄 ITC_MOD_tTest.R
    - 📄 ITC_MOD_ANOVA.R
    - 📄 ITC_MOD_LogisticRegression.R
    - 📄 ITC_FIG_HistogramSATByGender.R
    - 📄 ITC_FIG_ScatterplotSATByNumberCorrect.R
- 📁 Results
- 📁 Figures

Output

Independence of thinking bias and cognitive ability - recalculation of results

The report shows the recalculation of results from the replication study about the independence of thinking bias and cognitive ability, conducted by Baranski (2015).

Frequency tables and figures

The table shows the frequency of correct and incorrect solutions for the three card selection problems.

Distribution of responses

	P3	
	inc	co
P1, P2		
inc, inc	57	3
inc, co	3	14
co, inc	10	5
co, co	6	79

As can be seen in the following histogram, the average SAT scores seem to differ between genders. The scatterplot shows a slight increase in average SAT scores with more correct card selections. However, different variances between groups and one outlier seem to be evident from this view of the data.

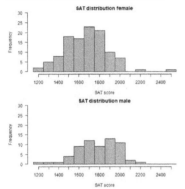

Model summaries

The following t-test summary is the recalculation of the test by Baranski (2015). The subsequent analyses of variance show an alternative analyses for the hypotheses. Finally, the logistic regression summaries respect the experimental meaning of the variables, setting the proportion of correct responses as the dependent variable and SAT scores as the independent variable.

Summary of the t-test.

Two Sample t-test data: sat by p1_pnotq_correct_relevelled t = 0.99484, df = 175, p-value = 0.3212 alternative hypothesis: true difference in means is not equal to 0 95 percent confidence interval: -31.50548 95.55087 sample estimates: mean in group correct mean in group incorrect 1735.714 1703.692

Summaries of ANOVA models for the three card selection tasks.

	Df	Sum Sq	Mean Sq	F value	Pr(>F)
p1_pnotq_correct	1	43393.47	43393.47	0.99	0.3212
Residuals	175	7672831.11	43844.75		

	Df	Sum Sq	Mean Sq	F value	Pr(>F)
p2_pnotq_correct	1	71987.05	71987.05	1.65	0.2009
Residuals	175	7644237.52	43681.36		

	Df	Sum Sq	Mean Sq	F value	Pr(>F)
p3_pnotq_correct	1	99805.32	99805.32	2.29	0.1317
Residuals	175	7616419.25	43522.40		

Summaries of logistic regression models (estimate and standard error).

	Dependent variable:	
	Problem 1 correct	All problems correct
	(1)	(2)
sat	0.001 (0.001)	0.001 (0.001)
Constant	-1.695 (1.289)	-1.811 (1.274)
Observations	177	177
Log Likelihood	-118.292	-120.856
Akaike Inf. Crit.	240.584	245.711
Note:		*p<0.1; **p<0.05; ***p<0.01

Figure 8.4 Example ITC results report.

Analysis

The first chunk of R code is called `Preparation` and it uses **source()** to include all the required script files consisting of calculations. We programmed these scripts in the previous chapters and they all contain inclusion guards to control for the creation of objects. They all return one object with results. The Rhtml report only calls these results objects or uses functions for results display in the subsequent code chunks. As can be seen, we separated calculation from results display throughout the analysis. In sum, the recalculation of the original *t*-test and the additional analyses cannot establish an association between cognitive ability and thinking bias.

8.2.3 Report generation

The package `knitr` helps us to make beautiful reports that you can hand in for the next team meeting. You can structure the Rhtml file with appropriate headings, prose, statistical tables, and figures. Combined with cascading style sheets (CSS, not covered here), `knitr` can further yield documents according to your own house style. Perhaps one team member knowledgeable in HTML and CSS creates an Rhtml document as a report template for the whole team and then all other members could create reports with a homogeneous visual appearance. Play around with the `knitr`, HTML, and CSS and further streamline your production of statistical results.

8.2.4 Key steps to create a table from scratch

The foregoing sections introduced many options to create tables with functions of different packages. Sometimes the number of options is overwhelming. Moreover, although some table functions have many parameters, they may not satisfy your particular needs. Consequently, you still have to tinker with the table in your word processor after creation. This is time-consuming and inefficient, especially if the numbers in your calculations change. Therefore, I would like to suggest a few key steps to take when you program a results table.

1. Draw a sketch of the table with paper and pencil. You will immediately see what you find important to display. Which rows and columns would you include? Which row and column labels would you employ? Later, the table functions in R will reveal how closely their output corresponds with your sketch. You can then decide where to spend time for fine-tuning.
2. Identify all objects in your R workspace which contain information for the table. Usually, descriptive statistics or model summaries go in a table for publication. Is there one object which holds all the necessary information, or is it required to aggregate information from different objects? Often, `cbind()`, `rbind()`, `matrix()`, or `data.frame()` provide good help for results aggregation.
3. Usually, the available objects in your workspace do not contain the particular row names, column names, or other labels that are required for a publication-ready table. During the analysis, one often works with abbreviations while in the publication we need written out labels. For example, the variable name t0_dep should become *Depression score at baseline* in the table. Sometimes, attributes with adequate labels for publication are available, but more often you need to create additional string objects with labels and pass them to the table-generating function.
4. Search for table-generating functions in different packages. In this chapter we already practised some good functions. For example, use the packages **R2HTML**, **sjPlot**, **xtable**, **htmltools**, **htmlTable**, **texreg**, or **stargazer**. For an important table, I usually try them all and then decide which table corresponds best with my sketch.
5. Do fine-tuning with the function of your choice. Study its parameter list for all options to change the appearance of the table. If necessary, use the additional string objects as further arguments for the table. For example, the default arguments for row and column labels usually need to be replaced with more comprehensible strings.

These key steps may take you from a raw sketch of your results table to the publication-ready table. A nice side effect of a programmed table is that you simply need to run the script for its creation again if the numbers or calculations of the analysis change.

8.3 Structuring the output of results

Up to now, some parts of your R code calculate results and other parts present the results. It is good advice to separate calculation from presentation. Furthermore, to present your results a few strategical considerations might help.

PROGRAMMING
ISSUE!

8.3.1 Separate calculation from presentation

Earlier, we talked about the division of the whole analysis in several scripts. To structure and connect them, we used the **source()** function and inclusion guards. In our example scripts we usually programmed one R object as output and then called it at the console or printed it in a report. One advantage of this strategy is separation of calculation from presentation. Reports and other outputs can easily use the same results objects by sourcing the respective script. Beside the console, such diverse outputs may be an HTML table, a figure, a conference presentation, or a paper manuscript. If we did not separate calculation from presentation, all presentation scripts would include the same calculations and thereby needlessly repeat code. Furthermore, changing calculations after some time in one place leaves outdated and wrong calculations lingering at other places in the code repository. So, it is better to maintain only one version of a particular calculation in only one place and then use **source()** to easily update all outputs to the actual and valid results. However, to calculate and present descriptive statistics sometimes requires only a few lines of code and it seems like needless complexity to use one script for calculations and another one for presentation. Conversely, as we have seen, results presentation can be very involved itself and may clutter statistical calculations. So, it is better to use self-contained scripts for results presentation. In sum, write scripts which create results objects and then source these scripts in the presentation scripts – separate calculation from presentation.

TEAMWORK!

8.3.2 The content of results tables

When working in a highly productive research team, you will constantly be calculating, presenting and publishing your results. That is, your job is not only to plan and interpret the statistics, but also to decide on an adequate display of which your audience can make sense. Besides the display of statistical results, your tables may profit from some more planning in your research team. Here are some further suggestions for you and your team to gather good content for your tables.

1   Enable recalculation of statistical results. Use information from your data that can be used to recalculate your inferential statistics and which offer the opportunity to calculate other inferential statistics that you did not include. For example, for all experimental conditions, variable means, standard deviations, and sample sizes let others recalculate your analysis of variance and other statistics. This information is also relevant if your study may enter the data pool for a meta analysis.

2   Be consistent across your own studies or studies of your team. Some research teams follow a research agenda which spans several studies. Often, all these studies have a

number of common variables or statistical methods. So, make sure to align the results of your own study with the other studies from your team regarding variables and statistical methods.

3 Search the literature for other important studies in your research domain and align with these studies. In several research domains researchers tackle a core set of issues or research questions. Often there is also a canon regarding which variables and analyses are relevant for empirical investigations. If you align your analyses and results presentation with this canon, you may be lucky and find citations of your work after some time.

4 Respect the author guidelines for results presentation from your target journal. This avoids much trouble in getting published. If the target journal want two-sided *t*-tests, make sure to provide these tests and not one-sided tests. If you disagree for good reason with this approach to statistical testing, then better search for another target journal. Similarly, also respect the guidelines for results preparation from professional associations from your research domain (e.g., the American Psychological Association (American Psychological Association, 2001) for psychologists).

This collection of hints is not comprehensive, but it provides some strategical considerations to the preparation of your results tables. It adds on to the more technical approach of the key steps to create a table (see above).

8.3.3 Convert the results report in other file formats using pandoc

HELPER SOFTWARE!

Many scientists write their papers in OpenOffice or Microsoft Word. For presentations, many use PowerPoint. However, R results are usually stored in plain text format or in HTML. We now use the universal and open source document converter pandoc to complete our workflow of results programming and produce results files suitable for OpenOffice, Word, or PowerPoint. Although pandoc is not R, we will automatically invoke the software in our R scripts.

Our first step is to download pandoc from https://pandoc.org/. I recommend downloading the program as a zip package and unpacking it close to your R working directory. It is a command line program without a graphical user interface; therefore, we may run it from our R scripts and need to take care of file paths of the files we will be converting. Saving the results in Word or OpenOffice format now works like this: we will first produce an Rhtml script consisting of HTML markup and R code. We then write a short auxiliary script to produce an HTML report with **knitr()** and then to convert this report to one of the office formats using pandoc.

Ch8_StatisticalReport.Rhtml

```
<!--## Statistical Report-->
<html>

<head>
<title>Statistical report</title>
</head>
```

```
<body>
<h1>Comparison of two independent samples</h1>
<p>This report presents a table, a t-test and a figure.</p>

<h2>The data set</h2>
<p>Take a look at the data.</p>
<!--begin.rcode
set.seed(12345)
d <- data.frame(A = gl(2, 10, labels = c("a1", "a2")),
                Y = round(rnorm(20), 2))
d
end.rcode-->

<h2>t-Test</h2>
<p>Compare the means of Y between the two levels of A.</p>
<!--begin.rcode
t.test(Y ~ A, d)
end.rcode-->

<h2>Bar diagram</h2>
<p>Bar diagram of the association between A and Y.</p>
<!--begin.rcode fig.width=7, fig.height=6
barplot(tapply(d$Y, d$A, mean))
end.rcode-->

</body>
</html>
```

This first script is an Rhtml report consisting of simple calculations with random data. Save this script in your working directory, but do not yet generate the report. Report generation is done in the following console script, together with file conversion.

Ch8_PandocStatisticalReport.R

```
## Convert statistical report with pandoc
library(knitr)

knit("Ch8_StatisticalReport.Rhtml")
system("pandoc.exe Ch8_StatisticalReport.html -s -o
Ch8_StatisticalReport.docx")
system("pandoc.exe Ch8_StatisticalReport.html -s -o
Ch8_StatisticalReport.odt")
```

Output

Statistical report

Comparison of two independent samples

This reports presents a table, a t-test and a figure.

The data set

Take a look at the data.

```
set.seed(12345)
d <- data.frame(A = gl(2, 10, labels = c("a1", "a2")),
                Y = round(rnorm(20), 2))
d
```

```
##      A     Y
## 1   a1  0.59
## 2   a1  0.71
## 3   a1 -0.11
## 4   a1 -0.45
## 5   a1  0.61
## 6   a1 -1.82
## 7   a1  0.63
## 8   a1 -0.28
## 9   a1 -0.28
## 10  a1 -0.92
## 11  a2 -0.12
## 12  a2  1.82
## 13  a2  0.37
## 14  a2  0.52
## 15  a2 -0.75
## 16  a2  0.82
## 17  a2 -0.89
## 18  a2 -0.33
## 19  a2  1.12
## 20  a2  0.30
```

t-Test

Compare the means of Y between the two levels of A.

```
t.test(Y ~ A, d)
```

```
##
## 	Welch Two Sample t-test
##
## data:  Y by A
## t = -1.1273, df = 17.979, p-value = 0.2744
## alternative hypothesis: true difference in means is not equal to 0
```

```
## 95 percent confidence interval:
##  -1.1970628  0.3610628
## sample estimates:
## mean in group a1 mean in group a2
##           -0.132            0.286
```

Bar plot

Bar plot of the association between A and Y.

```
barplot(tapply(d$Y, d$A, mean))
```

Figure 8.5 Statistical report converted to OpenOffice format with pandoc.

Analysis

The **knitr** library and its function **knit()** generates the statistical report as an HTML file. This could already be opened with a web browser. However, **system()** then obtains a string consisting of a command for the computer system to execute (e.g., for windows). The command calls the pandoc.exe program and supplies the just-generated HTML file as input and specifies a docx document as output. Note that if pandoc and the source document do not reside in the current working directory, you must spell out their paths completely. The second file conversion with pandoc generates an odt file suitable for OpenOffice (Figure 8.5). Locate the two files in your working directory and inspect their content. As can be seen, pandoc together with R give you the power to write results in different types of output files.

Some closing remarks about document conversion. Our example shows a blueprint for document conversion. Just try to fit it with the analysis for your own data or test how the converter works for different kinds of R objects. Perhaps tables generated with **stargazer** or **htmlTable** are worth a try. Our example can be modified to produce pdf documents, too. However, this still requires another software for the conversion, but you can check the comprehensive pandoc manual for how to do it. Moreover, we practised a simple example only. Pandoc gives you much control about formatting the output if you dive deeper into its functionality. Apart from R it is generally a powerful converter that runs from the command line of the operating system (e.g., Windows). Find more on document conversion here: www.r-statistics.com/2013/03/write-ms-word-document-using-r-with-as-little-overhead-as-possible/.

▬▬▬▬▬ What you have learned so far ▬▬▬▬▬

1   To store results in a text file
2   To create publication-ready tables with different add-on packages
3   To generate full reports of a statistical analysis
4   To separate calculation from presentation of results
5   The key steps to create a publication-ready table
6   To convert results files between different file formats

▬▬▬▬▬ Exercises ▬▬▬▬▬

Explore functions

1   Use the **MusicPreferencesSchool.csv** data set and create summary tables for descriptive statistics, an ANOVA, and a linear regression. Then practise table creation using functions of the packages **R2HTML**, **htmlTable**, **tables**, and **sjPlot**. Think about which of the functions suits you most.
2   Create one or more linear regression models with the **MusicPreferencesSchool.csv** data set. With these models compare the functions **stargazer()** and **htmlreg()** from the **stargazer** and **texreg** packages, respectively. Study the details of the parameter lists of both functions. Which parameters manipulate the out in a similar way?
3   Create a table with the **htmlTable** package and the function **htmlTable()**. Practise its CSS capabilities to modify the appearance of the table.

Solve tasks

1   Convert a data.frame or a model table of your choice to an xtable object and save it to a file using **xtable.print()**. Then, consulting the online documentation at http://pandoc.org, convert this file to a OpenOffice/LibreOffice *.odt file.
2   Write an R function **tomsword()** that takes as input arguments a filename and the output of xtable(..., type = "latex"), writes a .tex intermediary file, and then call pandoc to convert this to .docx.

Read code

1   The following code fragment uses the **MusicData.csv** data set to demonstrate how to write an HTML table in an HTML file. Describe in your own words what **htmlTable()** and **cat()** do.

```
## Descriptive statistics in an htmlTable

# Preparation
library(psych)
library(htmlTable)
d <- read.csv("MusicData.csv")

# Create table
descr <- describe(d[paste0("Ex", 1:20)])
HtmlDescriptives <- htmlTable(round(descr, digits = 2),
                              useViewer = FALSE)
cat(HtmlDescriptives,
    file = "HtmlDescriptives.html")
```

Apply in the real world

1   Separate calculation from results presentation in the analysis of your own data. To do this, prepare some scripts only for calculations and others for presentation (e.g., an Rhtml script to generate an HTML report). In the scripts devoted to calculations, store all results in objects. In the scripts for presentation, invoke the calculation scripts with **source()** and then program the display of the objects comprising the results.

Become a statistics programmer

1   The **knitr** package helps you to easily create HTML reports for your data analysis. The code chunks in each Rhtml file consist of some parameters to control the appearance of output in the final report. Study these parameters on the **knitr** reference card (Xie, 2021b), for example, to include or not include the source code in the report.

WBD   EXAMPLE!

1   Use the packages **stargazer** and **texreg** to create publication-ready tables with the regression analyses from the exercises in Chapter 7.
2   Create a full HTML report covering descriptive statistics, graphics, and regression models. Use the <p></p> element in HTML to add text to the report.

ITC   EXAMPLE!

1   Go back to the ANOVA for this example in the main text in Chapter 6. Use **xtable()** from the **xtable** package to create an HTML table for the ANOVA summary. Use **print.xtable()** with parameter setting type = "html".

(Continued)

2   Go back to the exercises for this example in Chapter 7 and again initialise the simple linear regression and the multiple regressions. Use **stargazer()** from the **stargazer** package to arrange the model summaries of all regressions in an HTML table.

EXAMPLE!   ## UPS

1   Use the package **xtable** to create HTML tables of the ANOVA results of Chapter 6. Use the functions **xtable()** and **print.xtable()** with the parameter setting type = "html".

EXAMPLE!   ## SCB

1   Go back to the SCB example in Chapter 8 in the main text. In the example, we created a table of regression results with **stargazer**. Recreate the HTML summary table using the **texreg** package with the function **htmlreg()**.

2   Create a full HTML report for the SCB example. Include descriptive statistics, graphics, ANOVA results, and regression results from the scripts in the main text of the foregoing chapters.

9

PREPARING GRAPHICS FOR PUBLICATION

━━━━━━━━━━━━━━━━━━━━━━━━━━━ Chapter overview ━━━━━━━━

This chapter shows R's great power to create graphics from scratch. You will practise the standard graphical utilities and other graphical systems R offers, like grid graphics or ggplot2. Programming features: graphics programming, graphic systems, 3D graphics.

9.1 Planning graphics for publication

Good graphics tell a story and highlight important results. So, graphics for publication need planning. In your research domain you and your colleagues know which graphics are common in the journals and which you need to convey your message. Sketch your graphics first and further obtain feedback for your design.

TEAMWORK! ### 9.1.1 Sketch your graphic

Designing and programming graphics from scratch uses R's full power. To practise this is pure fun! As you go along, with each example you will come up with many ideas of how to improve the display or which graphic you want to create yourself. The best way to start is possibly to focus on one key message from your results that the graphic should convey and then to quickly sketch this graphic. Prioritise the statistical information that it contains. Also think about the format: is the graphic for publication in a scientific journal or is it for a presentation? Is it restricted to black-and-white printing or can it use full colour. Perhaps you can even draw several sketches of one graphic and then select the best one. For example, profile diagrams are ideal to present multivariate data from different experimental groups. A profile for the grand mean vector and of profiles for the means of each group may look nice. So, sketch the profile diagram first. Play with the different proportions of plot region, figure region, and device region; test for the best orientation of axis annotation.

The literature of your research domain may be another source for inspiration. Skim through the recent issues of journals relevant in your field and see which types of graphics other researchers use. For example, do they include figures with the presentation of raw data, like in scatter plots, or is it usually aggregated data, like in bar diagrams? How do the journals include data presentation of qualitative data in comparison with quantitative data. Note how figures are annotated, how are the margins of the figure spaced, is a combination of main diagrams and auxiliary diagrams common? Use all this to refine your sketch. Another renowned standard of graphic design and layout for data display is Cleveland (1994).

PLANNING ISSUE! ### 9.1.2 Design of graphics

Another source of inspiration comes from the literature on graphic design: Duarte (2008) in her book about the creation of effective presentations gives a helpful introduction to some rules of data display. *Besides the general remark to keep things simple in data display, she distinguishes three conceptual layers that every graphic should satisfy in order to convey its message: background, data, and emphasis.* The background consists of the layout of the display, the coordinate system, axis annotation, and auxiliary lines in a diagram. The data is what we consider the main content of the diagram, say, a scatter plot, a bar chart, or mean profiles. The emphasis means selecting an aspect of the given display and manipulating it in order to catch the observer's attention, for example, to enlarge a single profile out of a bunch of profiles or to give data from a sample subgroup a special colour in a scatter plot. To use the three conceptual layers, we could use a high-level function to create the background and display the data, and then use a low-level graphic function to add emphasis to the figure. This takes only a few lines of code.

Ch9_PlotWithEmphasis.R

```
## High-level plot with emphasis
# Data creation
set.seed(12345)
X <- rep(1:10, each = 10)
Y <- X + rnorm(100, 0, 2)

# One high-level and one low-level function
plot(X,Y)
points(X[X > mean(X) & Y > mean(Y)],
       Y[X > mean(X) & Y > mean(Y)],
       pch = 19, col = "red", cex = 1.2)
```

Output

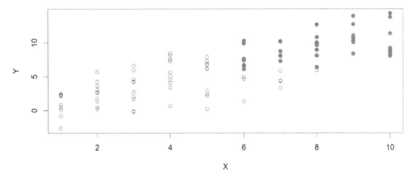

Figure 9.1 Diagram with emphasis.

Analysis

The script creates the variables X and Y first. It then uses **plot()** as a high-level function to draw the background and the data; and then it uses **points()** as a low-level function to highlight parts of the data. In this case, all data points that fall above the means of X and Y, respectively, are enlarged and coloured red.

In recent years the grammar of graphics appeared in the R community, mainly endorsed by Hadley Wickham and his development of the **ggplot2** library (Wickham, 2016). The grammar of graphics is an elaborated strategy to display data, and **ggplot2** encompasses this strategy in its syntax. Like the **lattice** library, it also builds on the **grid** graphics system introduced by Paul Murrell (2002). The grammar of graphics says how to map data to geometric objects like points, lines, or bars, and to their aesthetic properties like colour, shape, and size. The grammar offers a complex system and specifies the interplay between data, statistics, geometrical objects, and scales. It further allows us to think in layers of graphics; that is, we can overlay several iterations of mapping between data and aesthetical attributes of geometrical shapes. For example, to draw a scatter plot of raw data in the back and overlay it with a display of group means. The package **ggplot2** supports thinking in terms of the grammar and allows us to program accordingly. However, to work with layers of data and to use highlighting elements due to aesthetic manipulations is also possible with base graphics. Simply create a new figure and put all the graphical elements you need in it.

Ask your team or supervisor to provide feedback for your design. Criteria for such feedback may be whether the figure displays too much or too little data or whether the figure would fit in with the style of the target journal for publication. Update the design according to this feedback; in most cases they will improve and your colleagues will enjoy seeing their contributions to the final result.

9.2 Graphics from scratch

We will program a complete display using low-level graphics functions. Starting with an empty page, these functions add elements according to the painters model. We now also employ fonts in our graphics.

PLANNING
ISSUE!

9.2.1 Key steps to create a figure from scratch

Let us start with a list of key steps to create the figure:

1   calculate statistics for the display,
2   initialise the graphics file, for example png or pdf,
3   structure the device region,
4   set sizes of device, figure, and plot regions using **par()**,
5   initialise the first diagram with **plot.new()** and **plot.window()**, or use a high-level function,
6   add low-level elements to the plot,
7   repeat points 5 and 6 until all figure regions in the device are filled with a plot,
8   close the graphics file.

Keep these general steps in mind when we proceed through the details of the following blueprint for a plot.

Ch9_BlueprintPlot.R

```
## A blueprint for a plot from scratch
# Create data
set.seed(12345)
d <- data.frame(X <- rnorm(100))

# Create the plot
par(mar = c(4, 4, .5, .5), oma = c(0, 0, 0, 0))
plot.new()
plot.window(xlim = c(0, 100), ylim = c(-5, 5))
box()
axis(1)
```

```
axis(2)
mtext("Index", 1, line = 2)
mtext("Random value", 2, line = 2)
points(1:100, d$X)
```

Output

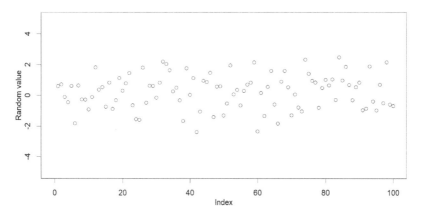

Figure 9.2   Output of a blueprint of a diagram from scratch.

Analysis

The code first sets parameters with **par()** to change sizes of plot and figure region; it then initialises the diagram with **plot.new()** and **plot.window()**. Finally, low-level graphic functions add elements to the diagram. It is helpful to run the code line by line to see the diagram evolve. Notice the different roles of **plot.new()** and **plot.window()**: the first substitutes a high-level function and simply initialises a new diagram without anything else, the second sets up the coordinate system of the diagram. Let us proceed to our study data and do a complex example where we use the **layout()** function and many other low-level functions.

9.2.2 UPS - Mean profiles   EXAMPLE!

The unpleasant sounds example is perfect for a demonstration of graphics from scratch. The following is a longer script, which uses many low-level graphic functions and prints its output directly in a png file. Our aim is to create mean profiles for each experimental group in all sound conditions.

UPS_FIG_MeanProfiles.R

```
## UPS - Mean profiles
source("Scripts/UPS_PRE_DataImport.R")

ith(UPS,
```

```
{
    # Preparation
    soundvariation <- interaction(sound, variation, sep = " ")
    LevelsOrder <- as.vector(matrix(1:28, nrow = 7, byrow = TRUE))
    levels(soundvariation) <- levels(soundvariation)[LevelsOrder]
    M <- tapply(rating, list(soundvariation, group, gender), mean)
    ftM <- ftable(M, col.vars = c(2, 3))
    M_AcrossGroupsAndGender <- tapply(rating, list(sound, variation),
    mean)
    M_ByGroupsAndGender <- tapply(rating, list(group, gender), mean)

    # Initialise diagram
    png(filename = "Figures/UPS_MeanProfiles.png",
        width = 1920, height = 1080,
        pointsize = 20)
    layout(matrix(c(1, 1, 1, 2,
                    1, 1, 1, 2,
                    1, 1, 1, 2,
                    1, 1, 1, 2,
                    1, 1, 1, 2,
                    3, 3, 3, 4), nrow = 6, byrow = TRUE))

    par(mar = c(0, 0, 3, .5), oma = c(1, 16, 0, 0))

    # Main diagram
    plot.new()
    plot.window(xlim = c(1,6), ylim = c(28, 1))
    box()
    segments(rep(1, 28), 1:28, rep(6, 28), lty = "dotted", col = "grey80",
lwd = 2)
    colors = paste0("grey", c(20, 40, 60, 80))
    lapply(as.list(1:4), function(X) lines(ftM[,X], 1:28,
                                           type = "b", pch = 16,
                                           col = colors[X],
                                           lty = X, lwd = 2,
                                           cex = 1.5))

    # Main diagram annotation
    mtext("Unpleasant sounds - Mean profiles by condition", 3, line =
    1, at = 1, adj = 0, cex = 1.5)
    mtext(levels(UPS$sound),
          side = 2, line = 9, at = c(1, 8, 15, 22), las = 1)
    mtext(levels(UPS$variation),
```

```
                   side = 2, line = 1, at = 1:28, las = 1)
        legend(1, 1, c("Female, origin known", "Male, origin known",
                       "Female, origin unknown", "Male, origin unknown"),
               cex = 1.5, border = "white",
               pch = 16, col = colors, lty = 1:4,
               title = "Condition", title.adj = 0.05)


        # Right diagram
        plot.new()
        plot.window(xlim = c(1,6), ylim = c(28, 1))
        box()
        axis(1, cex.axis = 1.2)
        segments(rep(1, 28), 1:28, rep(6, 28),
                 lty = "dotted", col = "grey80", lwd = 2)
        lines(as.numeric(M_AcrossGroupsAndGender), 1:28,
              type = "b", pch = 16, lwd = 2,
              cex = 1.5)
      mtext("Mean profile", side = 3, line = 1, at = 1, adj = 0, cex = 1.5)
        mtext("Rating", side = 1, line = 2)


        # Bottom diagram
        par(mar = c(3, 0, .5, .5))
        plot.new()
        plot.window(xlim = c(1,6), ylim = c(4.5, .5))
        box()
        axis(1, cex.axis = 1.2)
        segments(rep(1, 4), 1:4, rep(6, 4), lty = "dotted", col = "grey80",
        lwd = 2)
        points(as.numeric(M_ByGroupsAndGender), 1:4, cex = 1.5, col = "black",
ch = 16)
        mtext("Rating", side = 1, line = 2)


        # Bottom diagram annotation
        mtext(c("Female", "Male"),
              side = 2, line = 11, at = c(1, 3), las = 1)
        mtext(rep(c("Origin known", "Origin unknown"), 2),
              side = 2, line = 1, at = 1:4, las = 1)


        # Dummy diagram at the bottom right
        plot.new()
        # Close png graphics device
        dev.off()
    })
```

Output

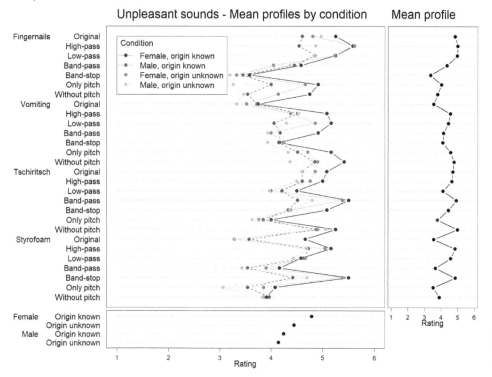

Figure 9.3 Unpleasant sounds mean profiles.

Analysis

Let us read through the script together. First, we include the script to load the data, in case the data frame is not already available. Then `with()` creates an environment for all of the subsequent calculations. It allows us to simply call the variables we need without indexing and it takes care that temporary objects are deleted after job completion. The script initially creates the factor soundvariation consisting of all combinations of levels of sound and variation. Then it calculates several means for the ratings respecting sound, variation, group, and gender and stores them in three objects: M, M_AcrossGroupsAndGender, and M_ByGroupsAndGender. We need these means for the diagrams. The script then calls **png()** to create a png file. It passes the desired filename and the sizes of the figure as parameters. The counterpart to this function is **dev.off()**, which closes the png file at the very end of the script. Next, **layout()** and **matrix()** structure the figure. In the six by four matrix it can be seen that the device region is divided into four figure regions, numbered from one to four. Figure region 1 goes to the top left and is three times as wide as region 2 to its right-hand side. Region 3 goes below and is only one fifth of the height of region 1. Region 4 is a dummy region just to complete the matrix. We will not use it but simply put an empty region here with **plot.new()**. Next, **par()** with its parameters mar and oma sets the size proportions of the device region, the figure regions, and the plot regions. After all this preparation, the script then draws the main diagram, that is, the big diagram in the top left corner of the device region. The functions **plot.new()** and **plot.window()** start the diagram, and all its visible elements are created by low-level graphics functions: **box()**, **segments()**, **lines()**, **mtext()**, and **legend()**. In the function calls, we use many parameter settings to create the diagram according to our wishes; in fact, a convenient high-level function would have done all this in the background. However, we have to do all these settings by hand. After all elements of the first diagram are in place, we proceed to the next diagram. It is in the top right corner and is also initialised with **plot.new()** and **plot.window()**. This diagram also uses the function **axis()** to create an x-axis at the bottom. Finally, the diagram at the bottom is created with the use of the same collection of low-level functions. After this diagram, **plot.new()** fills the fourth figure region with emptiness and then the figure is finished with **dev.off()** and the with environment is closed.

9.2.3 Using fonts in graphics

R is not that good in fonts. We can easily switch between serif, sans serif, and mono font family and also switch between font faces regular, bold, and italic. However, to use a certain arbitrary font, as is possible in a writing program, requires effort to set up. In the following two minimal examples I show R's basic font options and then how the **showtext** package supports you to use any font installed on your system.

Ch9_FontsInGraphics.R

```
## Using fonts in graphics
plot.new()
plot.window(xlim = c(0, 5), ylim = c(0, 5))
box()

text(x = 1:4, y = 4,
     labels = c("A", "beautiful", "graphic", "annotation"),
     font = 1:4,
     family = "serif")
text(x = 1:4, y = 3,
     labels = c("More", "words"),
     font = 1:4,
     family = "sans")
text(x = 1:4, y = 2,
     labels = c("Friendly", "words"),
     font = 1:4,
     family = "mono")
```

Output

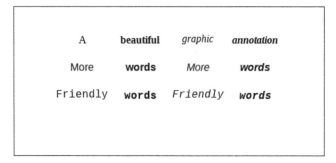

Figure 9.4 Using font styles in graphics.

Analysis

The output shows a few words in a dummy graphical display (Figure 9.4). We call the `text()` function three times and manipulate the font and family parameter to write three lines. The top line cycles through the four faces of a serif font. The middle line does the same for a sans serif font, and the bottom line for a mono type font.

To employ all fonts that your system knows, use the `showtext` package. It easily lets you include any font in your display. The following lines at the console show the available fonts. The paths and the file names returned by `font files()` figure as arguments for the `font add()` function in the subsequent script.

Ch9_SystemFonts.R

```
## Search for fonts in the system
library(showtext)

font_families()
font_paths()
dir(font_paths())
font_files()
```

Ch9_showtextExample.R

```
## Using fonts with the showtext package
library(showtext)

font_add("Bookman Old Style", "C:\\Windows\\Fonts\\BOOKOS.TTF")
font_add("Calibri",
         regular = "C:\\Windows\\Fonts\\calibri.ttf",
         bold = "C:\\Windows\\Fonts\\calibrib.ttf",
         italic = "C:\\Windows\\Fonts\\calibrii.ttf")

showtext_auto()
plot.new()
```

```
plot.window(xlim = c(0, 4), ylim = c(0, 4))
box()


text(x = 1, y = 3,
     labels = c("Serious looking Bookman Old Style"),
     family = "Bookman Old Style",
     adj = 0)
text(x = 1:3, y = 2,
     labels = c("Modern", "looking", "Calibri"),
     font = 1:3,
     family = "Calibri")
```

Output

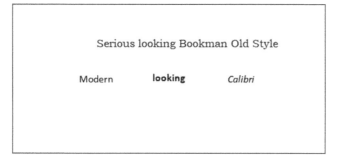

Figure 9.5   Changing fonts in graphics.

Analysis

The script includes the **showtext** package and then calls **font add()** to make Bookman Old Style and Calibri available for diagrams (Figure 9.5). The first parameter of **font add()** is an arbitrary name of how you want to call the font in the graphics function; the second parameter is the file path on your system, where to find the font in its regular version. The optional parameters bold and italic can take further font faces associated with a given font. Then, **showtext auto()** switches on the use of the newly added fonts. The **text()** function then obtains Bookman Old Style and Calibri as arguments for the family parameter. This way fonts are easy to use in graphics. To select a good and fitting font for your graphical display can make a difference in communicating your results. So, play around with it in your next data displays.

9.3 Graphic systems

The other graphic systems of R follow a different logic than base graphics. The systems are not necessarily compatible. Let us visit some of them, so that you can decide which one you prefer.

PLANNING
ISSUE!

9.3.1 Comparison of graphic systems

Prominent other graphic systems in R are `grid`, `ggplot2`, and `rgl`. These systems require one to think of graphics differently compared with base graphics. However, `ggplot2` is, in fact, based on grid. A graphic system is not merely a graphical library or graphic function within a library. Such libraries as `plotrix` (Lemon, 2006) or `gplots` (Warnes et al., 2020), or graphical functions within libraries like the `Hmisc` or the `psych` library, are not graphic systems because they all work with base graphics. Other libraries like `sjPlot`, which uses `ggplot2`, and `lattice`, which uses `grid`, are not graphic systems. Another graphic system uses different building blocks for the graphics. An example is `grid`, which uses so-called viewports as its basic layout elements. It does not comply with the distinction between plot, figure, and device region characteristic of base graphics. Viewports are simply rectangles that can be used for graphical display and that can occupy any position on the graphic device. A graphic device can contain any number of such viewports. Another graphic system is `rgl` graphics, which implements three-dimensional graphics. These graphics can be rotated and viewed from different angles and they can be animated. In the following sections, we will create graphics with `grid`, `ggplot2`, and `rgl`. Afterwards you can decide which system you prefer.

R LANGUAGE!

9.3.2 Grid graphics

Paul Murrell developed a graphic system for R, which is incompatible with base graphics and which endorses a different philosophy of creating a figure. His approach is described fully in Murrell (2016). The graphics device is no longer conceived of as structured in device, figure, and plot region. In grid graphics it is only the graphic device and rectangles on the device that hold content. Grid graphics are an alternative to base graphics. You have control over every aspect of the diagram, but you have to be familiar with some particular topics pertaining to the grid system: the use of graphical primitives, the control of viewports, coordinate systems, and graphical parameters.

Graphical primitives refer to the symbols to be drawn to represent the data points. Also the placement of text in a diagram requires a graphical primitive. The primitives are each called with their own function after the graphical display and the coordinate system has been set up. Such functions are **`grid.points()`**, **`grid.text()`**, **`grid.rect()`**, **`grid.segments()`**, **`grid.lines()`**, or **`grid.polygon()`** among others.

Graphical viewports are rectangular regions in the graphics device, which actually hold the data display. Any number of viewports in complex arrangements or even nestings can be placed on the graphics device. In a graphics device, which is considered the top-level viewport, all other viewports are arranged in a viewport tree. To manage viewports and to navigate in the viewport tree, use the functions **`pushViewport()`**, **`popViewport()`**, **`downViewport()`**, **`seekViewport()`**, and **`upViewport()`**. The main function to create a viewport is **`viewport()`** together with some parameters to specify its size and location.

A coordinate system is used to specify the location of graphical output relative to the current viewport. That is, although a viewport has an x- and y-axis as a coordinate system, different units like inch, centimetre, or lines can be used to place graphical elements. The function **`unit()`** is used for this, together with a numeric vector and a string specifying the unit.

Finally, graphical parameters help to change the appearance of the diagram. Among the graphical parameters are col, lwd, lty, cex, fontsize, and some others. Use **get.gpar()** to query all available parameters and their current settings.

The first example shows how to initialise a grid graphics display and how to print coloured data points. The script uses random data.

Ch9_gridExample.R

```
## grid - Basic example with random data
library(grid)

# Data creation
set.seed(12345)
n <- 20
A <- rep(1:3, each = n)
M <- matrix(c(7,   7,   8,
             10,  5,  11,
             14,  5,  10), 3, 3,
          dimnames = list(c("a1", "a2", "a3"),
                          c("X", "Y", "Z")))
d <- data.frame(A = factor(A),
               X = rnorm(length(A), rep(M[,"X"], each = n), 2),
               Y = rnorm(length(A), rep(M[,"Y"], each = n), 2),
               Z = rnorm(length(A), rep(M[,"Z"], each = n), 2))
rm(A, M, n)

# Create diagram
grid.newpage()
pushViewport(plotViewport(margins = c(5, 4, 2, 2)))
pushViewport(dataViewport(xData = d$X,
                          yData = d$Y))

# Display data and other content
grid.rect()
grid.xaxis()
grid.yaxis()
grid.points(d$X, d$Y,
           name = "dataSymbols",
           pch = 16,
           gp = gpar(col = as.numeric(d$A)*3))
grid.text("Variable X", y = unit(-3, "lines"))
grid.text("Variable Y", x = unit(-3, "lines"), rot = 90)
grid.text("Association between X and Y",
         x = unit(0.1, "npc"), y = unit(0.1, "npc"),
         just = c(0, 0))
```

Output

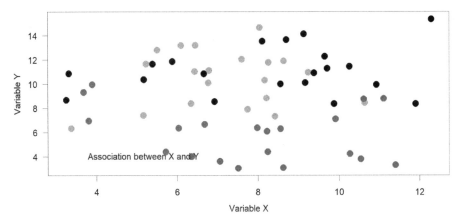

Figure 9.6 Basic example using the **grid** package.

Analysis

The script draws random data from the normal distribution first for three variables, X, Y, and Z and also includes a categorical variable with three levels, A. Then, **pushViewport()** begins the graphical display; however, it only initialises the data display without actually printing something. In this example, we use two convenience functions to create viewports for the data: **plotViewport()** and **dataViewport()**. The first one allows for the specification of the plot region according to the four margins that surround it. In this case the argument c(5, 4, 2, 2) creates a viewport comparable to the plot region of base graphics. The second viewport function, **dataViewport()**, is used to create a viewport with a coordinate system corresponding with the data ranges of the to-be-displayed variables X and Y. Note it would have been possible to create the plot region with one call of the **viewport()** function with appropriate arguments. However, the sequence of the two viewport functions can better be understood as, first, the positioning of a display and, second, the creation of a coordinate system for the data. Then, after creation of the diagram, we can display the data. The function calls **grid.rect()**, **grid.xaxis()**, and **grid.yaxis()** create a box marking the viewport boundaries and the x- and y-axis. Note that earlier in **dataViewport()** only the coordinate system was initialised, but nothing was displayed, and this coordinate system is now used in the two axis functions. Next, **grid.points()** displays the data. It is a simple scatter plot of the X and Y data vectors; however, the data points are coloured depending on the categorical variable A. The construction to set the colour shows how to generally set graphical parameters in grid. All graphical primitives in **grid** have a parameter gp, which stands for graphical parameter. This parameter accepts a gpar object returned by the **gpar()** function. This function takes a list of name-value pairs of graphical parameters and corresponding arguments. The help page of **gpar()** lists the available parameters.

Finally, we annotate the diagram with **grid.text()**. We name the x- and y-axis and we write a description of the diagram inside the plot region. Look how easily we can draw axis labels outside the boundaries of the viewport by specifying negative values for their positions. Moreover, the **unit()** function offers the convenience to specify the location of the strings very naturally in terms of lines, although the coordinate system of the viewport was specified earlier according to the X and Y data. Even in the placement of the description "Association between X and Y" we use the unit npc for our convenience to specify the position as a fraction of the width and height of the viewport, respectively. So, this text can be placed independently of the data. This entails the nice option that the **dataViewport()** with its xData and yData parameters can follow the data, while all annotation in the diagram can stay in place relative to the dimensionality of the viewport. This scatter plot is a basic example of how to create grid graphics from scratch. It illustrates important techniques related to this graphic system, which can be use in any complexity: (1) viewport creation, (2) pushing of viewports, (3) setting the coordinate system, (4) drawing of graphical primitives, (5) manipulating graphical parameters, and (6) measuring in different units to position graphical elements.

9.3.3 Handling viewports with grid graphics   R LANGUAGE!

Because the handling of a viewport is central to the **grid** system, and because it needs a little bit of understanding, let us discuss a small technical example. It shows how to create a nested structure of viewports and how to navigate between them using the name parameter. It further uses **grid.roundrect()** to show another beautiful graphical primitive – a rectangle with round corners.

Ch9_gridExample_HandlingViewports.R

```
## grid - handling viewports
library(grid)

pushViewport(viewport(x = .5, y = .5,
                      width = .8, height = .8,
                      name = "vp1"))
pushViewport(viewport(x = .2, y = .5,
                      width = .3, height = .4,
                      gp = gpar(fill = "grey80"),
                      just = "bottom",
                      name = "vp2_left"))
upViewport()
pushViewport(viewport(x = .7, y = .5,
                      width = .5, height = .4,
                      just = "top",
                      name = "vp3_right"))
upViewport()

grid.rect(gp = gpar(lty = "dashed"))
downViewport("vp2_left")
grid.rect()
```

```
grid.text("left viewport")
upViewport()
downViewport("vp3_right")
grid.roundrect(gp = gpar(col = "grey70"))
grid.text("right viewport")
```

Output

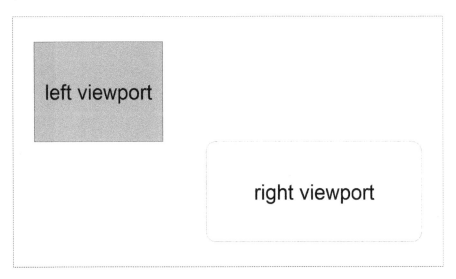

Figure 9.7 Basic output of graphical viewports with the `grid` package.

Analysis

The script creates and pushes three viewports and their names are vp1, vp2_left, vp3_right. We use the name parameter to set the names and they will be relevant when we navigate between viewports. After pushing vp2_left, this viewport has the focus for any subsequent graphics calls; so we use the **upViewport()** function to navigate back to vp1 and then push vp3_right. Otherwise, vp3_right would have been placed inside vp2_left. The ability of **grid** to navigate between viewports allows us to nicely structure the script in two parts: first, the creation and layout of viewports with **pushViewport()** and **viewport()** and, second, the output of content. After vp3_right is in place, we navigate back to vp1 and start placing graphical primitives. We mark the boundaries of vp1 with a rectangle with dashed line type. Notice again how we set the parameter lty with the **gpar()** function. Next, we move down in our viewport tree with **downViewport()**. To resolve the ambiguity in whether to move focus to vp2_left or vp3_right, the function takes the name of the desired viewport as an argument. We now border vp2_left with a rectangle and print some text inside it. Then, we use the sequence of **upViewport()** and **downViewport()** to move to vp3_right. This viewport obtains a rectangle with round corners and some text, too. The positions of the two viewports inside vp1 are determined by their x and y parameters. However, the just parameter determines which one of the four edges of the viewport is to be placed at the coordinates of x and y. That is, bottom means that the bottom edge is placed at the given y coordinate; top places the top edge at the y coordinate. By default, the horizontal adjustment places the centre of the viewport at the provided x coordinate. In sum, **grid** uses the general concept of a viewport as a rectangular region for holding some graphical content, whereas base graphics are bound to the differentiation between device, figure, and plot region. Use **grid** if you want to freely create a graphical display of any complexity.

9.3.4 Graphics with the ggplot2 package

R LANGUAGE!

Graphics with **ggplot2** follow the grammar of graphics developed by Wilkinson and Wills (2011) and further outlined by Hadley Wickham (2016). Each graphic is made of layers, which can be added on top of each other. There are two central functions **qplot()** and **ggplot()**. The first one is for quick diagrams, the second one for the fine control in layered displays. In fact, **qplot()** works like a high-level function in base graphics; that is, only one function call creates a complete diagram. It has several parameters, which allow for very different kinds of diagrams. Because the supply of the right data to the right parameters is central to using **ggplot2**, we gather a small table of the most important parameters.

Table 9.1 Parameters of the **ggplot()** function.

Parameter	Description
x	A variable that goes on the x-axis.
y	A variable that goes on the y-axis, usually dependent variable or other kind of empirical outcome.
data	The data frame which holds the variables used for the diagram.
color	The colour to use for the displayed data.
geom	Short form of geometric object. It is the type of geometric shape or object that represents the data in the diagram.
facet	Creates multi-panel layouts. Comparable with layout() of base graphics.

Let us now practise **qplot()** at the console with simulation data.

Ch9_ggplot2_Example.R

```
## Example ggplot2::qplot()
library(ggplot2)

# Data creation
set.seed(12345)
n <- 20
a <- rep(1:3, each = n)
d <- matrix(c(7,   7,   8,
              10,  5,   11,
              14,  5,   10), 3, 3,
            dimnames = list(c("a1", "a2", "a3"),
```

```
                                c("X", "Y", "Z")))
d <- data.frame(A = factor(A),
                X = rnorm(length(A), rep(M[,"X"], each = n), 2),
                Y = rnorm(length(A), rep(M[,"Y"], each = n), 2),
                Z = rnorm(length(A), rep(M[,"Z"], each = n), 2))


# Different uses of qplot()
qplot(x = X, y = Y, data = d)
qplot(X, Y, data = d, color = A)
plot(d$X, d$Y, col = as.numeric(d$A), pch = 16)

qplot(X, Y, data = d, geom = c("point", "smooth"))
qplot(A, X, data = d, geom = "boxplot")
qplot(A, X, data = d, geom = c("boxplot", "jitter"), color = A)

qplot(X, Y, data = d, facets = A ~ cut(Z, breaks = 2))
qplot(X, Y, data = d, facets = A ~ .)

myplot <- qplot(X, Y, data = d, facets = cut(Z, breaks = 2) ~ A)
myplot <- myplot + stat_smooth(method = "lm")
myplot
ggsave("myplot.png", myplot)
```

Output

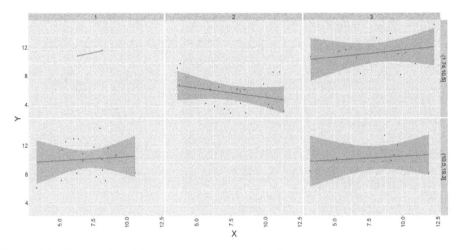

Figure 9.8 Scatter plot with six facets and regression lines.

Analysis

First, we include the **ggplot2** library and then create random data of three correlated quantitative variables, X, Y, and Z, and of one categorical variable with three levels, A. The data frame d comprises these variables. In the first function call of **qplot()**, the parameters x, y, and data look familiar; they tell ggplot2, which variable to use on the x-axis, which variable to use on the y-axis, and in which data frame to find the variables. This call creates the most basic bivariate scatter plot. Now, look at the resulting diagram; it is a simple scatter plot, but it has a grid in the background, which makes it beautiful and more readable. The next call of qplot() adds variable A as an argument to the colour parameter. For each data point, the level of variable A is now used to assign a colour to the point. The result is a scatter plot with nicely coloured points according to group. To compare, a similar call to the basic plot() function goes in the next line. This shows that almost the same result can be achieved with base graphics. The next diagrams show how ggplot2 works differently than base graphics. The next call of qplot() uses the geom parameter to specify which kind of graphical object or graphical representation the association between X and Y will assume: point and smooth. The first one now explicitly specifies the scatter plot, the latter one adds a fit line with a confidence band on top of the points. In general, we can specify any geom or groups of geoms that ggplot2 knows and pass them to this parameter; for example, density2d or quantile could also be used in this function call; then, ggplot2 will create the corresponding diagram. The next two calls of qplot() use the geoms box plot and jitter to create nice box plot illustrations of the relationship between the variables A and X. The next two calls of qplot() demonstrate multi-figure layouts with ggplot2 using the facets parameter. With this parameter, the graphical device can be split in rows and columns of diagrams according to the levels of one or two categorical variables. The parameter uses the tilde operator ~, which we know from formula notation, and you can think of its use in the same way here. However, here it tells the function which variable to use for the panel rows and which variable to use for the panel columns. If we use two categorical variables to form the panel layout, in this case A and Z, split into two levels, and we write facets = A ~ cut(Z, breaks = 2) as an argument to qplot(), then the categories of A will define the rows of the layout and the categories of Z will define the columns of the layout. If a layout of panels in only one row or one column should be created, then ~ is used with only one variable; for example, facets = A ~ . for one column of panels and facets = . ~ A for one row of panels. The remaining lines of code create the object myplot and show how elements can be added to a diagram using the + operator. In this case a faceted display of the scatter plot of X and Y receives linear model fit lines with the stat_smooth() function. The object holding the diagram can now be displayed simply by calling its name. Finally, ggsave() saves the diagram with the file format given in the suffix of the file name.

In base graphics we usually worry much about the structure and layout of the diagram. For example: do I need a little bit more margin to the left? Do I need to extend the x-axis limits, so that the legend fits in the plot area without covering data points? And how do I homogenise the appearance of different diagrams in multi-figure displays? In contrast, ggplot2 only requires that you think about the variables in your design and how they map to aesthetics. The rest of figure creation is taken care of by ggplot2. In consequence, you can quickly analyse your data and produce beautiful diagrams.

9.3.5 UPS – ggplot2 display respecting (almost) all variables EXAMPLE!

In the example unpleasant sounds there is a short list of experimental factors which can all affect the pleasantness rating for the sounds; these are group, gender, age, sound, and variation. Let us create a display in which we consider all factors of influence and create a diagram rich in information which could serve in a team discussion.

UPS_FIG_ggplot_ScatterplotDesign.R

```
## UPS - Scatter plot of the complete design
source("Scripts/UPS_PRE_DataImport.R")

library(ggplot2)

ggplot(data = UPS, mapping = aes(x = sound, y = rating, color = age)) +
  geom_point() +
  geom_jitter()

with(UPS,
     {
       UPS$age_cut <- cut(UPS$age,
                          breaks = seq(10,80,10),
                          labels = c("10-20","21-30",
                                     "31-40","41-50",
                                     "51-60","61-70",
                                     "71-80"))
       ggplot(data = UPS, mapping = aes(x = sound,
                                        y = rating,
                                        color = variation)) +
         geom_point() +
         geom_jitter() +
         facet_grid(age_cut ~ group)
     })
```

Output

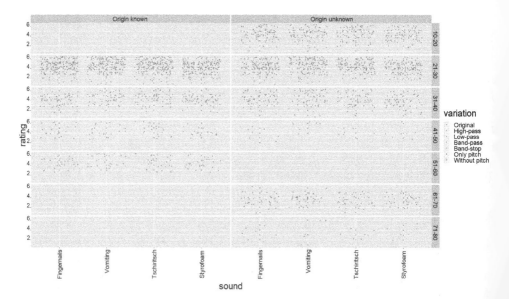

Figure 9.9 Unpleasant sounds scatter plot using all variables.

Analysis

In the script we use two calls of the main function **ggplot()** to create two figures. The first figure illustrates how we basically want to generate the diagram: a scatter plot of jittered points of the association between the independent variable sound and the dependent variable rating including, moreover, the association between rating and age. We achieve this with the initial call of **ggplot()** followed by + **geom_point()** and + **geom_jitter()** to define which geom to use. In **ggplot()** we supply the two parameters data and mapping with arguments. First, we supply the name of the data frame. Second, we tell R with **aes()** how the variables should appear in the diagram: sound as the independent variable goes to the x-axis, the rating as the dependent variable goes to the y-axis, and the many different levels of age will be mapped as many different colours in the diagram. After these specifications, we can simply call **geom_point()** and **geom_jitter()** without arguments, because everything regarding the diagram has been specified already. The parameters in this place could be used to change defaults of the functions, though this is not necessary here.

Let us now turn to the next **ggplot()** call, which is embedded in an environment created with **with()**. What we want to achieve with this more complex diagram is to include more independent variables and other factors of influence in the figure. However, creation of a new categorical variable age_cut comes first. We use **with()** here, so that the variable will disappear after diagram creation. In **ggplot()** the aesthetic mapping again receives three arguments for x, y, and color. However, now we represent the sound variations as colours in the figure. And the geoms are again point and jitter. However, we now include two more categorical variables to create a grid layout of diagrams according to the categories of the newly created age_cut variable and of group. To achieve the layout, we use **facet_grid()** and supply the age and group variable in formula notation.

9.3.6 Three-dimensional graphics with the rgl package

R LANGUAGE!

No other graphics package can compare with the **rgl** package (Murdoch & Adler, 2021) because this package creates three-dimensional (3D) diagrams. Furthermore, it allows user interaction and animation. The package's origins date back to the work of Murdoch and Adler (2007). It also uses the distinction between high-level and low-level graphics functions and maintains many of the well known terms: in many cases it simply adds 3d as a suffix to the function names for the 2D analogues. For example, there is a function **plot3d()** for 3D scatter plots and there are low-level functions like **points3d()**, **lines3d()**, or **segments3d()**. In the following, we will practise the basics of 3D diagramming with three blueprint scripts: the first demonstrates 3D high-level functions and animates the diagram, the second creates a diagram from scratch analogous to our earlier 2D example, and the third script introduces 3D multi-diagram layouts and highlights some differences in handling the graphics device compared with standard graphics. Start now with the first script. Execute it at the console line by line and feel like a magician.

Ch9_Blueprint_HighLevel_rgl.R

```
# Blueprint high-level rgl diagram with animation
library(rgl)
set.seed(12345)
```

```
A <-  rep(1:2, each = 5)
X <- rnorm(10)
Y <- rnorm(10)
Z <- rnorm(10)

plot3d(X, Y, Z, col = A, type = "s")

f <- spin3d(axis = c(0, 0, 1), rpm = 10)
play3d(f, duration = 5)
```

Output

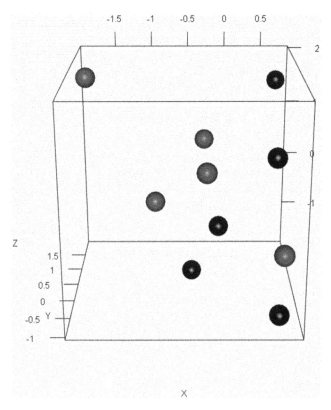

Figure 9.10 A 3D scatter plot created with `rgl`.

Analysis

The core of 3D diagramming is to use **plot3d()** and specify three numerical variables of the same length, X, Y, and Z. To use this function, library(rgl) loads the package first. Then, we create four data vectors, one categorical and three quantitative. The categorical variable A helps us to demonstrate how easy it is to display points in 3D space grouped by the categories of another variable. So, **plot3d()** takes three variables sampled from the normal distribution as the first three arguments. With the colour parameter `col`, it takes the categorical variable `A` and so distinguishes between groups. Importantly, although `A` is categorical it needs to be passed as a numerical variable in this place. With the execution of **plot3d()** the 3D diagramming is already accomplished and you can inspect the beautiful result. Click with the left mouse button inside the diagram and drag the diagram around to view it from different angles; furthermore, right click on the diagram and drag the mouse up and down to zoom in and out on the diagram. Now, execute the remaining functions of the script at the console. This accomplishes the animation. The function **spin3d()** creates a function holding the information of how to animate the diagram and **play3d()** specifies the duration of the animation. There is also a similar function **movie3d()**, which writes the animation into a movie file that you can play in other applications on your system. Isn't it fun that so much is possible with so little programming?

The next code blueprint focuses on low-level functions. It uses several such functions in a reasonable order; so, you can simply adapt them to your own data.

Ch9_Blueprint_rgl_fromScratch.R

```
## Blueprint rgl diagram from scratch
library(rgl)
set.seed(12345)

n <- 20
A <- rep(1:3, each = n)
M <- matrix(c(7,   7,   8,
              10,  5,   11,
              14,  5,   10), 3, 3,
            dimnames = list(c("a1", "a2", "a3"),
                            c("X", "Y", "Z")))
d <- data.frame(A,
                X = rnorm(length(A), rep(M[,"X"], each = n), 2),
                Y = rnorm(length(A), rep(M[,"Y"], each = n), 2),
                Z = rnorm(length(A), rep(M[,"Z"], each = n), 2))
rm(A)

with(d,
     {
         open3d()
         decorate3d(xlim = c(0, 20),
                    ylim = c(0, 20),
                    zlim = c(0, 20),
                    box = FALSE,
                    axes = FALSE)
         axis3d("x", at = seq(0, 20, 5), pos = 0)
         axis3d("y", at = seq(0, 20, 5), pos = 0)
         axis3d("z", at = seq(0, 20, 5), pos = 0)
         grid3d("Z", col = "black")
```

```
points3d(X, Y, Z, col = A)

M <- lm(Z ~ X * Y)
c <- coef(M)
XX <- 1:20; YY <- 1:20
dd <- data.frame(expand.grid(X = XX, Y = YY))
dd$"X:Y" <- dd$X * dd$Y
pdd <- predict(M, newdata = dd)
mpdd <- matrix(pdd, 20, 20)
surface3d(1:20, 1:20, mpdd, col = "lightcyan", alpha = .5)
})
```

Output

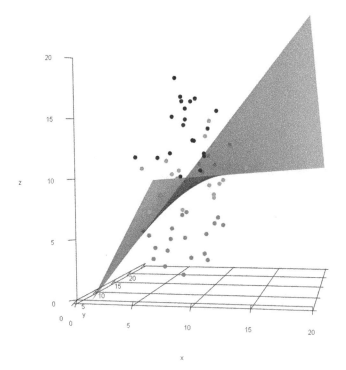

Figure 9.11 Output of a 3D diagram from scratch using the **rgl** package.

Analysis

The script starts with the creation of a data frame, which consisting of one categorical vari-able A with three levels and three quantitative variables X, Y, and Z. The matrix M is used to simulate some effect of the levels of A on the quantitative variables. After creation of the data frame we use an environment created with **with()** to make all variables of d easily accessible. In the 3D graphics of **rgl**, **open3d()** creates an empty graphics device. Then, **decorate3d()** sets up the coordinates of the device with the limits of the three scales. This is similar to the combination of **plot.new()** and **plot.window()** in standard graphics. However, **decorate3d()** already does printing; that is, in this case it places the axes labels in the diagram. In fact, the function could even create axes and a box around the diagram, but we prefer to do this by hand. So, we set the respective parameters to FALSE and use three calls of **axis3d()** for the X, Y, and Z axes, respectively. An interesting function, which adds to the 3D impression of the diagram, is **grid3d()**. It adds reference planes as visual orientation to the scene. The argument Z for the parameter side specifies that the grid is to be placed orthogonal to the z-axis, that is, in the plane defined by the x- and y-axis. Until now, we set up the plot window without any data display. The function **points3d()** now does this. It simply creates a 3D scatter plot with the points coloured according to their level of the variable A. Finally, we draw the surface for the prediction of Z by the predictors X and Y and their interaction. In fact, this is one big advantage of 3D diagramming. We not only have regression lines but regression planes or surfaces. To print the surface, we need information from the associated regression model, so we call **lm()** and supply the three quantitative variables. Then we use some functions associated with linear predictions like **expand.grid()** and **predict()** to create a data column of predicted values. We have to arrange this data column in matrix form and then hand it over to **surface3d()**. The function **surface3d()** needs a matrix, and the rows and columns of this matrix stand for all the value combinations of the X- and Y-coordinates. The numeric entries of the matrix give the Z value at which to print the surface for the given X- and Y-coordinates. Now, take this blueprint of rgl functions and use some variables of your own data to create a 3D diagram.

The third rgl script introduces the layout of several 3D diagrams on one device. It is similar to **layout()** in 2D, but there are differences, too. The following script uses random data from three quantitative variables and displays the variables in four panels of the graphics device.

Ch9_Blueprint_rgl_layout.R

```
## Blueprint rgl layout with four diagrams
library(rgl)
set.seed(12345)

d <- as.data.frame(mvtnorm::rmvnorm(100,
                                    sigma = matrix(c(3, 2, 2,
                                                     2, 3, 1,
                                                     2, 1, 3),
                                                     3, 3)))
names(d) <- c("X", "Y", "Z")

with(d,
    {
        layout3d(matrix(1:4, 2, 2))
        subsceneList()
        currentSubscene3d()
        plot3d(X, Y, Z, type = "s")
        useSubscene3d(subsceneList()[2])
        lines3d(X, Y, Z)
```

```
useSubscene3d(subsceneList()[3])
triangles3d(X[1:99], Y[1:99], Z[1:99], col = 1:33)
useSubscene3d(subsceneList()[4])
aspect3d(2, .2, 1)
quads3d(X, Y, Z, col = 1:25)
bg3d("light grey")
box3d()
})
```

Output

Figure 9.12 Layout of four 3D diagrams using the `rgl` package.

Analysis

For random sampling of the data frame d, we use **`mvtnorm::rmvnorm()`**. The function takes a theoretical covariance matrix sigma as an argument and returns correlated data. The function **`with()`** creates the environment for the 3D diagrams. Inside, **`layout3d()`** and **`matrix()`** create a layout of four subscenes on the graphics device. A call of **`subsceneList()`** shows that the subscenes are numbered and the function **`useSubscene3d()`** can be used to switch between them. We start diagramming at the top left panel and use a simple high-level graphic function. To proceed to the next panel, we use **`next3d()`** and then use **`lines3d()`** as a low-level graphic function for diagramming. Note, unlike the low-level functions of standard graphics, **`lines3d()`** also acts like a high-level function because it has the capacity to initialise the diagram like **`plot.new()`** and to set up the coordinate system like **`plot.window()`**. To proceed to the next panel we again use **`next3d()`**. Now, **`triangles3d()`** creates the triangles version of our data. Finally, we use the data to diagram quadrangles. The lower right panel also demonstrates three other functions: **`aspect3d()`** changes the aspect ratios between the three axes of the diagram; that is, a unit on the x-axis is to be printed two times bigger than a unit on the z-axis and a unit on the y-axis is to be printed five times smaller than a unit on the z-axis. Finally, **`bg3d()`** creates a background colour and **`box3d()`** creates a box around the diagram.

Let us move back to how we proceeded through the panels of the device; we used **next3d()** for this. However, other functions like **subsceneList()**, **currentSubscene3d()**, **useSub-scene3d()** also make switching between panels easy. Since all panels are numbered, we could use **useSubscene3d()** to switch back and forth between panels.

The **rgl** package adds 3D diagramming to the capabilities of R. You can start with high-level functions and produce impressive diagrams with only one line of code; you can then refine your diagrams with low-level functions or even start your diagram from scratch. Together we covered the basics of **rgl** usage, that is, high-level functions, the creation of a diagramming blueprint from scratch and the layout of several diagrams on one device. However, there is more to play with in **rgl**: manipulate the material properties of data points, experiment with illumination of the scene, or create animations of a journey through the data points and store this in a movie file. Then, present your output in the next team meeting or even use it in a conference presentation. Finally, search the web for further examples to show you how far you can go with 3D diagrams and the power of **rgl**.

9.3.7 Closing remarks about graphic systems

R LANGUAGE!

This chapter introduced powerful tools in R to produce any graphic. You now can decide which graphic system works best for your requirements and then simply create graphic after graphic in this system. In every one of the described graphics systems you will discover even more power compared with what we talked about. For example, **rgl** can also use illumination and animation of the scene, **ggplot2** also knows animation in conjunction with further packages, **grid** sustains a powerful object model of its graphical primitives which allows you to add, edit, or remove graphical elements during runtime of your statistical program. So, there are still many things for you to discover.

═══════════ **What you have learned so far** ═══════════

1 To plan your graphics
2 To combine high-level and low-level functions
3 To create base graphics from scratch with low-level functions
4 To manipulate fonts in graphics
5 To use the viewport approach to graphics of the **grid** system
6 To produce layered graphics according to the grammar of graphics using **ggplot2**
7 To produce three-dimensional graphics with the package **rgl**

═══════════ **Exercises** ═══════════

Explore functions

In this chapter we used **par()**, **plot.new()**, **plot.window()**, **box()**, and **axis()** to create a blueprint for a graphic from scratch. Practise different layouts for your next diagrams with these functions. Use **box()** and dummy data (e.g., plot(1:10)) to

(Continued)

visualise the layout. Pay close attention to the initialisation of the x- and y-axis with `plot.window()`.

2   Use the `MusicData.csv` data set to practise the **`plot3d()`** function from the **`rgl`** package. For the x, y, and z coordinates select three of the 20 rating variables relating to the evaluation of musical examples. Add animation to your display with **`spin3d()`** and **`play3d()`**.

Solve tasks

1   Use the `MusicData.csv` data set to program a multivariate mean profile from scratch. Use the blueprint and example from section 9.2 from the main text to get started. The means of the 20 variables should be arranged in a vertical profile.

2   Extend the mean profile from the foregoing task by adding a table with descriptives. Use **`text()`** to add mean, standard deviation, and sample size on the left-hand side of the diagram, next to the y-axis.

Apply in the real world

1   Think about your own research project. Which diagram would be nice to present at the next team meeting or conference. Try to gather information in the different graphics systems as to whether a high-level function is available which gives a good overview of the data. Create an initial diagram using a high-level function and then augment the figure with low-level functions. For example, think about highlighting of data points with colour or a different symbol, or add a regression line to a scatter plot.

Become a statistics programmer

1   Think about common graphical displays in your research field. What do other researchers present on conferences and in research papers? Gather a collection of printouts of graphics you like and keep them available for your brainstorming about your own graphics.

2   Play around with the diagram function and practise the supply of data vectors to graphical parameters. Start with the function call `plot(1:100, pch = 1:25, col = 1:100, cex = c(.5, 1, 2, 3))`. Try to pass vectors of data of different length to other graphical parameters as well. Think about the principle of recycling of vectors in R, which is common in many situations in R. Then think about how to apply this principle with real data. For example, create a scatter plot of two quantitative variables and use different colours for the data points depending on the respective level in a categorical variable.

Read code

1   The following code creates box plots with additional data points. It uses the data set `MusicPreferencesSchool.csv`. The mean rating for Song 1 and 2 figures as the dependent variable, the factor `Musician` as the independent variable. Describe in your own words how high-level and low-level functions create the figure. Why is it good to use **`jitter()`** here?

```
## Box plots with data jitter


d <- read.csv("MusicPreferencesSchool.csv")
```

```
with(d,
     {
          MeanPref <- (PreferenceSong1 + PreferenceSong2) / 2
          boxplot(MeanPref ~ Musician,
                  xlab = "Musician",
                  ylab = "Mean preference")
          points(jitter(as.numeric(Musician),
                        amount = .2),
                 MeanPref,
                 pch = 16)
     })
```

2   The following script creates histograms for several variables. Each histogram also shows error bars. Describe in your own words why it is good to define a function first and then use the function for each variable.

```
## Histogram with error bars


d <- read.csv("MusicData.csv")


# Define function
HistogramErrorBars <- function(X = rnorm(100),
                               ymax = 30,
                               ...)
{
        M <- mean(X, na.rm = TRUE)
        SE <- sd(X, na.rm = TRUE) / sqrt(length(X))

        hist(X,
             breaks = .5:6.5,
             ylim = c(-2, ymax + 1),
             yaxt = "n", ...)
        axis(2, at = seq(0, ymax + 1, 5), las = 1)
        points(c(M - SE, M, M + SE),
               rep(-1, 3),
               pch = c("[", "|", "]"),
               type = "o")
}


# Call function
YMAX <- max(apply(d[paste0("Ex", 1:20)], 2, table))
layout(matrix(1:20, nrow = 4))
par(mar = c(2.5, 3, 2, .5))
lapply(paste0("Ex", 1:20), function(X)
{
        print(X)
```

(Continued)

```
HistogramErrorBars(d[[X]],
                   ymax = YMAX,
                   main = X)
})
```

EXAMPLE! **WBD**

1 Which diagram would be most informative regarding the primary hypothesis of an effectiveness of the web-based depression intervention? Sketch the desired diagram on paper. Now try to create the diagram from scratch using base graphics. Use the functions **plot.new()**, **plot.window()**, and **par()** to set up the graphic and then fill the device with the help of low-level functions.

EXAMPLE! **ITC**

1 Go back to the ITC examples in Chapter 5 showing the SAT distributions split by gender. Recreate this display using **grid** graphics and **ggplot2**.

EXAMPLE! **UPS**

1 Go back to the UPS example in the main text in Chapter 9. Simplify the **ggplot2** display to show only the association between the variables sound and rating. Exclude all other variables from the display.

EXAMPLE! **SCB**

1 In Chapter 5, we created a matrix of scatter plots to display the associations between the variables of numbers of conflict - the dependent variables. Recreate parts of this display from scratch using **grid** graphics and viewports. A matrix of diagrams using two or three variables may suffice.

10

DATA SIMULATION

Chapter overview

This chapter introduces simulation as a powerful technique to create and evaluate experimental designs. R helps us to opt for the best design. It provides us with large numbers of simulation runs to compare the efficiency of different analyses as a supplement for statistical power analysis. Programming features: random number generation, simulation reproducibility, vector and matrix capabilities for the analysis of multiple data sets, display of simulation results.

10.1 The idea of data simulation

Let us first familiarise ourselves with the idea of data simulation and with its benefits. As researchers and designers of empirical studies we will find simulation valuable for the determination of sample size.

One of R's powers is data simulation. In the design of empirical studies this helps us to compare between designs and to determine the required sample size. We can compare and evaluate different designs before spending resources to gather the actual data. Furthermore, if we include the statistical test of the main analysis in the simulation, we will learn how many subjects to include to give a certain effect size a good probability of yielding a significant result. For example, a clinical study may test the effectiveness of a new treatment using physical exercise against pain. Of course, the experiment includes a control condition; however, there are also several staff to conduct the exercises with patients. If the treatment is new and not already tested, it is questionable whether its effect can measure up with the data nuisance that results from employing the random effects of study staff and perhaps study sites. In data simulation, besides the supposed intervention effect, the further effects on pain level can be included. The simulation could reveal how strong the treatment effect must be or how many subjects to include in the experiment to ensure that the assumed treatment effect has any chance to yield a significant result. In fact, this example illustrates the use of simulation to assess statistical test power of an effect. Statistical test power is the probability in a statistical test of finding a significant result, given a certain statistical effect size. The effect size is usually the true difference between the means of treatment and control condition, measured in units of the standard deviation. An established effect size is Cohen's d (Cohen, 1988), but there are more effect sizes to know (Kirk, 2012).

We always want a high statistical test power in our experiments and one of the determinants of it, besides effect size, is sample size. With data simulation, we would program thousands of replications of our experiment using different specifications of sample size and compare the proportions of significant results between different specifications. This way, we may find a sample size that at the same time complies with our resources and ensures high test power. However, we may also reach the conclusion that our resources are not sufficient to attain the required sample size that the simulation suggests. Then, perhaps, stronger experimental control of the data collection or statistical control by inclusion of covariates may allow us to specify a higher effect size in the simulation. This may, in turn, allow for a reduction of sample size while preserving power. In the clinical example, stronger experimental control may ensue after standardised training of the staff conducting the new treatment and the control condition. Stronger statistical control would follow from the inclusion of the baseline measurements of pain in the design and data simulation.

But how does simulation in R work? An R script receives the design specifications in a handful of code lines. Usually, these specifications are population means, population variances and covariances, and sample sizes. We derive these quantities from our specification of all elements that comprise the experimental design, that is, independent variables, dependent variables, covariates, and nuisance variables (Lipsey, 1990). In the clinical example, we may know estimates of the population means of the control condition or the means of treatments similar to our intervention. With the specifications, R generates random data with one of the random data generation functions. Often, `rnorm()` suffices for this, but `rmvnorm()` of the `mvtnorm` package is better for the simulation of correlated data (Genz et al., 2021). With these functions we create thousands of data sets and store them in an R object. Then, we calculate thousands of statistical tests and count the number of significant results. The proportion of significant results will correspond with the statistical test power that the design implies.

In general, simulations reveal the behaviour of experimental and statistical designs, given that the assumptions are reasonable. Tests of the design become possible without allocation of many experimental resources, for example, subjects or testing material. You can even run simulation scenarios to make some point with the stakeholders about certain necessities of the experimental design that you consider important. You can persuade people to agree with your plans for empirical investigation by demonstrating how various obstacles of experimentation can affect the data quality and the possible conclusions from the experiment. In the clinical example, the data simulation could demonstrate that a certain random factor like study site should be kept constant or restricted to a few levels only, in order to give a small intervention effect high power. This may even provide good reason to increase experimental control at the expense of generalisation of the results. With data simulation you can master the experimental design before collection of real data.

10.2 Data simulation in R

This section introduces four ways to simulate data. We first sample a data vector from the normal distribution. We continue with simultaneous sampling from more than one normal distribution. Then, we sample correlated data. Finally, we program a large number of simulation runs and use this for statistical power analysis.

10.2.1 Sampling from the normal distribution   R LANGUAGE!

Let us begin with a straightforward example and sample 200 data points from the normal distribution. Then, we test this sample for significance against a criterion value.

Ch10_RandomDataNormalDistribution.R

```
## Random data from the normal distribution
set.seed(12345)

# Generate data
mu <- 50.8; sigma <- 5; n = 200
x <- rnorm(n, mu, sigma)
print(x, digits = 4)

# Figures
layout(matrix(1:2,1))
qnorm(x); hist(x)

  Test H1: mu > 50
 ritM <- qnorm(p = .05,
            mean = 50,
```

```
              sd = sigma/sqrt(n),
              lower.tail = FALSE)
ifelse (mean(x) > CritM,
        "Result significant",
        "Result not significant")
```

Analysis

The part about data generation includes three specifications, namely expected mean, expected standard deviation, and sample size. Afterwards, **rnorm()** draws *n* values at random from the normal distribution. Subsequently, a Q-Q diagram and a histogram help to assess whether the random sampling was successful. With a big sample size, these displays should look very convincing that the data was indeed drawn from a normal distribution. The next part of the script does the z-test line by line in that it uses **qnorm()** to select a critical value from the normal distribution with mean 50 and standard error `5/sqrt(200)`. This cuts off an area at the upper tail of the distribution with a probability of at most 5%. Thus, **qnorm()** takes a probability and a specification of a distribution as parameters and returns a value of the continuum of possible values, which cuts of the desired probability at the upper or lower tail of the distribution. Using the critical value, the script compares the empirical mean value with the critical mean value and then prints out one of two possible results, depending on the comparison of means. This simulation simply shows the creation and usage of random numbers.

R LANGUAGE! ## 10.2.2 Simultaneous sampling from different distributions

With **rnorm()**, we can also sample from different distributions. The function takes vectors of more than one mean and variance as arguments and then simulates complex designs of independent measurements. Here is an example; it creates the data first and then runs a statistical test.

Ch10_SimulationDifferentDistributions.R

```
## Simulation from different normal distributions
set.seed(12345)

# Specifications
J = 6
n = 10
mu      <- c(25, 15, 23, 20, 15, 22)
sigma   <- c(1, 2, 5, 2, 4, 6)

# Create data frame
RandomData <- data.frame(Subj = 1:(J*n),
                         A = factor(rep(1:J, times = n)),
                         y = round(rnorm(J*n, mean = mu, sd = sigma)))

# Check distributions
boxplot(y ~ A, RandomData)
```

Output

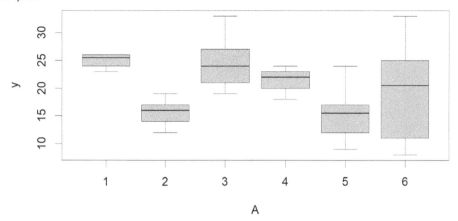

Figure 10.1 Box plot of empirical distributions sampled from six different normal distributions.

Analysis

The difference compared with the foregoing script is that we specify more than one value for `mu` and `sigma`, respectively. We then pass on these objects to **`rnorm()`**. The function now samples `J` times `n` numbers – the number of groups times the group size. With each subsequent number **`rnorm()`** proceeds to the next entries of `mu` and `sigma`. If the function reaches the end of these two objects, it starts over again. This is called recycling and it is used often in R programming. In fact, the switching between distributions corresponds with the structure of factor `A`, so we can later distinguish to which group each data value belongs. We close with a box plot to confirm that **`rnorm()`** used different distributions for the sampling (Figure 10.1).

10.2.3 Multivariate data simulation

R LANGUAGE!

To sample multivariate data the package **mvtnorm**, created by Torsten Hothorn, provides the **`rmvnorm()`** function (Genz et al., 2021). A big advantage over **`rnorm()`** is that it can easily produce correlated data. In regression analysis, repeated measures designs, or in the analysis of covariance, we need data of this kind. The parameter list of **`rmvnorm()`** is similar to **`rnorm()`**; however, means of a multivariate data vector have to be specified and a covariance matrix with variances and covariances of all variates.

Ch10_rmvnorm_Example.R

```
## Using rmvnorm()
library(mvtnorm)
set.seed(12345)

  <- 100
u <- c(50, 52, 50, 48)
```

```
sigma <- matrix(c(5, 2, 2, 1,
                  2, 4, 2, 2,
                  2, 2, 4, 3,
                  1, 2, 3, 5), nrow = 4)

d <- rmvnorm(N, mu, sigma)

cov(d)
```

Analysis

This example first attaches the **mvtnorm** package to the search path. Then it creates three arguments for **rmvnorm()**: the sample size, the mean vector, and sigma, which is the symmetrical covariance matrix used for sampling. Then **rmvnorm()** produces the data frame. We use cov(d) to inspect how close the observed covariances correspond with the theoretical ones. Notice that **rmvnorm()** and **rnorm()** work a little bit different: **rmvnorm()** uses a mean vector argument as the means of different variables. In contrast, **rnorm()** uses a mean vector for recycling; that is, it uses the means in the same output vector. For example, in rnorm(50, c(10, 100), 5), the sampling switches back and forth between mu = 10 and mu = 100 for each subsequent data value.

R LANGUAGE! ## 10.2.4 Repeated data simulations

The simulation approach most often uses large numbers of data sets and not only one to find answers to a design problem. So, the last example of this section shows how to specify parameters for a simulation and then run the simulation several times.

Ch10_RepeatedSimulations.R

```
## Running repeated simulations
set.seed(12345)

# Specifications
J = 2
n = 50
mu      <- c(50, 53)
sigma   <- 5

# Simulation
runs <- 1000
RandomData <- lapply(1:runs,
                function(X)
                {
                    data.frame(Subj = 1:(J*n),
                            A = factor(rep(1:J, times = n)),
                            y = round(rnorm(J*n, mean = mu, sd = sigma)))
                })
```

```
# Analysis
pvalues <- unlist(lapply(RandomData,
                  function(X) t.test(y ~ A, X)$p.value))
table(ifelse(pvalues <= .05, "significant", "non-significant"))
```

Output

```
> table(ifelse(pvalues <= .05, "significant", "non-significant"))

non-significant      significant
           139              861
```

Analysis

The script starts with specifications for the simulation. In the simulation section we declare runs as the number of data simulations. Then **lapply()** allows us to repeatedly execute an anonymous function, which creates a data frame with random data. This way we obtain a large number of data sets and put them in the object RandomData. Because RandomData is a list, we use **lapply()** again for data analysis. We specify another anonymous function, which calculates a *t*-test for all data frames stored in RandomData. However, we store only the *p*-values of theses tests in pvalues. Lastly, we count the number of significant and non-significant results and find out that 861 out of 1000 tests reached significance. We conclude that the *t*-test for our design specifications had a statistical test power of about 86%. In sum, with this example we practised a minimal simulation example for a statistical power analysis using data simulation.

10.2.5 Making a data simulation reproducible   R LANGUAGE!

Statistical simulations work with random data. However, everybody has heard somewhere that computers cannot actually create random data, because they can only do things that somebody programmed. However, for data simulation it suffices if the computer-simulated data appears and is indistinguishable from data that was created with a true random process, like throwing a dice. In fact, computers start at a certain point or number and follow an algorithm to produce the next number, and starting from the next number to produce another next number and so on. As researchers doing data simulation we can take advantage of this procedure: if we supply the computer with the number to start the selection of random numbers, we can exactly reproduce the computer-generated sequence of random numbers. We can even pass on the starting number to our colleagues and other researchers and they will be able to generate exactly the same random sequence. This makes teaching of statistics and demonstrations with random data much easier. In fact, you may have noticed that you can produce all random data sets used in this book yourself, when you run the scripts yourself. If it were truly random, this should not be possible. However, the **set.seed()** function does the job of setting the first number for the random data generators in R like **rnorm()** or **sample()**. So, use **set.seed()** with an arbitrary number before using a random data function. Then, the particular combination of the call of **set.seed()** and the random data function always produces the same random data. Let us try it at the console.

Ch10_ReproducibleRandomData.R

```
## Reproducible random data with set.seed()
set.seed(12345)
sample(1:10)

set.seed(12345)
sample(1:10)
sample(1:10)

set.seed(12345)
sample(1:10)
rnorm(3)
rnorm(3)

set.seed(12345)
sample(1:10)
rnorm(3)
```

Output

```
> set.seed(12345)
> sample(1:10)
 [1]  3  8  2  5 10  9  7  6  4  1
> set.seed(12345)
> sample(1:10)
 [1]  3  8  2  5 10  9  7  6  4  1
> sample(1:10)
 [1] 10  1  8  7  6  9  4  2  3  5
> set.seed(12345)
> sample(1:10)
 [1]  3  8  2  5 10  9  7  6  4  1
> rnorm(3)
 [1] -0.2841597 -0.9193220 -0.1162478
> rnorm(3)
 [1] 1.8173120 0.3706279 0.5202165
> set.seed(12345)
> sample(1:10)
 [1]  3  8  2  5 10  9  7  6  4  1
> rnorm(3)
 [1] -0.2841597 -0.9193220 -0.1162478
```

Analysis

The lines illustrate reproducible data simulation with **set.seed()**, **sample()**, and **rnorm()**. We always use the same arbitrary number in **set.seed()** to start data generation and every sample(1:10) that follows it creates the same sequence of numbers. If another call of `sample(1:10)` follows, it produces another sequence, because it uses a new starting point that came out of the preceding `sample()`. This also works if different random data functions are combined, like `sample()` and `rnorm()`. And with any combination of random data functions, we can always go back to the original starting point with `set.seed()` to let them produce exactly the same mass of random numbers. In sum, if you ever want to demonstrate something with random data, use `set.seed()` before data generation to make it reproducible.

10.2.6 Using the linear model equation to simulate data

STATISTICAL ISSUE!

We now simulate data for a more realistic experimental design, which we can often use in everyday work. The design consists of a dependent variable Y and two independent variables A and B, with A comprising $J = 2$ levels a_1 and a_2, and B comprising $K = 3$ levels b_1, b_2 and b_3. A, B and the interaction AB figures as factors with fixed effects. The design furthermore consists of a random error term. Here, the following model of the completely randomised factorial design applies (Kirk, 2012):

$$Y_{ijk} = \mu + \alpha_j + \beta_k + (\alpha\beta)_{jk} + \epsilon_{i(jk)}$$

To simulate data with this design, we only need to apply the formula. That is, we specify the expected grand mean μ, the effects α_j and β_k and $(\alpha\beta)_{jk}$, and sample the error term or random component $\epsilon_{i(jk)}$ from the normal distribution. Consequently, all the components on the right-hand side of the linear model need a data column with a number of entries corresponding with the sample size N, respectively. The sum of these vectors produces the dependent variable. For the analysis, we then pretend to forget our model specifications and retain only the information to which combination of factor levels of A and B each measurement of Y belongs. With the analysis of the data vector Y with descriptive and inferential statistics we statistically rediscover the modelled effects. The following script simulates the data for this design and then applies the analysis.

Ch10_SimulationExperimentalDesign.R

```
## Simulation of an experimental design
set.seed(12345)
library(psych)

# Specify simulation
N <- 120
Mu <- rep(5, N)
Alpha <- rep(c(-1,1), each = N/2)
Beta <- rep(c(-.5, .3, .2), times = 2, each = N/6)
AlphaBeta <- rep(c(0, 0, -.1, .6, -.5, 0), each = N/6)
Epsilon <- rnorm(N, 0, 3)

  Model equation
  <- Mu + Alpha + Beta + AlphaBeta + Epsilon
```

```
# Data frame creation
d <- data.frame(Y,
                A = factor(Alpha, labels = c("a1", "a2")),
                B = factor(Beta, labels = c("b1", "b2", "b3")))

# Statistical analysis
describeBy(d$Y,
           group = list(d$A, d$B),
           mat = TRUE)[c("n", "mean", "sd")]
m <- aov(Y ~ A * B, data = d)
summary(m)
model.tables(m)
```

Output

```
> # Statistical analysis
> describeBy(d$Y,
+              group = list(d$A, d$B),
+              mat = TRUE)[c("n", "mean", "sd")]
     n     mean        sd
X11 20 3.729550 2.501806
X12 20 6.796380 3.453288
X13 20 4.415529 3.890672
X14 20 6.352413 2.953296
X15 20 5.511561 3.626853
X16 20 7.024937 3.293857

> m <- aov(Y ~ A * B, data = d)

> summary(m)
            Df Sum Sq Mean Sq F value   Pr(>F)
A            1  141.6  141.57  12.861 0.000495 ***
B            2   24.1   12.05   1.094 0.338203
A:B          2   12.9    6.45   0.586 0.558300
Residuals  114 1254.9   11.01
---
Signif. codes:  0 '***' 0.001 '**' 0.01 '*' 0.05 '.' 0.1 ' ' 1

> model.tables(m)
Tables of effects

 A
A
    a1      a2
-1.0862  1.0862
```

```
B
B
       b1        b2        b3
-0.3754  -0.2544   0.6299


  A:B
     B
A    b1        b2        b3
  a1 -0.4472   0.1177   0.3295
  a2  0.4472  -0.1177  -0.3295
```

Analysis

The script consists of three parts: (1) creation of a dependent variable Y as the sum of systematic effects and a random component, (2) creation of a data frame with the dependent and independent variables, and (3) the statistical analysis. The first part simply follows the definition of a design model of data as the sum of effects. Many statistics books include these models, for example Kirk (2012), Winer et al. (1991), or Hays (1994). In this part of the script, we play god knowing the eternal truth about how the variables A and B relate to Y. Each of the $N = 120$ subjects belongs to one of the conditions of A and B and thus has some true effect on these variables and their interaction. We store these effects in the vectors Alpha, Beta, and AlphaBeta. Up to this point nothing is random – only fixed effects. However, personal characteristics, non-constant data collection, and other accidental empirical conditions impose error on the systematic effects, which is represented by Epsilon, the random component of the model. Summation of systematic components and a random component for each subject produces the simulated vector for the dependent variable Y. In the next part of the script, however, we forget the eternal truth and simply retain the dependent variable Y. We turn the variables consisting of the numeric effects into the factors A and B, which consist only of the information regarding the group to which each subject belongs. The aim now is to empirically assess the effects of A, B and their interaction regarding Y. With the simulated data frame, we can calculate descriptive statistics to compare between groups and an analysis of variance to assess whether to conclude that there is a systematic influence of A and B on Y. This simple case of design simulation shows how we assign numbers as effects of the experimental design, add a random component, and then sum these sources of variance to form the simulated data.

10.3 Simulation applications

We use random data for teaching and in the preparation of research to compare between different designs of an empirical study. In research this helps us to decide for a design with high statistical test power.

10.3.1 Comparison between a completely randomised design and a randomised block design

STATISTICAL ISSUE!

Let us use simulation to compare between experimental designs. Perhaps our plan is to compare three pills against headache in a randomised experimental design: the first pill is a well established one and acts as a control condition. The second one is a recent innovation but it is expensive to produce. The third one is also an innovation but much cheaper to produce than the second one. We now turn our ideas about the effectiveness of the pills in parameter settings for our dependent variable headache improvement: we set the effects of each pill in the linear model as $\alpha = (-2, 1, 1)$ and set a grand mean of $\mu = 10$. To arrive at these numbers, we need to study earlier research using one or more of the three pills and using the dependent variable. We further need ideas about the variance of the data from earlier research. We may also know from earlier research that a certain quantifiable physiological factor partly determines the dependent variable. But this factor is not easy to measure. So, we ask whether blocking or stratification for this factor in the experimental design enhances statistical test power for the headache pills intervention. In this case, blocking means measuring the physiological factor of each participant before group allocation and then forming several groups of three participants with equal or similar measurements along the continuum of the physiological measure. These groups are called blocks. Within each block, a random permutation of the three participants is distributed among the three experimental groups. This way, each of the three conditions shows almost the same distribution regarding the physiological measure, and the association between the blocking variable and the dependent variable can enter as an effect in the experimental design model. In effect, this procedure usually makes it easier for the intervention effect to reach significance if there is a considerable block effect. To find out whether blocking is meaningful in a given design, we simulate the experimental design as a completely randomised design including only the pills intervention, and we simulate a randomised block design including the pills intervention and blocking for the physiological factor. We run a large number of these simulations and check which design produces more significant results, indicating higher statistical test power.

Ch10_ComparisonANOVARandomizedBlock.R

```
## Comparison of ANOVA and randomised block design
set.seed(12345)


# Specify simulation
N = 120                              # number of subjects
J = 3                                # number of groups
n = N/J                              # group size
Mu <- rep(10, N)                     # grand mean
Alpha <- rep(c(-2, 1, 1), each = n)  # group effect
Pi <-  c(M = 0, S = 3)               # random block effect
Epsilon <- c(M = 0, S = 5)           # random component


# Data frame creation
runs = 100
RandomData <- lapply(1:runs,
                     function(X)
                     {
                       data.frame(A = factor(rep(c("a1", "a2", "a3"),
                                                 each = n)),
                                  Block = factor(rep(paste("Block", 1:n),
```

```
                                         times = J)),
                        Y = Mu + Alpha +
                            rep(rnorm(n, Pi["M"], Pi["S"]),
                                times = J) +
                            rnorm(N, Epsilon["M"], Epsilon["S"]))
                 })

# Analysis
pvalues <- t(sapply(RandomData,
                function(X)
                {
                    list(CR = summary(aov(Y ~ A, X))[[1]][["Pr(>F)"]][1],
                        RB = summary(aov(Y ~ A + Error(Block),
                                    X))[[2]][[1]][["Pr(>F)"]][[1]])
                }))

# Results
plot(pvalues[, "CR"], pvalues[, "RB"],
    pch = 16, xlim = c(0, 1), ylim = c(0, 1))
abline(h = .05)
abline(v = .05)
table(ifelse(pvalues[, "CR"] <= .05, "significant", "non-significant"))
table(ifelse(pvalues[, "RB"] <= .05, "significant", "non-significant"))
```

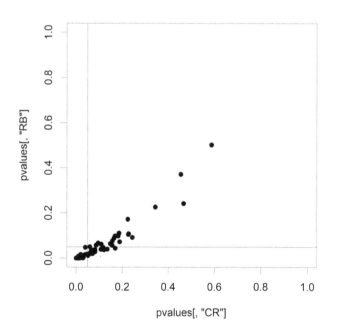

Figure 10.2 Comparison of p-values for a completely randomised design and randomised block design.

Output

```
> # Results
> plot(pvalues[, "CR"], pvalues[, "RB"],
+        pch = 16, xlim = c(0, 1), ylim = c(0, 1))
> abline(h = .05)
> abline(v = .05)
> table(ifelse(pvalues[, "CR"] <= .05, "significant", "non-significant"))

non-significant     significant
           36              64
> table(ifelse(pvalues[, "RB"] <= .05, "significant", "non-significant"))

non-significant     significant
           20              80
```

Analysis

The script first obtains the simulation settings. Suppose that we are restricted to at most $N = 120$ participants divided into $J = 3$ groups. We further implement the intervention effect together with the grand mean for the dependent variable. Furthermore, we specify the mean and standard deviation of the random block effect Pi, which we will later sample from the normal distribution. We finally set Epsilon as the variability of the dependent variable. The next section of the script creates the data frame using **lapply()** and an anonymous function. Inside the function we create a data frame with the grouping and blocking variable and with the design model equation to calculate Y. The function **lapply()** allows us to create 100 data frames and store them in the object RandomData. The next part of the script calculates the analysis using **sapply()** and another anonymous function. The function calculates a completely randomised ANOVA and a randomised block ANOVA for each data frame stored in RandomData. These analyses need only supply the p-value for the intervention effect, so we employ a complex expression with several indexing operations using **[[]]** to keep only the p-values from the ANOVA summaries. Finally, we display the results of 100 completely randomised and randomised block analyses, respectively. The display of p-value pairs reveals that more randomised block analyses reach significance than completely randomised analyses. The subsequent tables confirm this impression. To conclude, given the same specifications for the intervention effect in all analyses, the randomised block design yields more significant results. The randomised block design in this simulation, thus, has a higher statistical test power then the completely randomised design.

Comparison of statistical designs before data collection is an economic way to check whether your study set-up has a reasonable chance to yield a significant result if your research hypothesis is true. Suppose your hypothesis is indeed true (nobody will ever know), this helps you to plan an experiment that has a high probability of supporting your claim.

EXAMPLE! ## 10.3.2 SCB – Data simulation

For the study of student reconciliators against bullying at school, simulation data may help us to test the statistical analyses before we obtain the funding for real data. The following script defines a function to create a random data frame as the potential study data. We can then repeatedly use this function to create many data sets.

SCB_FUN_DataSimulation.R

```
## SCB - Data simulation function
if(!exists("SCB_DataSimulation"))
{
  library(mvtnorm)
  SCB_DataSimulation <- function(N = 180,
                                 Mu = 20,
                                 A_effect = c(0, 0, 0),
                                 B_effect = c(0, 0, 0),
                                 Mu_X1 = 5,
                                 Mu_X2 = 15,
                                 Sigma = diag(c(2, 5, rep(5, 12))))
  {
    # Initialise data frame
    d <- data.frame(ID = 1:N,
                    A = rep(c("No intervention",
                             "Enhanced teacher attendance",
                             "Conflict reconciliators"), each = 60),
                    B = rep(c("Primary school",
                             "Secondary school",
                             "Further education"), 3, each = 20),
                    Mu = rep(Mu, N),
                    A_effect = rep(A_effect, each = 60),
                    B_effect = rep(B_effect, times = 3, each = 20))

    # Generate correlated random data
    Variates <- c("X1", "X2",
                  "Y1C1", "Y2C1", "Y3C1",
                  "Y1C2", "Y2C2", "Y3C2",
                  "Y1C3", "Y2C3", "Y3C3",
                  "Y1C4", "Y2C4", "Y3C4")
    d[Variates] <- as.data.frame(rmvnorm(N, mean = c(Mu_X1, Mu_X2, rep(0,12)),
                                         sigma = Sigma))

    # Generate dependent variables with linear model equation
    d[c("Y1C1", "Y2C1", "Y3C1",
        "Y1C2", "Y2C2", "Y3C2")] <- d$Mu + d$B_effect + d[c("Y1C1", "Y2C1",
    Y3C1",
                                                           "Y1C2", "Y2C2", "Y3C2")]
    d[c("Y1C3", "Y2C3", "Y3C3",
        "Y1C4", "Y2C4", "Y3C4")] <- d$Mu + d$A_effect + d$B_effect + d[c("Y1C3",
    Y2C3", "Y3C3",
                                                                        "Y1C4",
    Y2C4", "Y3C4")]
```

```
    d[c("Mu", "A_effect", "B_effect")] <- NULL
    return(d)

  }

}
```

SCB_SIM_DataSimulation_NoEffect.R

```
## SCB - Analysis of variance by simulation
source("Script/SCB_FUN_DataSimulation.R")
set.seed(12345)

p_values <- replicate(200,
                 {
                    d <- SCB_DataSimulation()
                    m <- anova(lm(Y3C1 ~ A, data = d))
                    m$"Pr(>F)"[1]
                 })
prop.table(table(cut(p_values, breaks = c(0, .05, 1))))
```

Output

```
> prop.table(table(cut(p_values, breaks = c(0, .05, 1))))

(0,0.05]  (0.05,1]
   0.03      0.97
```

Analysis

The first script defines the function **SCB DataSimulation()** to generate one data frame with the planned study design. The second script calls the function 200 times using **replicate()** and without any parameter adjustments to check how many significant F-tests of the study intervention we would get if there was actually no intervention effect present. The call set.seed(12345) initialises the random data generator for the creation of pseudo-random numbers. This allows for the exact recreation of the data in repeated runs of the script. The script creates a data frame with two fixed factors, A and B, representing the treatment condition and the school, respectively. Next, it creates the theoretical parameters of all distributions of the dependent variables. Furthermore, **rmvnorm()** from the **mvtnorm** package generates the data frame according to the defined correlation structure. Finally, the script aggregates the data frame on the basis of the design model. The second script just replicates the function **SCB DataSimulation()** and calculates an analysis of variance with each data set. It then displays the proportions of significant and non-significant results in a table.

EXAMPLE! ## 10.3.3 SCB - Power of the analysis of covariance

It is well known that selection of a good covariate in the analysis of variance can greatly enhance the statistical test power of the assumed intervention effect. The covariate should correlate by about $\rho = .6$ with the dependent variable (Kirk, 2012). In the following, we use our data generation function **SCB_DataSimulation()** to insert such a high correlation between bullying on the schoolyard after the intervention took place with the covariate of condition of the school building. We then compare whether the effect of conflict reconciliators more often reaches significance in the *F*-test of an analysis of covariance then in a normal analysis of variance.

SCB_SIM_DataSimulation_PowerANCOVA.R

```
## SCB - Power of analysis of covariance by simulation
## Simulation with small effect of factor A and covariate

source("Script/SCB_FUN_DataSimulation.R")
set.seed(12345)

# Set effects and associations
InterventionEffect <- c(.6, -.2, -.4)
CovMat <- diag(c(2, 5, rep(5, 12)))
Covariance <- 2
CovMat[lower.tri(CovMat)] <- Covariance
CovMat[upper.tri(CovMat)] <- Covariance
CovMat[cbind(c(1, 2), c(2, 1))] <- 1

# Generate study replications
SimStat <- t(replicate(100,
                   {
                       d <- SCB_DataSimulation(A_effect = InterventionEffect,
                                                 Sigma = CovMat)
                     m_simple <- anova(lm(Y1C3 ~ A, data = d))
                     m_covariate <- anova(lm(Y1C3 ~ A + X1, data = d))
                     c(Means = with(d, tapply(Y1C3, A, mean)),
                         r   =  with(d,  by(cbind(X1,  Y1C3),  A,  function(x)
                         {cor(x$X1, x$Y1C3)})),
                         p_simple = m_simple$"Pr(>F)"[1],
                         p_covariate = m_covariate$"Pr(>F)"[1])
                   }))
SimStat <- as.data.frame(SimStat)

#View(SimStat)

# Display results
plot(SimStat$p_simple,
     SimStat$p_covariate,
     xlim = c(0, 1),
     ylim = c(0, 1))
abline(h = .05)
abline(v = .05)
```

```
# Table of significant p-values
with(SimStat,
    {
      p_breaks <- c(0, .05, 1)
      addmargins(prop.table(table(cut(p_simple, breaks = p_breaks),
                               cut(p_covariate, breaks = p_breaks),
                               dnn = c("p_simple", "p_covariate")))))
    })
```

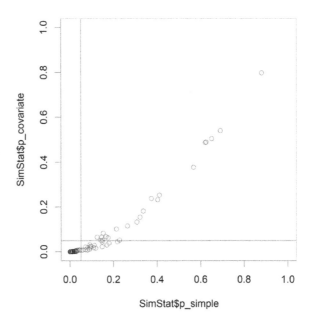

Figure 10.3 Comparison of p-values of F-tests in ANOVA and ANCOVA.

Output

```
> # Table of significant p-values
> with(SimStat,
+       {
+         p_breaks <- c(0, .05, 1)
+         addmargins(prop.table(table(cut(p_simple, breaks = p_breaks),
+                                  cut(p_covariate, breaks = p_breaks),
+                                  dnn = c("p_simple", "p_covariate")))))
+       })
            p_covariate
p_simple    (0,0.05] (0.05,1]  Sum
  (0,0.05]      0.59     0.00 0.59
  (0.05,1]      0.19     0.22 0.41
  Sum           0.78     0.22 1.00
```

Analysis

The script first sets the parameters for the simulation. Besides the intervention effect this means specifying the covariance matrix with the associations between the covariates and the dependent variables. Then, **replicate()** runs 100 data simulations and returns simulation statistics in the SimStat object. The functions **plot()** and **table()** take the *p*-values from this object and let us compare between the designs. The diagram shows that the analysis of covariance yields more significant results for the intervention effect than the analysis of variance. The cross table of *p*-values confirms this impression. With the proportions of *p*-values below $\alpha = 0.05$, the analysis of covariance indicates a statistical test power of almost 80% whereas the analysis of variance indicates only about 60%.

10.3.4 SCB - Data set for the analysis in this book   EXAMPLE!

In the others chapters, we calculated many examples with the SCB data set. The following script generated the data frame for these analyses.

SCB_SIM_DataSimulation_Example.R

```
## SCB – Example data
source("Script/SCB_FUN_DataSimulation.R")
set.seed(12345)

# Set effects and associations
InterventionEffect <- c(.6, -.2, -.4)
SchoolEffect <- c(.4, .3, -.7)
CovMat <- diag(c(2, 5, rep(5, 12)))
Covariance <- 2
CovMat[lower.tri(CovMat)] <- Covariance
CovMat[upper.tri(CovMat)] <- Covariance
CovMat[cbind(c(1, 2), c(2, 1))] <- 1

# Generate random data frame
SCB <- SCB_DataSimulation(A_effect = InterventionEffect,
                    B_effect = SchoolEffect,
                    Sigma = CovMat)
SCB <- data.frame(SCB[c("ID", "A", "B")],
               round(SCB[c("X1", "X2",
                        "Y1C1", "Y2C1", "Y3C1",
                        "Y1C2", "Y2C2", "Y3C2",
                        "Y1C3", "Y2C3", "Y3C3",
                        "Y1C4", "Y2C4", "Y3C4")]))

# Export data frame
write.csv(SCB, file = "./Data/ExperimentConflictReconciliationData.csv")
```

Analysis

Although an earlier script created several data sets, this script creates only one. It is the data set used in the other chapters of this book. It sets the intervention and school effects to some arbitrary values, and also uses a high covariance between the covariates and the dependent variables. The function **write.csv()** writes the data frame in a file. All other scripts of the student reconciliator example import and use this data set.

━━━━━━━━━ **What you have learned so far** ━━━━━━━━━

1   To create a data simulation with random numbers
2   To simultaneously sample from different distributions
3   To program repeated runs of a data simulation
4   To compare the statistical test power between statistical designs
5   To make data simulation reproducible

━━━━━━━━━ **Exercises** ━━━━━━━━━

Read code

1   Read and run the following script to simulate correlated data. How do we take advantage of the idea of factor analysis in the simulation? Closely observe the structure of Sigma, the empirical correlations, and the matrices of factor loadings.

```
## Simulation of correlated data

library(mvtnorm)
library(psych)

# Simulation
set.seed(11345)
Lambda <- matrix(c(.6, .1, .1,
                   .6, .1, .1,
                   .6, .1, .6,
                   .1, .6, .6,
                   .1, .6, .1,
                   .1, .6, .1,
                   .1, .6, .1,
                   .1, .6, .1),
                 nrow = 8,
                 byrow = TRUE)
Sigma <- Lambda %*% t(Lambda) + diag(rep(.3, nrow(Lambda)))
d <- mvtnorm::rmvnorm(n = 100, sigma = Sigma)
```

```
# Display correlations
pairs.panels(d)

# Factor analysis
f1 <- factanal(d, 2)
f2 <- factanal(d, 3)
f3 <- factanal(d, 4)

loadings(f1)
loadings(f2)
loadings(f3)
```

11

PUTTING IT ALL TOGETHER: THE STRUCTURE OF A STATISTICAL ANALYSIS

━━━━━━━━━━━━━━━━━━━━━━━━━━━━━ Chapter overview ━━━━━━━━━━

Sometimes, after years, we go back to one of our statistical analyses with big worries as to how all these function calls fit together. In this chapter, we will find remedy for this. We will review practical advice on programming style and on the structure of the analysis. This helps us to maintain the analysis and, further, to enhance and extend it after time has passed.

11.1 Programming style

Let us first discuss two aspects relating to coding style: One involves generic suggestions for good style; the other includes elements of R to maintain good style and to write readable and robust code.

11.1.1 Programming style in R

We already discussed strategies to program R – for example, to use one script for each task in the analysis, to use inclusion guards, or to separate calculation from presentation. There are some more helpful rules for a good programming style which make the analysis easier. Your code is easier to read, errors are easier to find, maintenance and extension of the code are easier, and, most important, you can share code in your research team. Let us visit some of the rules suggested by Goodliffe (2007), which are widely accepted among programmers.

- "Write your code to be read. By humans. Easily" (Goodliffe, 2007, p. 59). The days of the lone statistics programmer are gone. Today, the labour of analysis is divided in the research team. So, write readable code and share it.
- Clarity of the code is more important than its brevity. Some strive for the shortest possible expression to do a complex task. Sometimes this makes code unintelligible for others. If the task is complex, use intermediate steps to reach your goal and name the intermediate objects so that others can guess their content.
- Use a consistent and conventional presentation of the code. Read code written by others and infer the conventions of code presentation. If required, add your own rules for how to structure the code. Then, stick to these rules.
- Write self-documenting code. If you consistently use a good naming convention, write simple lines of code, and let each script do one meaningful thing, you will not need many comments to tell the reader what the code does – the code will document itself. Do not use comments to say in other words what the code does. Reserve comments to say why something is done. "Only add comments if you can't improve the clarity of the code in any other way." (Goodliffe, 2007, p. 65).
- Not in the list by Goodliffe (2007), but helpful anyway: refactor your scripts. Refactoring is an activity to optimise the code and make it more intelligible without changing the code's output (Martin, 2009). For example, to replace for loops with a vectorised solution, to better structure scripts by separating variable declarations from variable manipulation, to better structure the separation of results calculation and results presentation. Refactoring makes your scripts appear more professional. They will be easier to read and comprehend (even for yourself). You can also better maintain and extend the analysis and share the scripts with others.

11.1.2 Elements of the R language to maintain good style

It is one thing to be consistent with generic style guidelines; it is another to use the R language to maintain good style. Next, we will recall some helpful elements of R.

- *Often use* `with()` *to structure the analysis.* It offers a triple benefit: first, it allows for easy access to variables in a data frame. Second, it keeps the workspace tidy, because temporary

objects are removed after completion of the calculations. Third, it also visually groups calculations which belong together and makes the code easier to read.

- Split complex tasks in several scripts and link the scripts with **source()**.
- Use inclusion guards to create objects only once in an analysis session.
- For recurring tasks, use **function()** to program your own analysis functions.
- Use attributes to enrich your data frames with additional information. For example, this
could be variable labels, value labels, or item wordings from questionnaires.
- Do not call functions inside other function calls too often. *Functions can be called inside
the calls of other functions in R. This often yields very complex expressions. Consequently,
some suggest writing a separate line for each function call to get clean code and to use
intermediate objects to store the output of functions and then pass these objects to the next
functions.* For example, in x <- seq(1, 5, 2) and then rep(x, each = 4), x would be
such an intermediate object. This is good practise instead of writing several function calls in
one line. *However, sometimes when you hack in code at the console just to see if some idea
works, then several function calls in one line may suffice.* This makes it even more important
to distinguish between, on the one hand, an intelligible R script as the result of planning and,
on the other hand, a quick and dirty approach to quickly supply results. At the console I often
use several functions in one line; when storing the operations in scripts I try to rearrange the
code to make it readable.

11.2 Management of a complete statistical analysis

Good programming style is one thing; the other is how to synthesise the complete analysis –
how to structure scripts and keep the analysis open for refinement.

11.2.1 To manage the statistical analysis is vital to the project

PLANNING
ISSUE!

For a research project there usually is no one-and-only valid analysis. *In reality, at some point in
your project, you, your team, and your supervisor have to decide which analysis to apply to your data.
In good empirical projects you decide this and even publish this decision, before gathering the data.*
Otherwise you would be susceptible to selecting the type of analysis which produces significant results or the strongest effects. However, when you obtained the data, you then prepare
the data so that it can be used for this analysis, check the prerequisites for the analysis, do
the analysis, and then publish it. When published, you are accountable for the analysis and
you should be able to replicate and explain every step of it from the raw data to the published
table with the summary of the analysis. So, our strategy to structure the analysis needs to
account for this. However, as I said before, there is no one-and-only data analysis for a project
of a noteworthy size. More commonly, there are several ideas that you have in mind or which
your team members suggested to you in the last meeting; some ideas that support, refine,
or complement your main analysis. *Nothing should prevent you from exploring your data from
such different perspectives. However, in the mass of scripts that you produced for the analysis, your
main analysis must remain identifiable and replicable up to all the decimal numbers of your published results.* Consider, for example, that one of the supplementary analyses requires another
grouping of factor levels compared with the main analysis, which you easily program in one
of the data preparation scripts after you published the main results. If this data preparation

script is part of your main analysis, then the former valid and published analysis becomes corrupted and could become non-replicable by you. Even worse, you may not even notice the corruption of the main analysis. And if some more time passes you may even forget about at what place in your scripts you changed the something that marked the beginning of a non-replicable analysis. If somebody kindly asks you to share your analysis scripts or if you need to recalculate it, you find out that your main analysis virtually does not exist any more and that there is now a big question mark between the raw data and the published results table. *Consequently, the structure of the analysis needs to accomplish two things: (1) to contain the main analysis to allow for easy replication and (2) to allow for extension of the analysis.*

PROGRAMMING
ISSUE!

11.2.2 General rules for the structure of R scripts

Before we go to the scripting rules, let us list some dangers that can occur if scripts are generated without care.

- Important and accurate code may be conflated by badly written, outdated and inaccurate code. We tend to leave half-completed analyses in the scripts because they often contain a good approach or a fancy R expression that we like. However, if this code does the wrong calculations, remove it. Otherwise, it wastes your time with every code reading session. Furthermore, important code may not be retrievable in a mass of scripts with wrong code.
- Code may be doubled at several places in the analysis. Updates in one place may not be carried out in other copies of the code. A certain variable transformation or other data preparation may be required in several analyses. It is easy to copy-paste the respective code where it is needed. If you change the code in one place, however, different versions of code under the same name enter your scripts. Better to put the data preparation in one script and invoke this script, where necessary.
- Scripts grow too long. Review of a long script is very hard and prone to errors. It is nearly impossible to delete a suspected superficial line of code, because it may be that its output is required hundreds of lines later. Commenting out code instead of deletion is no true remedy, because it makes the code even more unintelligible. In fact, the commenting facilities are made for comments and are no safeguard against the careless deletion of code.

There may be more dangers in bad scripting; however, we now turn to some rules to simultaneously maintain integrity and extensibility in the statistical analysis:

- Let a script create at most one object. Then use an inclusion guard for this object to prevent its repeated generation in an analysis session. Furthermore, match the file name of the script with object name of the object created. However, if the script creates more than one object, collect them in a list, which is again a single object as output. Use each object name only once in an analysis session. Following this basic advice, you will easily keep track of all objects in an analysis session.
- A script should not permanently manipulate an object it has not created itself. If scripts manipulate objects for which they are not responsible, you will not notice undesired changes in objects while running the analysis. It is another way of saying that each script must stick to its purpose. An exception for this is preparatory scripts, which usually add new variables to a data frame.

- A script must always run completely. Thereby, it sources all its preparatory scripts and so creates the data and objects as needed. This way you may start each analysis session with any script and run it without error. Furthermore, in writing new code you will not be distracted by code written in the preliminary scripts.
- *Create a header section for all scripts.* This header should include the date of creation and modification, version number, author, aim, and a rough description of the script's contents. This enables a quick grasp of the main purpose and contents of the script for others, and even for you, if you go back to the script after a few years. Use comments to create this section. Beware, however, describing the contents of the script in too much detail in the header section or in any other comment. These descriptions become outdated if you change the code later and possibly nobody notices that they are outdated.
- *Use code comments to state the aim or purpose of the code.* Do not simply restate in everyday language what the code does. This may sound counterintuitive, because human-readable prose appears to be much more accessible than programming code. However, there is a phenomenon that as you write, rewrite, and amend your code, there is usually little feeling rewriting and amending the comments, too. If the job is done, everything else wastes time. So, trivial code comments that simply restate what happens, quickly become outdated as the code develops. Reading the comments at some later time, thus, might mislead you or others. They will not only be worthless, they will be harmful to the intelligibility of the analysis. The phenomenon is well known as ROT in programming: redundant, outdated, or trivial (Liberty, 2001). One of the strategical cornerstones of agile programming is to reduce documentation (Goodliffe, 2007). The strategy includes focusing on the creation of readable and intelligible code for the acquainted reader. This means a reader of the code who is basically familiar with the R syntax should be able read through the code and understand what happens. The strategy builds on naming conventions for the objects in the code; for example, to create long enough names for objects which speak for themselves. This strategy has a team advantage: good scripts can be reviewed in team meetings, where everybody can suggest amendments. Thereby, it spreads good code writing habits and also familiarises R newbies with the language and the house rules for programming a statistical analysis. In sum, do not write ROT comments in your code; instead, write readable code.
- Use *inclusion guards* when creating objects or defining functions. This ensures that even if a script is sourced two or more times, it runs only one time. If, for example, a load-data script is followed by a calculate-scores script, which attaches new variables to the loaded data set, then a later call of the load-data script would overwrite the data frame and, thus, erase the previously attached variables. It is preferable to load the data frame only once and that an inclusion guard prevents repeated loading. Chapter 3 shows how to use them.

11.2.3 Structuring the complete analysis

PROGRAMMING
ISSUE!

Designing a statistical analysis is like designing other computer software. Software programmers consider the design of software as even more important than the actual coding (Goodliffe, 2007). Similarly, in statistical analysis you also need careful planning. In fact, you will plan code scripts and how they connect to form the complete analysis. So, identify the central activities of the analysis and then prepare at least one script for each activity. These can be data acquisition, data preparation, descriptive statistics, figure preparation, inferential statistics, preparation of tables, and preparation of the results report. Our emphasis in planning the grand design of the analysis is the same as in professional software programming (Goodliffe, 2007).

If, for example, you need three figures for the results of your manuscript, then create at least three scripts, one for each figure. You could do it all in one script, but experience shows that initial plans for figures seldom prevail, but change with each team discussion about your results. So, in a script which creates three figures, one may receive much attention and many refinements and modifications. Usually the code for such figures grows line by line with every revision. It would not look nice if another abandoned figure still remains in the script. So, throw the abandoned figure away or store it in another script. Name the scripts to reflect their content.

All steps of the analysis should be scripted. Data acquisition, variable transformations, case deletions, statistical calculations, and results output should all reside in R code. This way, all steps of the analysis are ready for team discussion and critical revision. Nowadays, the scripts also fit as supplements to journal publications and other researchers can review your analyses, too. Furthermore, scripting automatically creates a protocol for all steps of the analysis, which you can use to rethink your analysis even months later.

Maintain a fixed organisation for all scripts. Statistical analysis comprises different tasks that reoccur in any of your projects. There will always be a phase of data acquisition. Almost every project yields results tables and figures. Additionally, many projects allow for the distinction between data description and inferential statistics. So, if you maintain such a structure, your team members can expect certain scripts to be available, which facilitates code sharing and reuse. For example, if there is one script which calculates a dependent variable with raw data, all analysis scripts that use this variable can include the calculation and do not need to calculate themselves.

Prefer the writing of new code instead of changing existing code. Sometimes you are not satisfied with a given analysis and you change its code. However, two months later your team agrees that the initial version of the analysis was, in fact, better than the later one. If you changed the code, you have to reconstruct and rewrite the earlier analysis, and recreation of a thing you know you already had is a terrible job. Therefore, it would have been better to create new code every time a new idea for analysis comes up. It is simple then to rerun earlier versions of an analysis. Moreover, do not corrupt valid existing code with changes. In doing this you would actually abandon an old valid analysis, which you may have published already. So, to extend the analysis, write new code rather then changing existing code for a new analysis. If you have a bar diagram and now you want a scatter plot, write a new script for the scatter plot. Do not change the bar diagram script if it is valid; simply write a scatter plot script.

Distinguish between different phases of the analysis. Such phases are data acquisition, preparation of the data frame for the analysis, exploratory analysis, graphics, model calculations, and results output. Almost all my analyses comprise them. For the scripts of each phase, use meaningful prefixes in their names. Otherwise, if you have more than 20 or so scripts in the analysis, their structure becomes incomprehensible. Meaningful prefixes may enhance the structure of the analysis. Table 11.1 shows example prefixes. The filenames PRE_ImportData_20210401.R, PRE_CalculateScores_20210401.R, FUN_EnhancedHistogram_20210401.R, and MOD_Linear Regression_20210401.R all suggest what they do at first sight. Furthermore, use a fixed folder structure for your analysis to store all relevant files. I prefer to use four folders in each analysis: Data, Scripts, Results, Figures. So, the FIG scripts usually write their output in the figures folder and the DES scripts write their output in the results folder.

Table 11.1 Prefixes of script names.

Prefix	Phase of analysis
PRE	Preparation
FUN	Function definition
DES	Descriptive statistics
MOD	Models and inferential statistics
OUT	Results output
FIG	Graphical output of results
CON	Analysis session at the R console
SIM	Data simulation

draw the structure of your analysis. Use boxes to draw the scripts and objects which your analysis requires. Connect the boxes with arrows to reflect their dependencies (Figure 11.1). Software programmers often use the so-called Unified Modeling Language (UML) to visualise the details of the information that software components contain and how these components interact to eventually yield the working software (Rumbaugh, Jacobson, & Booch, 2005). For data analysis, we should learn from this approach and also visualise the structure of our analyses. This way we better find inconsistencies and can illustrate the analysis structure in our team.

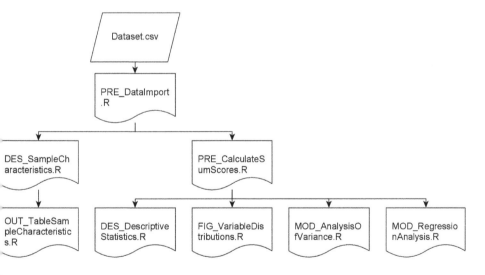

Figure 11.1 Example structure of a statistical analysis.

What kind of visualisation of the structure you chose depends on the characteristics of your project. Bigger projects usually come with a fixed data analysis plan that is published already. The visualisation needs to account for this analysis plan. Other projects have a more explor-atory character and require the visualisation of a number of different analyses of the data.

Still, other projects need many transformations of the raw data, and statistical and graphical checks to secure data quality. It may be important for projects about the creation or validation of psychometric scales. The visualisation also may reflect these steps. Finally, projects may include data that changes in regular intervals, perhaps the monthly sales numbers or patients in a hospital. So, we may also want to visualise the structure of monthly or quarterly statistical reports.

Put the results in an HTML report. In Chapter 8 we created reports of our analyses. To set up an HTML report as an Rhtml file, that is, to combine HTML markup language and R code, is another way to structure the statistical analysis. The HTML markup gives structure to the analysis and at the same time allows for a description of the results in normal language. In fact, I suggest using the Rhtml language mainly for structuring purposes for the final report and not for statistical calculations. Program the results in R scripts and then use **source()** with these scripts to include them in the Rhtml file. Use R code in the Rhtml file simply to display results, for example, by calling a table with **stargazer** or **xtable**.

EXAMPLE! ## 11.2.4 WBD – Structure of the analysis

Consider the following example structure of our statistical analysis about positive imagery-based cognitive bias modification in depression. We display all the scripts that comprise the analysis (Figure 11.2).

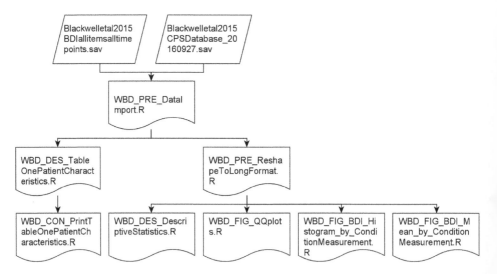

Figure 11.2 Structure of the analysis scripts of the example web-based cognitive bias modification for depression.

What is relevant with this structure is that it comprises the complete analysis. Your analysis is broken down into several small tasks and each script accomplishes one of these tasks. There is a script for data import, another one for the recoding of variables and the calculation of scores, and so on. Scripts later in the analysis include some calls of **source()** to invoke earlier scripts to supply the data. For example, the script calculating descriptive statistics not only needs the data import script but also the script which reshapes the data from wide format to long format. As can be seen, several of the analysis scripts use the data in long format.

11.3 Where to go from here

We now move on beyond our own data and tap data from the research community. There are vast online resources of available data sets, all easily imported in R. Finally, to conclude, we step back from the data and look at the big picture of how data and computing power enable the progress of science.

11.3.1 Obtaining data from online repositories

PLANNING ISSUE!

There are endless options for analysis of original data. *R is the right tool to acquire and ana-lyse thousands of data sets, freely available on the internet.* Example topics are health, educa-tion, environment, and business. You will surely find data from your own domain, as well. So, search the repositories and get the data. Table 11.2 provides an overview of open data resources. Go to the websites and skim through the masses of data. *Several data servers hosted by the departments of national governments provide access to plain data files in csv format.* Usually, each data set has an online address that you can call from the R console. Simply copy the web address of a data file and paste it in the **read.csv()** function. For example, a data set relat-ing to a school survey about crime and safety, hosted by the National Center for Education Statistics of the US Department of Education, can be directly imported.

```
d       <-      read.csv("http://nces.ed.gov/surveys/ssocs/data/txt/1999_00_
ssocs-cd.txt")
utils::View(d)
str(d)
```

After loading the data from the data repository, it is available for calculations, but check the website for other related information about the data. Including such a web address in the script, however, makes it rely on the remote data source and changes in the address of the data set require changing the call of **read.csv()**, too.

Table 11.2 Open data web resources.

Institution	Description	Web address
GOVDATA (Germany)	Administrative data for Germany.	www.govdata.de/
UK Data Archive (UK)	Hosts research data in the social sciences and humanities. Offers much advice about how to prepare and archive research data.	www.data-archive.ac.uk/
UK Data Service	Hosts data sets structured by topics.	www.ukdataservice.ac.uk/

(Continued)

Table 11.2 (Continued)

DATA.GOV (USA)	Repository of the U.S. Government's open data. Structured by topics such as health, education, climate, or business. Contains more than 180,000 data sets, supplied by more than 170 organisations.	www.data.gov/
HealthData.gov	More than 2500 health-related data sets. Website managed by the U.S. Department of Health & Human Services.	www.healthdata.gov/
eurostat (Europe)	Data hosted by the statistical office of the European Union. Provides aggregated data and offers restricted access to micro data. Data tables structured by topics like population and social conditions, environment, or energy. Interactive data tables.	ec.europa.eu/eurostat/web/main/home
dataplanet	Big archive of international data sets sorted by subject or country.	dataplanet.sagepub.com/
Journal of Open Psychology Data	Free access to data that accompanies published articles. Peer review, does not host the data but links to data repositories.	openpsychologydata.metajnl.com/
OpenScienceFramework	Hosts many academic projects. Not only data but project resources.	osf.io
Zenodo	Open data repository developed by researchers.	zenodo.org

The data available on the internet distinguishes between raw data and aggregated data. Raw data is also called micro data on some sites; it is data of single subjects' responses collected in a data frame. Aggregated data comes in prefabricated tables that the vendor found to convey a useful message. Download of aggregated data is usually free on the internet, whereas the download of raw data usually requires a user account. Another distinction relates to the kind of data provider. Most interesting to researchers working in the domains of psychology, health, education, or political science seem to be data from governmental agencies, initiatives sponsored from different sources, libraries, and scientific journals. Beware of whether the content went through peer review prior to publication.

Some journals nowadays, like Annals of Internal Medicine, occasionally request authors to provide data and even the statistical scripts together with the publication. The benefit of this is to make not only the study reproducible but also the calculation of results and the applied statistical methods. It requires more work by you to furnish the data set and the scripts for others to see. However, in this book we practised the means to put data and scripts in good form. Our hard work as researchers to create high-quality data naturally suggests sharing the data with colleagues or with the research community. So, not only one researcher may run statistical analyses on a given data set, but anyone who is competent to do so. This is the idea of open data. Perhaps you may be willing to confer your data to the scientific community as well, using one of the suggested repositories.

11.3.2 Object-oriented programming in R

R LANGUAGE!

In R everything is an object. However, users of R only occasionally think of R as an object-oriented programming language. So, what aspect of this is important for statistical analyses? When do we have to define our own classes? As users, we usually only experience the part of object-oriented programming related to object creation and function use. For example, we create an ANOVA object and then query this object for information using functions like `summary()` or `model.matrix()`. The other side of object-oriented programming is to define new classes from scratch, which accomplish tasks that nobody programmed before. Perhaps there is an exotic statistical analysis which is not available in R at present. If it is important for you, you could implement the analysis yourself using object-oriented programming in R. If you succeed, you can even put it in a new R package and supply it to the scientific community. To continue this idea, you should enquire how S3 and S4 classes work in R. This is described in Matloff (2011).

11.3.3 The scientific enterprise

PLANNING ISSUE!

As researchers using empirical methods, we take part in the currents and trends of science. We follow the current research paradigm (Kuhn, 1962) and somehow have the feeling of what is good and not-so-good science, especially as there are no concerns regarding the epistemic and ontological frameworks that imbue us. We even promote some criteria of what a good scientific program should accomplish. For example, the falsification of a theory tentatively is one of the core criteria, if not the only one, of any idea to be scientific (Popper, 1994). *We should not fall into the trap of thinking that our current research is not determined by epistemological trends and is simply rational.* Some of the biggest insights into nature in the past have been accomplished under very different circumstances where people did not have our computing capacity but only had paper and pencil. *Even the meaning of observational data and how it relates to theory is not a settled issue that was available at all times, and it is still not settled today.* That's history; however, with our statistical software today we have the idea that everything in science is only a matter of precision and that, given enough resources, we can reach any precision we want and therefore any degree of truth that we want. However, data and tons of data is one thing; the idea of what it is that yields the data is another. The Ptolemaic world view in astronomy could serve us for all time to structure our observations of the stars. In fact, Copernicus' revolution was in-part supplied with the richness of the observational data of former generations; but it was nevertheless a bold revolution in thought to set the Earth in motion, which was nowhere given in the data. In fact, and this is just a guess, since it is well known that the Ptolemaic view collapsed partly because of the high complexity of the epicycle system to describe the motion of the planets, perhaps if the people in those days had had the computing resources of our time, this complexity would not have been experienced as so severe. It could have been managed by simply giving more power to the computer. Supposedly, the computing capacity could perhaps even disguise the fundamental problems in theory. If we think of modern approaches to the analysis of data, the linear modelling capacities of R are so vast and offer so many parameters of adjustment and optimisation that sometimes the issue of how to demonstrate a psychological cause and effect is trivialised in a mass of correlations that we sometimes collect from our samples. In fact, the increasing power of numerical calculations does not necessarily correspond with advances in theory. So, reflect more about what a statistical software like R can do for you and what could be the joy of a theoretical endeavour in your research.

What you have learned so far

1   To plan a complete statistical analysis in R from scratch
2   To divide statistical tasks over different scripts
3   To extend an analysis without corrupting it
4   To obtain data from online repositories

Exercises

Apply in the real world

1   Start a new analysis of data available to you.
2   Set up project folders like data, scripts, results, and figures.
3   Prepare a template for a script that contains elements that each script is going to have. These can be a header section, an inclusion guard, and one or more calls of `source()`.
4   Decide for a naming convention for your file names and object names.
5   Prepare a bunch of scripts that will definitely be part of the analysis and use the script template to initialise them. Such scripts could be for data import, variable transformations, descriptive statistics, statistical modelling, results output, and diagrams. Preferably, each script should have one object as output.
6   Go to one of your existing analyses and improve its structure.
7   Decompose scripts that produce too much output in several scripts, each producing only one object.
8   Use inclusion guards if scripts have new objects as output.
9   Try to impose a structure that scripts later in the analysis invoke earlier scripts with `source()`. For example, each script doing some analysis could invoke the data import and variable transformation script.
10 Go back to one of your former analyses or publications. Try to recalculate your (published) results with old or new statistical scripts up to the last decimal point. Were you successful?
11 Ask your supervisor or colleagues to provide you with one of their recent publications consisting of a data analysis. Also obtain the corresponding data set. Try to recalculate the exact results of the published paper with the given data using R.

REFERENCES

Agresti, A. (2002). *Categorical data analysis* (2nd ed.). Hoboken, NJ: John Wiley & Sons.

Agresti, A. (2007). *An introduction to categorical data analysis (Wiley Series in Probability and Statistics)*. Hoboken, NJ: John Wiley & Sons, Inc.

Ahmed, E., & Braithwaite, V. (2006). Forgiveness, reconciliation, and shame: Three key variables in reducing school bullying. *Journal of Social Issues, 62*, 347–370. doi:10.1111/j.1540-4560.2006.00454.x

American Psychological Association (APA) (2001). *Publication manual of the American Psychological Association* (5th ed.). Washington, DC: American Psychological Association.

Aquino, J. (2021). descr: Descriptive Statistics. Retrieved from https://CRAN.R-project.org/package=descr.

Baath, R. (2012). The state of naming conventions in R. *R Journal, 4*(2), 74–75.

Bates, D., Mächler, M., Bolker, B., & Walker, S. (2015). Fitting linear mixed-effects models using lme4. *Journal of Statistical Software, 67*(1), 1–48. doi:10.18637/jss.v067.i01

Behrens, J. T. (1997). Principles and procedures of exploratory data analysis. *Psychological Methods, 2*(2), 131.

Blackwell, S. E., Browning, M., Mathews, A., Pictet, A., Welch, J., Davies, J., ... Holmes, E. A. (2015). Positive imagery-based cognitive bias modification as a web-based treatment tool for depressed adults: A randomized controlled trial. *Clinical Psychological Science, 3*(1), 91–111. doi:10.1177/2167702614560746

Casella, G., & Berger, R. L. (2002). *Statistical inference* (2nd ed.). Pacific Grove, CA: Duxbury/Thomson Learning.

Ceyhan, E., & Goad, C. L. (2009). A comparison of analysis of covariate-adjusted residuals and analysis of covariance. *Communications in Statistics – Simulation and Computation, 38*(10), 2019–2038. doi:10.1080/03610910903243687

Christensen, R. H. B. (2019). Ordinal-regression models for ordinal data. Retrieved from https://CRAN.R-project.org/package=ordinal.

Cleveland, W. S. (1994). *Visualizing data*. Summit, NJ: Hobart Press.

Cohen, J. (1988). *Statistical power analysis for the behavioral sciences* (2nd ed.). Hillsdale, NJ: Erlbaum.

Cohen, J. (1990). Things I have learned (so far). *American Psychologist, 45*(12), 1304–1312.

Crawley, M. J. (2005). *Statistics. An introduction using R*. Chichester: Wiley.

Csardi, G., & Nepusz, T. (2006). The igraph software package for complex network research. Retrieved from http://igraph.org.

Dahl, D. B., Scott, D., Roosen, C., Magnusson, A., & Swinton, J. (2019). xtable: Export Tables to LaTeX or HTML. Retrieved from https://CRAN.R-project.org/package=xtable

Defossez, A., Ansarinia, M., Clocher, B., Schmück, E., Schrater, P., & Cardoso-Leite, P. (2020). The structure of behavioral data. arXiv:2012.12583

Draper, N., & Smith, H. (1981). *Applied regression analysis. Series in probability and mathematical statistics*. New York: Wiley.

Duarte, N. (2008). *slide:ology: The art and science of creating great presentations*. Beijing: O'Reilly.

Elff, M. (2021). memisc: Management of survey data and presentation of analysis results. Retrieved from https://CRAN.R-project.org/package=memisc

Fox, J. (2005). The R Commander: A basic-statistics graphical user interface to R. *Journal of Statistical Software, 14*(9), 1–42.

Fox, J., & Weisberg, S. (2019). *An R companion to applied regression* (3rd ed.). Thousand Oaks, CA: Sage.

Gagolewski, M. (2020). R package stringi: Character string processing facilities. Retrieved from www.gagolewski.com/software/stringi/

Genz, A., Bretz, F., Miwa, T., Mi, X., Leisch, F., Scheipl, F., & Hothorn, T. (2021). mvtnorm: Multivariate normal and t distributions. Retrieved from https://CRAN.R-project.org/package=mvtnorm

Goodliffe, P. (2007). *Code craft. The practise of writing excellent code.* San Francisco: No Starch Press.

Gordon, M., Gragg, S., & Konings, P. (2021). htmlTable: Advanced Tables for Markdown/HTML. Retrieved from https://CRAN.R-project.org/package=htmlTable

Harrell, F. E. (2015). *Regression modeling strategies* (2nd ed.). Cham: Springer.

Harrell Jr, F. E. (2021a). Hmisc: Harrell miscellaneous. Retrieved from https://CRAN.R-project.org/package=Hmisc

Harrell Jr, F. E. (2021b). rms: Regression modeling strategies. Retrieved from https://CRAN.R-project.org/package=rms

Hays, W. L. (1994). *Statistics* (5th ed.). Fort Worth: Harcourt Brace College Publishers.

Hlavac, M. (2018). stargazer: Well-formatted regression and summary statistics tables. Retrieved from https://CRAN.R-project.org/package=stargazer

Horstmann, C. S. (2007). *Big Java* (3rd ed.). Hoboken, NJ: Wiley.

Hosmer, D. W., & Lemeshow, S. (2000). *Applied logistic regression* (2nd ed.). New York, NY: Wiley.

Hothorn, T., Bretz, F., & Westfall, P. (2008). Simultaneous inference in general parametric models. *Biometrical Journal, 50*(3), 346–363.

Hrynaszkiewicz, I., Norton, M. L., Vickers, A. J., & Altman, D. G. (2010). Preparing raw clinical data for publication: guidance for journal editors, authors, and peer reviewers. *BMJ (Clinical Research ed.), 340*, c181. doi:10.1136/bmj.c181

James, D., & Hornik, K. (2020). chron: Chronological objects which can handle dates and times. Retrieved from https://CRAN.R-project.org/package=chron

Kelley, K. (2007). Confidence intervals for standardized effect sizes: Theory, application, and implementation. *Journal of Statistical Software, 20*(8), 1–24.

Kelley, K. (2021). MBESS: The MBESS R package. Retrieved from https://CRAN.R-project.org/package=MBESS

Kirk, R. E. (2012). *Experimental design: Procedures for the Behavioral Sciences* (4th ed.). Sage Publications.

Kuhn, T. S. (1962). *The structure of scientific revolutions.* Chicago: University of Chicago Press.

Lecoutre, E. (2003). The R2HTML package. *R News, 3*(3), 33–36.

Leifeld, P. (2013). texreg: Conversion of statistical model output in R to LaTeX and HTML tables. *Journal of Statistical Software, 55*(8), 1–24.

Lemon, J. (2006). Plotrix: A package in the red light district of R. *R-News, 6*(4), 8–12.

Liberty, J. (2001). *Sams teach yourself C++ in 21 days* (4th ed.). Indianapolis, IN: SAMS.

Lipsey, M. W. (1990). *Design sensitivity.* Newbury Park, CA: Sage.

Loftus, G. R. (1996). Psychology will be a much better science when we change the way we analyse data. *Current Directions in Psychological Science, 5*, 161–171.

Lovelace, R. (2014). Consistent naming conventions in R. Retrieved from www.r-bloggers. com/2014/07/consistent-naming-conventions-in-r/. 2014-07-15

Lüdecke, D. (2020). sjPlot: Data visualization for statistics in social science. Retrieved from https://CRAN.R-project.org/package=sjPlot

Lüdecke, D. (2021). sjstats: Statistical functions for regression models (Version 0.18.1). Retrieved from https://CRAN.R-project.org/package=sjstats

Martin, R. C. (2009). *Clean code*. Upper Saddle River, NJ: Prentice Hall.

Masson, M. E. J., & Loftus, G. R. (2003). Using confidence intervals for graphically based data interpretation. *Canadian Journal of Experimental Psychology*, *57*(3), 203–220.

Matloff, N. (2011). *The art of R programming: A tour of statistical software design*. San Francisco, CA: No Starch Press.

Meyer, D., Zeileis, A., & Hornik, K. (2020). vcd: Visualizing categorical data. R package version 1.4-7.

Murdoch, D. (2020). tables: Formula-driven table generation. Retrieved from https://CRAN.R-project.org/package=tables

Murdoch, D., & Adler, D. (2007). useR! – International R User Conference. Ames, IA.

Murdoch, D., & Adler, D. (2021). rgl: 3D visualization using OpenGL. Retrieved from https://CRAN.R-project.org/package=rgl

Murrell, P. (2002). The grid graphics package. *R News*, *2*(2), 14–19.

Murrell, P. (2016). *R graphics*. Boca Raton, FL: CRC Press.

Open Science Collaboration (2015a). Estimating the reproducibility of psychological science. *Science*, *349*, 6251. doi:10.1126/science.aac4716

Open Science Collaboration (2015b). Reproducibility project: Psychology. Retrieved from https://osf.io/ezcuj/. 2015.

Pinheiro, J., Bates, D., DebRoy, S., Sarkar, D., & R Core Team (2021). nlme: Linear and nonlinear mixed effects models. Retrieved from https://CRAN.R-project.org/package=nlme

Popper, K. R. (1994). *Logik der Forschung (10. Aufl.)*. Tübingen: Mohr.

Reuter, C., & Oehler, M. (2011). Psychoacoustics of chalkboard squeaking. *Journal of the Acoustical Society of America*, *130*(4), 2545.

Reuter, C., Oehler, M., & Mühlhans, J. (2014). Physiological and acoustical correlates of unpleasant sounds. Joint Conference ICMPC13-APSCOM5, 4–8 August, 2014, Seoul, Korea.

Revelle, W. (2021). psych: Procedures for psychological, psychometric, and personality research. Retrieved from https://CRAN.R-project.org/package=psych

Rubin, D. B. (1976). Inference and missing data. *Biometrika*, *63*(3), 581–592.

Rubin, D. B. (1987). *Multiple imputation for nonresponse in surveys*. New York, NY: Wiley.

Rumbaugh, J., Jacobson, I., & Booch, G. (2005). *The Unified Modeling Language reference manual* (2nd ed.). Boston, MA: Addison-Wesley.

Sarkar, D. (2008). *Lattice: Multivariate data visualization with R*. New York: Springer.

Serlin, R. C., & Harwell, M. R. (2004). More powerful tests of predictor subsets in regression analysis under nonnormality. *Psychological Methods*, *9*(4), 492–509.

Stanovich, K. E., & West, R. F. (2008). On the relative independence of thinking biases and cognitive ability. *Journal of Personality and Social Psychology*, *94*(4), 672–695. doi:10.1037/0022-3514.94.4.672

Stevens, S. S. (1946). On the theory of scales of measurement. *Science*, *103*, 677–680. doi:10.1126/science.103.2684.677

Su, Y.-S., Gelman, A., Hill, J., & Yajima, M. (2011). Multiple imputation with diagnostics (mi) in R: Opening windows into the black box. *Journal of Statistical Software*, *45*(2), 1–31. doi:10.18637/jss.v045.i02

Swinton, J. (2019). The xtable gallery. Retrieved from https://cran.r-project.org/web/packages/xtable/vignettes/xtableGallery.pdf. 2019-04-21

Thomas, M. A. (2019). Mathematization, not measurement: A critique of Stevens' scales of measurement. *Journal of Methods and Measurement in the Social Sciences, 10*(2), 76–94. doi:10.2458/v10i2.23785

Timm, N. H. (1975). *Multivariate analysis with applications in education and psychology*. Monterey, CA: Brooks/Cole.

Tukey, J. W. (1997). *Exploratory data analysis*. Reading, MA: Addison-Wesley.

U.S. Preventive Services Task Force. (2009). Screening for depression in adults: U.S. preventive services task force recommendation statement. *Ann Intern Med, 151*(11), 784–792. doi:10.7326/0003-4819-151-11-200912010-00006

Van Buuren, S., & Groothuis-Oudshoorn, K. (2011). MICE: Multivariate imputation by chained equations in R. *Journal of Statistical Software, 45*(3), 1–67.

Warnes, G. R., Bolker, B., Gorjanc, G., Grothendieck, G., Korosec, A., Lumley, T., … Rogers, J. (2017). gdata: Various R programming tools for data manipulation. Retrieved from https://CRAN.R-project.org/package=gdata

Warnes, G. R., Bolke, B., Bonebakker, L., Gentleman, R., Huber, W., Liaw, A., … Galili, T. (2020). gplots: Various R programming tools for plotting data. Retrieved from https://CRAN.R-project.org/package=gplots

Wason, P. C. (1966). Reasoning. In B. Foss (ed.). *Reasoning*. Harmonsworth: Penguin.

Wickham, H. (2016). *ggplot2: Elegant graphics for data analysis*. New York: Springer-Verlag.

Wickham, H. (2019). stringr: Simple, consistent wrappers for common string operations. Retrieved from https://CRAN.R-project.org/package=stringr

Wilcox, R. (2012). *Introduction to robust estimation and hypothesis testing*. New York: Academic Press.

Wilkinson, L.Wills, G. (Ed.) (2011). *The grammar of graphics* (2nd ed., softcover reprint of the hardcover 2nd ed., 2005). New York, NY: Springer.

Winer, B. J., Brown, D. R., & Michels, K. M. (1991). *Statistical principles in experimental design*. Blacklick, OH: McGraw-Hill Humanities/Social Sciences/Languages.

Xie, Y. (2015). *Dynamic documents with R and knitr* (2nd ed.). Boca Raton, FL: Chapman and Hall/CRC.

Xie, Y. (2021a). knitr: A general-purpose package for dynamic report generation in R. Retrieved from https://yihui.org/knitr/

Xie, Y. (2021b). knitr reference card. Retrieved from https://cran.r-project.org/web/packages/knitr/vignettes/knitr-refcard.pdf. 2021-09-29

Yoshida, K., & Bartel, A. (2021). tableone: Create 'Table 1' to describe baseline characteristics with or without propensity score weights. Retrieved from https://CRAN.R-project.org/package=tableone

INDEX